CHARACOIDS of the world

t.f.h.

by dr. jacques gery

Cover: *Hyphessobrycon serpae.* Photo by H.J. Richter.

ISBN 0-87666-458-3

t.f.h.

Printed in the U.S.A.

Distributed in the U.S.A. by T.F.H. Publications, Inc., 211 West Sylvania Avenue, P.O. Box 27, Neptune City, N.J. 07753; in England by T.F.H. (Gt. Britain) Ltd., 13 Nutley Lane, Reigate, Surrey; in Canada to the book store and library trade by Clarke, Irwin & Company, Clarwin House, 791 St. Clair Avenue West, Toronto 10, Ontario; in Canada to the pet trade by Rolf C. Hagen Ltd., 3225 Sartelon Street, Montreal 382, Quebec; in Southeast Asia by Y.W. Ong, 9 Lorong 36 Geylang, Singapore 14; in Australia and the south Pacific by Pet Imports Pty. Ltd., P.O. Box 149, Brookvale 2100, N.S.W., Australia. Published by T.F.H. Publications, Inc. Ltd., The British Crown Colony of Hong Kong.

PREFACE

This book, devoted to the most diversified group of American and African fresh-water fishes, has two goals. First, to provide the general reader, who is an aquarist more often than not, with as many good illustrations of aquarium type characoid fishes as possible. This kind of pictorial guide will enable him, it is hoped, to identify at first sight a good portion of the aquarium species. At the same time, the non-specialized reader will get an idea of the position of the species in the classification (genus, family and so on) and, if he wants, can refer to the text for more systematic and, occasionally, biological data, using the index of Latin names at the end of the volume.

The second aim of the book is to give to zoologists as well as to advanced aquarists a scientific account of our present knowledge of the group, which has not been reviewed as a whole for half a century. This account has been made as complete and reliable as possible, with a minimum of words and a bibliography reduced to a few recent works, but it is by no means to be taken as a true revision of the suborder.

In order to classify and to describe all, or almost all, known species and enable curators of museum collections and other interested specialists to identify them with reasonable accuracy, dichotomous tables, also called identification keys or dichotomous keys, are presented along with a discussion of the various species and their particular characteristics.

The author is aware that such a system causes a kind of psychological block to certain persons. These readers are asked to content themselves with the illustrations and general statements regarding the groups, which should be quite sufficient in most cases. Others, it is hoped, will make the simple effort to enter the field of "yes or no" reasoning, which is basically the computer's, as well as our own brain's, way of thinking. To master the use of a key, which only takes a few minutes, the principle is as follows: one is asked, step by step (a, b, c, etc.) a question to which the answer is either yes or no. If the organism to be identified has such or such a feature (answer yes), one goes to the next step. If not (answer no), one goes to the corresponding branch of the alternative (represented by aa, bb, cc, etc.), which is more or less remote in the key, until the identification is achieved.

Actually, this will probably not always be the case, and the reader must be prepared for some false leads. The impossibility of identifying a particular species may be due to the following: certain keys are surely imperfect, suffering from some defect which can only be detected after several tests; certain exceptions to the presence or absence of a character may cause difficulties, although they should be mentioned in the key when known; and the whole group is still imperfectly known and it may be, chiefly if the species is small and comes from some remote area, that the unidentifiable specimen has never been scientifically described (a so-called new species). The characoids now number more than 1200 species (including 200 or so African ones), and, based on past experience, in every large collecting sample 5-10% of the species can be expected to be new.

ACKNOWLEDGMENTS

The author wishes to extend his thanks to the persons who have greatly contributed to the book:

Dr. Warren Burgess, who had the difficult task of making the author's language readable by English-speaking people, and the people whose remarkable photographs provide the indispensable illustrations, principally Dr. H. R. Axelrod and the late Harald Schultz, in whose memory this book is dedicated.

Table of Contents

INTRODUCTION

Once upon a time, an African fresh-water fish decided to become a characoid.

This happened long ago, in the Mesozoic (70-175 millions of years?) when conditions were favorable. Volcanism was intense, working profound modifications of the earth. Our fish lived in a mythical land, called Gondwanaland, formed by the united continents of Africa and South America. They were soon to be separated, owing to a gigantic break which had just started, forming a chain of great lakes. A similar process is actually in action in Africa today; the new break starts with the Red Sea and is prolonged by the Rift Valley forming the Great Lakes of Africa (Tanganyika, Malawi, Victoria, etc.).

Such lakes are the most suitable places to allow what is called explosive evolution to occur because they offer the greatest ecological opportunities to the species. These species are able to occupy a multitude of empty spots such as rocks, sand, grass-beds, bottom, surface, and so on, or potential so-called ecological niches. Another possible cause of rapid and active speciation in Great Lakes, as contrasted with the "normal" evolution in adjacent rivers, may be the lack of important fish predators (such as *Hydrocynus* or *Lates* in certain African lakes) which, when present, restrain the diversification of the species.

The ambitious fish mentioned at the beginning of the story (which is not a fairy-tale but a hypothesis based on certain well established facts), was probably a species resembling what we now call a gonorynchiform. That small group of partly fresh-water fishes, mostly African, with *Kneria, Cromeria* and others, seems indeed to have some "pre-ostariophysine" features.

An ostariophysine fish (characoids were apparently the first ostariophysans, although certain authors believe that the catfishes are more ancient) is characterized by at least two features: a chain of ossicles, called Weber's ossicles, uniting the gas-bladder with the internal ear and permitting the hearing of high-frequency sounds; and a chemical alarm system, telling the schooling fish that one of its congeners has been attacked by a predator. These two adaptations are supposed to be useful to fishes living in schools in fresh-water biotopes (which is the case in most characoids and cyprinoids).

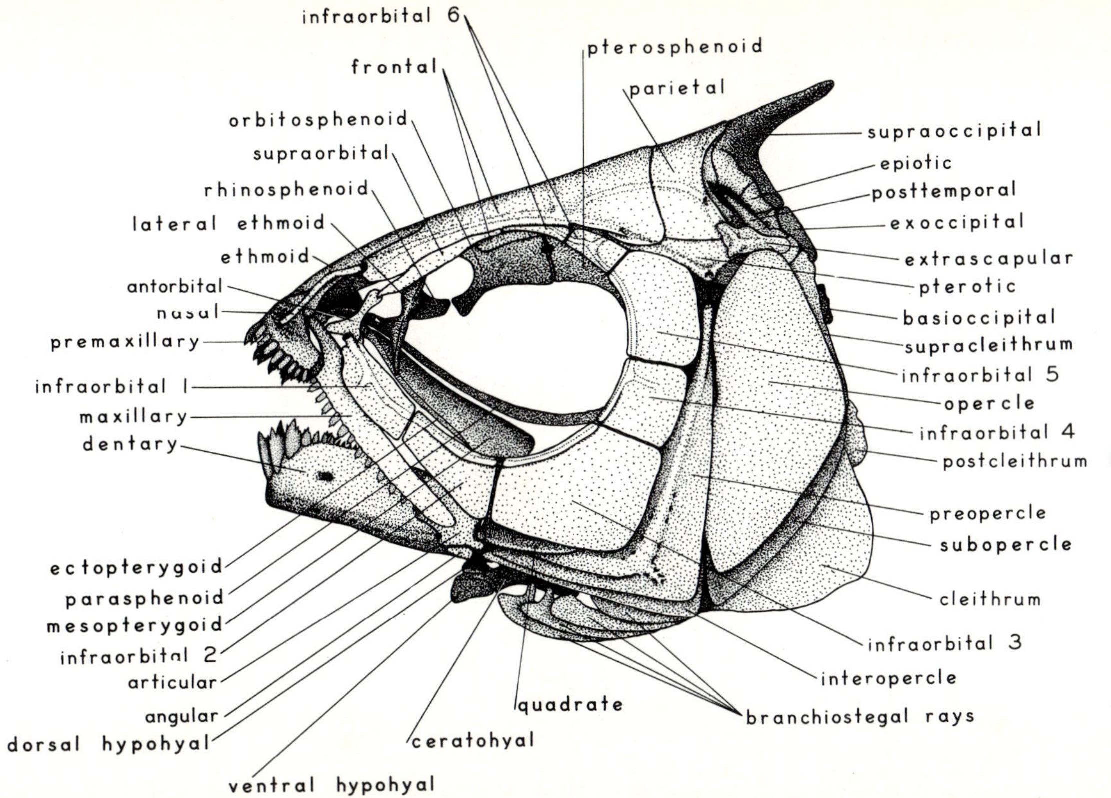

Lateral view of the skull of a generalized characoid (after Weitzman, 1962).

Recent investigations have shown that most, if not all, of the *Kneria*-like recent fishes do possess rudiments of these structures. They have specialized cervical vertebrae (the first three) as if they were in the process of differentiating ossicles (which is achieved in the characoids), and they also have a typical fright reaction if a wounded fish of their kind (or a cyprinid) is introduced into a school.

With all the above cited conditions brought together, our *Kneria*-like hypothetical procharacoid was roughly in the same position as a cichlid was, rather recently, in a great African lake, that occupied all potential (empty or not completely exploited) niches and gave rise to the famous, extremely diversified, recent fauna of the Great Lakes. It is likely that the procharacoid literally underwent explosive radiation with such a spread of adaptations that there are no other such examples in the vertebrates except for the Australian marsupials.

This evolution was probably possible thanks to the general organization, with several generalized features, which characterizes a characoid. *In this group, there is usually an adipose fin and, almost always, well developed scales. Teeth are always present, at least in the post-larva. There is no rostral bone, no preethmoid, and the prevomer is not toothed and is directly articulated with the palatine head and lateral ethmoid. Finally, the Weberian ossicles are complete, with the tripus fixed on the body of the third vertebra by means of a vibrating lamella, in contrast to all other ostariophysans.*

Once the continents were separated, the above defined characoids evolved, for their own part, on each side of the newly-formed Atlantic Ocean. It was very probably impossible for these strictly so-called primary fresh-water fishes to survive in marine water, except for short periods of time or short distances.

It is believed (but not by all zoologists) that in Africa they soon gave rise to the cyprinids, of which the primitive stock could have been *Barilius*-like fishes (?), which in turn radiated and became ardent competitors of their own parents (the descendants of the procharacoids) in the very same biotopes. These barbs emigrated to Asia and North America by the northern route, while the characoids apparently never did (*Astyanax fasciatus* reached Texas by a different route, the southern one through Central America).

In America, the characoids gave rise to the highly specialized gymnotoids after the separation of the continents. Gymnotoids never developed in Africa, where their niches are occupied by knife-fishes and mormyroids.

Neither the gymnotoids nor the catfishes (of unknown origin), specialized very early, were such competitors as were the barbs in Africa. The American (Neotropical) characoids were thus free to diversify in almost all possible directions. In addition, the drift of the continents had caused the

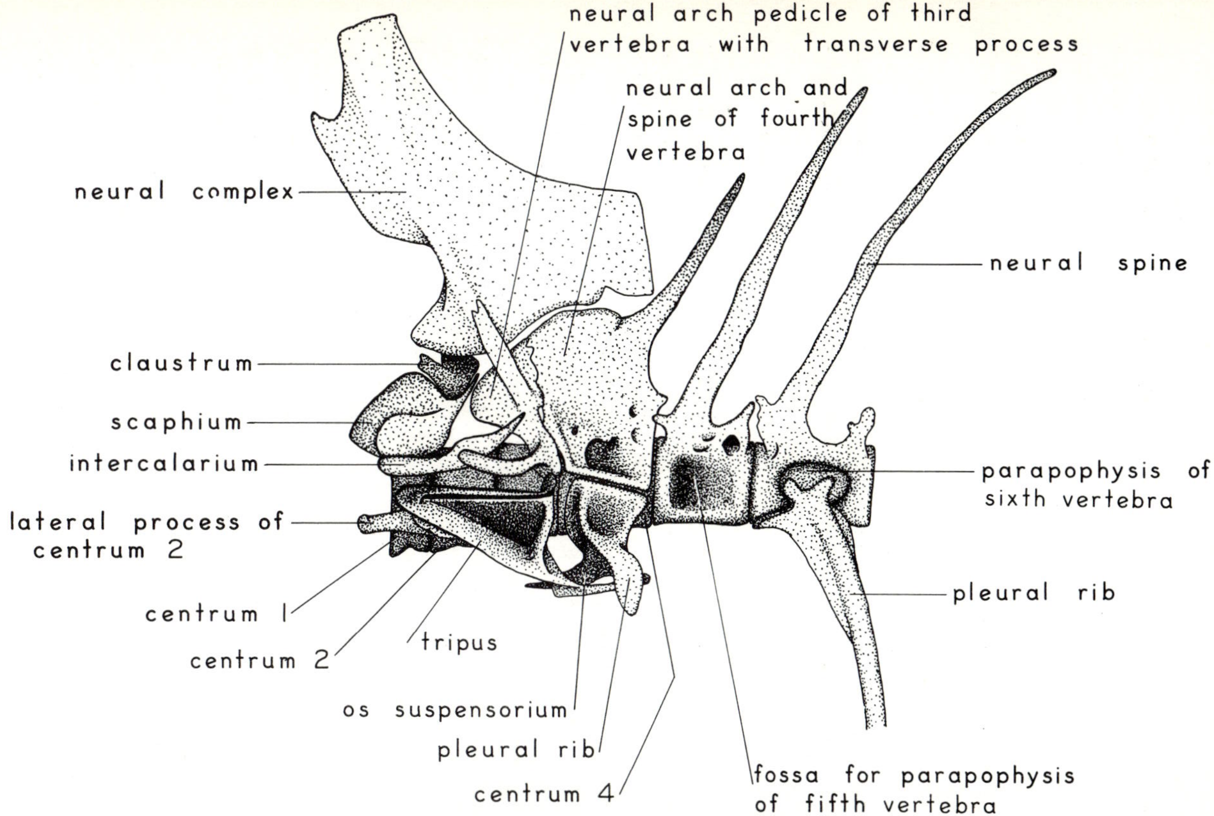

Weberian apparatus of a generalized characoid (after Weitzman, 1962).

upheaval of the Andes, as if the western coast of South America had abutted against a static block in its course to the west. This in turn caused the formation of the Amazon basin, the greatest of the world, which offered the best opportunities to a young, opportunistic group of fishes.

These two reasons explain why the Neotropical characoid fauna represents, in number of species, about 80% of the total, and the African ("Ethiopian") fauna consequently only 20%, with 11 and 3 families respectively, according to certain modern authors.

On each side of the Atlantic Ocean, similar ecological pressures determined a parallel evolution. Certain evolutionary convergences are worth mentioning:

The African pike-characoid *Hepsetus* has its counterpart in the American *Acestrorhynchus*.

The large African alestids (except *Hydrocynus*) strongly parallel the American Bryconinae, whereas the small alestids are scarcely discernable from the numerous American tetras.

Hydrocynus, a modification of an alestid toward predation, has close similarities with the Neotropical *Hoplias* or *Erythrinus*.

Most of the Citharinidae are convergent with certain Hemiodidae and Curimatidae (two rather closely related American families), whereas some others, very small and bottom-living fishes (*Nannocharax* and allies), are strikingly similar to the Neotropical Characidiidae.

In the above cited examples, only knowledge of the origin, or the study of anatomical details, can tell the non-specialist whether he is dealing with an African or an American line, which diverged from each other so many millions of years ago. *Hepsetus*, the most primitive characoid according to T. Roberts (forming the family Hepsetidae), has an accessory ectopterygoid bone and a canal into the supraopercle. These characters are not shared by any Neotropical characoid, even the Ctenoluciidae which are apparently the closest group. The citharinids have ctenoid scales (except for one genus), numerous ventral rays, pharyngeal "bags" and other features which do not exist (at least together) in American families. Finally, the alestids differ from their closest American relatives, the Characidae, in having no rhinosphenoid, a bony tube enclosing the olfactory nerve, a toothless, paddle-like maxilla, and other characters.

It has been pointed out above that the evolution of the characoids, chiefly of the Neotropical ones, resembles that of the more recent cichlids from certain African lakes. Both groups, each with a different time scale however, entered new biotopes and exploited all of their possibilities.

Indeed, most of the adaptations enumerated by Fryer and Iles (1972) concerning the African cichlids are found in the characoids. They practice commensalism, mimicry, shoaling (also called schooling), migration, burrowing in sand, and so on. They have developed three means of communication, namely sounds, chemicals, and recognition marks (mostly on humeral region and caudal peduncle, as well as on dorsal and adipose fins). They have, in certain cases, elaborated or evolved breeding habits such as the care of the young as practiced by the piranhas. Finally, they have developed an array of feeding habits, which is largely comparable to that of the cichlids, including scale eating or fin biting, as well as other adaptations which are almost unique among teleosts, such as flight (hatchet-fishes) or group-predation (piranhas).

Owing to the millions of years of evolution at the disposal of the characoids, these adaptations occurred more than once and in different phylogenetic lines. This resulted in an extremely confusing complexity of groups. It is often difficult to tell from a character shared by different forms the role of true phylogeny (common origin) and of convergency (different origin). That is how a generalized body-shape, like that of a salmon or a trout for example, is to be seen at the same time in *Brycon, Salminus* and *Astyanax,* which originated from three different evolutionary lines (although in the same family).

Evolution of a predatory existence, that is predation mostly on fishes, appeared at least eight times in completely different lines: *Hepsetus, Hydrocynus* and certain Citharinidae in Africa, *Ctenolucius* et al., *Erythrinus* et al., *Acestrorhynchus* et al. and several other Characinae, not forgetting the piranhas (Serrasalmidae) in South America.

The plant-eaters belong to at least three families, namely Citharinidae, Anostomidae and Curimatidae. But some members of other groups, for example *Deuterodon* in the Tetragonopterinae, and several others, also developed a herbivorous habit. The same is true for the limnivorous species, found in the Citharinidae as well as in the Curimatidae, the "Aufwuchs"-eaters, which nibble food from rocks or fallen trees (two or three families), the plankton-feeders (at least two families), and the specialized forms referred to above. Scale-eaters appear in two families, Characidae and Serrasalmidae, and the fin-biters also in two families, Citharinidae and (probably) Serrasalmidae.

Several species are adapted to life in river torrents. They belong to three families, Citharinidae (with *Nannocharax* et al.), Characidiidae (with *Characidium* et al.) and Hemiodidae (with *Parodon* et al.).

A ventral keel appeared in several subfamilies of the Characidae (*Triportheus, Gilbertolus, Gnathocharax, Pseudocorynopoma* and so on), and also in certain Curimatidae, as well as in the Serrasalmidae, culminating, in the Gasteropelecidae, in a complex of adaptations permitting flight a few inches above the water surface.

Finally, sexual dimorphism is evident in the Characidae as well as the

Alestidae, Serrasalmidae, Crenuchidae and even in one curimatid genus, whereas care of the young appeared in at least three lines, certain Erythrinidae and Serrasalmidae, and certain Pyrrhulininae in the family Lebiasinidae.

The above enumeration of their intricate adaptations helps to explain why the systematics of the characoids has long been, and still is, a difficult task. They were recognized as a homogeneous group only in 1844 by two German ichthyologists, Müller and Troschel. The Viennese Kner and his student Steindachner (between 1858 and 1915) followed by accurately describing a number of species without paying much attention to their classification. It remained for one of the best ichthyologists of all times, Carl H. Eigenmann, to undertake, in the U.S., a natural classification of the South American groups. His principal monographs were issued between 1910 and 1927. He was followed by his student, George S. Myers, the still very active father of the "characologists," who in turn handed the torch down to his own students, J. Böhlke, S. Weitzman and several others, forming a chain of interest which is presently still growing.

Recently a series of anatomical studies by S. Weitzman, T. Roberts, and others helped specify the systematic position of several critical groups, which led to the recognition of many families within the suborder (a division which could never have been accepted in Eigenmann's time) and revealed the importance, as well as the complexity, of the characoids, presently considered as a key-group among the teleosts.

The first authoritative division of the characoids into several families (16) was done by S. Weitzman, in Greenwood et al. (1966). The arrangement of the families in the present book takes into account that classification, as modified by Géry (1972), with a few alterations from the 1972 order.

As compared with Weitzman's classification, the present one is as follows:

S. WEITZMAN in GREENWOOD et al. (16 families)	PRESENT ARRANGEMENT (14 families)
(1) CHARACIDAE	(2) ALESTIDAE; (6) CRENUCHIDAE; (7) CHARACIDIIDAE; (13) SERRASALMIDAE; (14) CHARACIDAE
(2) ERYTHRINIDAE	(4) ERYTHRINIDAE
(3) CTENOLUCIIDAE	(5) CTENOLUCIIDAE
(4) HEPSETIDAE	(1) HEPSETIDAE
(5) CYNODONTIDAE	RHAPHIODONTINAE, a subfamily of the CHARACIDAE
(6) LEBIASINIDAE	(8) LEBIASINIDAE
(7) PARODONTIDAE	PARODONTINAE, a subfamily of the HEMIODIDAE
(8) GASTEROPELECIDAE	(12) GASTEROPELECIDAE
(9) PROCHILODONTIDAE	PROCHILODINAE, a subfamily of the CURIMATIDAE

(10) CURIMATIDAE	(11) CURIMATIDAE (including PROCHILODINAE, CHILODINAE and ANODINAE)
(11) ANOSTOMIDAE	(9) ANOSTOMIDAE
(12) HEMIODONTIDAE	(10) HEMIODIDAE (including PARODONTINAE)
(13) CHILODONTIDAE	CHILODINAE, a subfamily of the CURIMATIDAE
(14) DISTICHODONTIDAE	DISTICHODINAE, a subfamily of the CITHARINIDAE
(15) CITHARINIDAE	(3) CITHARINIDAE (including DISTICHODINAE and ICHTHYBORINAE)
(16) ICHTHYBORIDAE	ICHTHYBORINAE, a subfamily of the CITHARINIDAE.

The evolutionary importance of the group will be conspicuous, it is hoped, in the body of the book, which will deal successively with the African and American characoids, ending, in a third part, with the most successful Neotropical group, the Characidae.

Part I
The African Families

KEY TO THE AFRICAN FAMILIES★

a. Scales cycloid; teeth usually well developed, not movable; joint between two sides of mandible with some interlocking device, movable; no pharyngeal "bags".
  b. Predaceous fishes armed with strong canines; maxilla entirely toothed; an accessory ectopterygoid bone; a large supraopercle bearing a lateral line canal ... HEPSETIDAE
  bb. Omnivorous fishes usually with incisiform and molariform teeth, rarely piscivorous and with canines (as in the subfamily Hydrocyninae, with a single genus); maxilla never toothed; no accessory ectopterygoid bone; no supraopercle ... ALESTIDAE (2 subfamilies)

aa. Scales ctenoid (except in *Citharinus*, which is a bream-like fish very different from *Hepsetus* or from an alestid); teeth usually small, often movable; joint between two sides of mandible simple, sometimes ankylosed; pharyngeal "bags" present ... CITHARINIDAE (3 subfamilies)

Family Hepsetidae

African Pike-like Characoids

Hepsetus odoe, formerly called *Sarcodaces odoe*, is the single representative of the family. Its external resemblance with an *Acestrorhynchus*, by convergent evolution, is striking. However, an examination of the finer structures of its skull reveals a number of differences. *Hepsetus*, according to Roberts, is more closely related to another South American group, the Ctenoluciidae, but with many more primitive features (accessory ectopterygoid bone, flattened cranial roof, large canal-bearing supraopercle, and so on) which make it probably the most primitive characoid. The African pike lives in almost all rivers of the tropical area, with the exception of the Nile basin. It is a fish-predator, well armed with strong, caniniform teeth, similar to those of *Hoplias, Acestrorhynchus* and the like. The maxillary bone is

★ This key, like all other keys in the book, is for identification purposes only. It is not intended to reflect relationships.

entirely toothed as in most predators. According to the author's own experience in Africa, it is a good game-fish, behaving like a European pike, putting up a good fight. It reaches 30 cm in length and is edible. Its color is brownish with indistinct bars on the body and small spots on the fins; its adipose fin is bright orange and black, apparently forming a recognition signal.

Svensson, as well as Johnels, has reported that *Hepsetus* spawns somewhat in the manner of the anabantids. A nest of floating foam is made by the spawning pair and the eggs (which are afterwards guarded by both parents or one parent only) are deposited in it.

Family Alestidae

African Tetras

The Alestidae are the African counterparts of the Characidae, and there is little doubt that the two groups have the same origin. One example will show how similar they are. A new genus and species *(Pellegrinina heterolepis)* was once described as a new African form close to *Alestes* and *Brycinus*, because the specimen studied was thought to come from "some part of West Africa." A more critical evaluation revealed afterward that the label on the bottle was incorrect and that the fish was a well-known South American *Chalceus*.

Despite the general resemblance between the two groups (which led several ichthyologists to recognize only one family from both sides of the Atlantic Ocean), the Alestidae have the following characteristics in contrast to those of the Characidae (at least those characids which most resemble the alestids, namely the Bryconinae): no rhinosphenoid (a small bone in front of the orbitosphenoid, present in *Brycon*); olfactory nerve enclosed in a bony tube formed by the orbitosphenoid and the lateral ethmoid; lateral line, when present, very low, the number of scales between lateral line and ventral fin never exceeding 3 (usually 2). In the Alestinae, the maxillary bone (more or less covered by the first infraorbitals) is toothless, usually paddle-like with a rounded contour and a thin pedicel (like a music note). The premaxillary bone has a characteristic pedicel-like posterior process extending to the maxilla, which is rather firmly united to it; the inner premaxillary teeth are invariably four in number (apparently a primitive number, also to be found in the American Hemibryconini), and molariform (except in the small species), usually with the crushing face excavated, the posterior border rounded, multi-cuspidate, and the anterior one with two prominent cusps or points. On the mandible there is usually a pair of small conical teeth behind the main series.

In the Hydrocyninae, which are ichthyophagous predators derived from some *Alestes*, the powerful jaws are armed with a single series of very acute caniniform teeth; it is believed that these teeth correspond to the outer series of the alestins. Another peculiarity of this group is the type of middle joint between the two parts of the mandible. It is a "bolted hinge," whereas that of the alestins is said to be "a fluted tenon." Otherwise, as concerns the general organization, a *Hydrocynus* does not differ essentially from an *Alestes*. Accordingly, the family Alestidae will be split into two groups at the subfamilial level:

a. Teeth in 2 rows on the upper jaw (three rows in one genus), the inner row usually composed of molariform teeth; omnivorous species ... ALESTINAE

aa. Teeth in 1 row on both jaws, caniniform; ichthyophagous species ... HYDROCYNINAE

SUBFAMILY ALESTINAE

African Tetras

Quite artificially, one may divide the alestins into two groups, Alestini and Petersini, which could be called tribes. The Petersini look like the young of the Alestini and constitute what are properly called the African tetras. They differ from the Alestini not only by their small size (frequently in correlation with the reduction of the lateral line), but also in other "regressive" characters. The inner teeth of the upper jaw are no longer excavated and molariform, but have retained a single, pluricuspidate cutting edge, and the inner mandibulary conical teeth are often lacking. Some alestins, the so-called *longipinnis*-group, form a sort of transition between the two groups, the species being small and the molariform structure of their teeth scarcely evident.

Tribe Alestini

The present author recognizes 3 genera diagnosed by the following key:

a. An adipose lid covering at least part of eye; always a fontanel; scales with radiating "radii" which are never anastomosed.

 b. Premaxillary teeth triserial; 3 scales between lateral line and ventral fin; dorsal fin well in front of midbody, filamentous in adult male; body deep (depth 2.8-3.5 in standard length) ... *Bryconaethiops*

 bb. Premaxillary teeth biserial; $1\frac{1}{2}$-2 scales between lateral line and ventral fin; dorsal fin at midbody or even behind, not filamentous; body elongate (depth 3.25-5 in standard length) ... *Alestes*

aa. No adipose lid; usually no fontanel, except in a few species *(B. leuciscus, B. longipinnis, B. chaperi, B. opisthotaenia)*; "radii" of the scales usually anastomosed ... *Brycinus*

GENUS *BRYCONAETHIOPS*

This genus includes a few species which change considerably during growth, certain of the adults being scarcely identifiable with the young. By the time they reach 200 mm in length they are usually very deep, very dark (sometimes almost black), and the filaments of the dorsal fin of the males reach the caudal fin. The group is at once recognized by the 3 series of teeth on the upper jaw (it is somewhat like modified *Alestes*, just as *Creagrutus*, in the American tetras, looks like a modified *Hemibrycon*), but is urgently in need of a revision at the species level.

The two or three species, *B. microstoma*, *B. boulengeri* and perhaps *B. macrops* (which, if valid, is much rarer), are mostly from the Congo basin (including Ubanghi) although some species are found in Cameroon and Gaboon.

Bryconaethiops sp. Photo by Dr. J. Gery.

GENUS *ALESTES*

A few species, namely *Alestes dentex*, *A. baremose*, *A. macrophthalmus*, *A. liebrechtsii* and *A. stuhlmanni*, have rather recently been separated from the bulk of the *Alestes*-like species on the basis of anatomical details such as adipose lid, fontanels, etc. They are usually large (up to 45 cm), elongate species inhabiting large rivers. Some are of economic importance. *Alestes dentex* and *A. baremose*, from the Nile, Niger, Tchad, Senegal, etc., are sibling species, being separable only by the number of gill-rakers.

Alestes macrophthalmus is a southern form from the Congo and Gaboon, *A. liebrechtsii* is restricted to the Congo basin, and the small *A. stuhlmanni* seems to be an endemic form from East Africa.

Except for some red color on the caudal fin in certain forms, species of *Alestes* are usually uniform silvery, without characteristic markings.

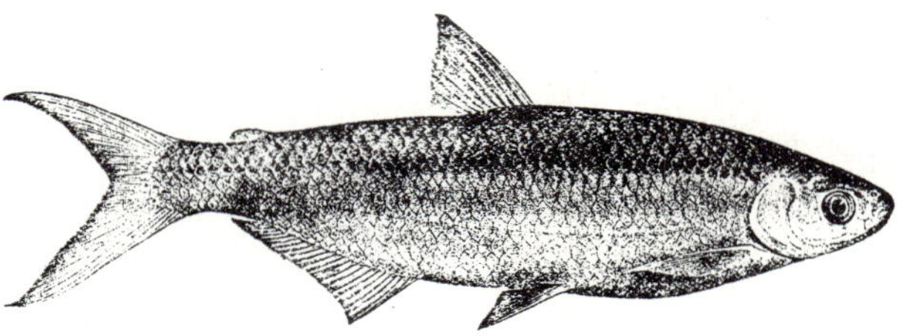

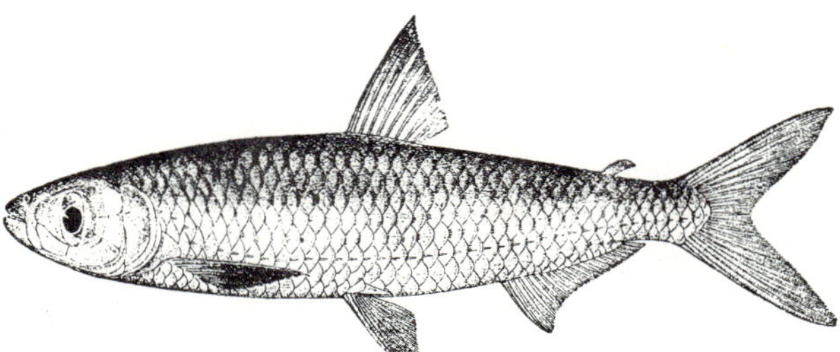

Alestes baremose and *A. liebrechtsii* (photos by Dr. J. Gery); *A. dentex* (after Boulenger, 1909); *A. macrophthalmus*, holotype (after Boulenger, 1909).

GENUS *BRYCINUS*

This genus is an aggregate of species in which one may recognize five groups. Some of these could, perhaps, be ranked as subgenera, although overlapping of characters is frequent; the others are quite artificial assemblages. The *Brycinus macrolepidotus* - group (1), which contains the type species of the genus *Brycinus, B. macrolepidotus*, and a few other species, is characterized by the large size (up to 50cm), straight dorsal profile, dorsal fin far behind mid-body, and overhanging snout. The other four groups, which have dissimilar characters, may be recognized as follows:

a. Interorbital usually less than 3 times in length of head; caudal peduncle high (its depth 1.0-1.5 in its length).
  b. 16-28 gill-rakers on lower arch (fontanel always absent; scales usually with anastomosed radii) . . . (2) *Brycinus nurse-imberi*-group
    c. Dorsal fin slightly in front of mid-body . . . *B. nurse*
    cc. Dorsal fin at mid-body or slightly behind . . . *B. imberi*-group (6-7 spp.)
  bb. 11-18 gill-rakers on lower arch (fontanel present or absent; scales with or without anastomosed radii).
    d. Body depth 3.5-3.66 in standard length; 4 scales between dorsal fin and lateral line; scales without anastomosed radii; no fontanel; dorsal fin usually behind mid-body . . . (3) *B. rutilus*-group (2 spp. ?)
    dd. Body depth 2.3-3.5 in standard length; 5-6 scales between dorsal fin and lateral line; scales with anastomosed radii; sometimes a fontanel; dorsal fin usually at mid-body or in front of mid-body . . . (4) *B. longipinnis*-group (8-9 spp.)
aa. Interorbital 3.0-3.5 times in length of head; depth of caudal peduncle 1.5-2.0 in its length (fontanel always absent; scales without anastomosed radii) . . . (5) *B. humilis*-group (4 spp. ?)

(1) The *Brycinus macrolepidotus*-group

The large species belonging to the nominal genus *Brycinus* have a distinctive shape, the straight snout (which continues the long, also straight, predorsal line) overhanging the mandible. The smallest of the six included species, *B. brevis* (225 mm) from Nigeria and the Gold Coast, is also the deepest (body depth 2.75 to 3.0 in standard length), with few lateral line scales (21-23), and is a uniform color. *Brycinus carmesinus* is very close to it.

The four other species are more elongate (body depth 3.5 to 4.0 in the standard length). Two of them, *B. grandisquamis* from the Congo basin and *B. batesii* from South Cameroon, attain 26 cm in length and can be easily recognized thanks to a conspicuous spot on the end of the caudal peduncle. They have, respectively, 23-27 and 28-29 lateral line scales, and 13-14 versus 15-16 anal rays.

Brycinus sp., juvenile. Photo by G. Timmerman.

Brycinus macrolepidotus, the type species of the genus, inhabits a large part of Africa from the Nile to the Congo. It usually has a broad dark longitudinal band. *Brycinus rhodopleura* is its representative in Lake Tanganyika, differing only in the number of lateral line scales (22-26 scales in *B. macrolepidotus* versus 28-29 in *B. rhodopleura*). They reach, respectively, 42 and 30 cm in length.

(2) The *Brycinus nurse-imberi*-group

Brycinus nurse (type species of the genus *Brachyalestes*, not used much nowadays) and the *Brycinus imberi*-group (with *B. imberi, B. jacksoni, B. affinis, B. taeniurus, B. kingsleyae* and *B. bimaculatus*), are medium-sized (100-230 mm) alestins with the dorsal fin at mid-body or thereabouts and 16-28 gill-rakers (usually less than 18 in the other groups).

Brycinus nurse is the only species in this group to have the dorsal fin slightly in advance of mid-body, and it is also the largest (230 mm); it has the same geographical distribution as *B. macrolepidotus*, plus Lakes Victoria and Rudolf. The caudal fin and part of the ventral and anal fins are usually red; humeral and caudal spots are present.

Species of the *B. imberi*-group strictly speaking have the dorsal fin inserted at mid-body or slightly behind. *B. kingsleyae* and *B. taeniurus* from South Cameroon, Gaboon and Chiloango are immediately recognizable by the dark lateral band beginning behind the dorsal fin level and ending on the middle caudal rays. *Brycinus kingsleyae* has a humeral spot (lacking in *B. taeniurus*) and a shorter anal fin (12-13 branched rays versus 15-17).

B. imberi (of which *curtus, bequaerti* and *jacksoni* are possibly synonyms) from Cameroon to Kasai, *B. bimaculatus* from the upper Congo, and *B. affinis,* an endemic species from East Africa, are mostly distinguished by their color pattern. *B. imberi* usually has a conspicuous humeral spot and a broad caudal spot ending on the caudal rays; *B. bimaculatus* has the same caudal spot but it is restricted to the caudal peduncle and also has a lateral spot much farther back (at the level of the end of the dorsal fin); *B. affinis* has the same humeral spot as *B. imberi,* but has no caudal spot at all.

(3) The *Brycinus rutilus*-group

This small group differs from the preceding one mostly in having fewer gill-rakers and from the next one by the structure of the scales and the more

Brycinus bimaculatus, holotype (after Boulenger, 1899) and *B. nurse* (photo by Dr. J. Gery).

elongate body. *B. rutilus* is a large form (up to 32 cm) with the fins sometimes bright vermilion (hence its name). It inhabits the same regions in West Africa as *B. macrolepidotus*, but is nowhere captured with it (a vicarious species). *B. stolatus* from the Congo is a small (?), poorly known species without color marks.

(4) The *Brycinus longipinnis*-group

This quite unnatural group forms a sort of transition with the small alestins called tribe Petersini. Most of the species are small. They sometimes have a fontanel and, at least when young, their inner premaxillary teeth are more compressed than in typical *Brycinus*. However, a close examination reveals a pair of small "shoulders" in front of the largest teeth, which is almost the only character which distinguishes them from the next tribe.

B. leuciscus, from the Senegal and Niger basins (where it is called *tineni* and is of economic importance), has long been confounded with the young of *B. nurse*, but it has yellow fins instead of red. Another species, more recently described, is also very close. This is *B. dageti* from the Tchad basin, which differs from *B. nurse* in being mature at half the size of the latter species, as well as by the number of gill-rakers. It seems to be the sibling of *B. nurse* and might well be classified together with *B. leuciscus*, within the *B. nurse*-group.

The *B. longipinnis*-group strictly speaking is in need of revision (like other *Brycinus* groups). The following key will aid in distinguishing the species:

a. Not more than 30 lateral line scales (18-21 branched anal rays; dorsal fin filamentous in male; caudal band present) . . . *B. longipinnis, B. chaperi* and an undescribed form from the Ivory Coast (West Africa, plus Equatorial Africa for *B. chaperi* only ?)

aa. 32-37 lateral line scales.

b. 16-21 branched anal rays.

c. An elongate caudal spot; 16-17 branched anal rays; 33 lateral line scales . . . *B. sadleri* (endemic species from Lake Victoria)

cc. Caudal spot broad or absent; 17-21 branched anal rays; 33-37 lateral line scales.

d. A broad caudal spot; dorsal rays filamentous in male, marked with a black bar; 33-35 lateral line scales . . . *B. intermedius* (South Cameroon)

dd. Body and fins plain; 37 lateral line scales . . . *B. tessmanni* (South Cameroon)

bb. 24-26 branched anal rays (body and fins plain; 32 lateral line scales) . . . *B. tholloni* (Congo basin)

B. longipinnis and *B. chaperi*, probably sympatric in West Africa, are the best known alestins among aquarists. The two species are difficult to tell apart, especially when one has to deal with young individuals. As a rule, *B. chaperi* has slightly more lateral line scales than *B. longipinnis* (28-30 versus 26-27), as well as a longer snout, shorter maxillary bone, broader interorbital and deeper caudal peduncle. Its dorsal fin is also inserted farther back than that of *B. longipinnis*, being at mid-body or slightly behind (versus slightly in front of mid-body).

(5) The *Brycinus humilis*-group

The four species *B. humilis, B. abeli, B. lateralis* and *B. opisthotaenia* are also very close to certain species of *Micralestes*. They are small (maximum size 150 mm), with few scales (24-34) and few branched anal rays (12-17).

Brycinus opisthotaenia, from South Cameroon and Gaboon, is the largest species and has retained many *Brycinus macrolepidotus* characters. It is remarkable in having contact tubercles (similar to those on the snout and

Brycinus humilis, holotype (after Boulenger, 1909) and *B. rutilus* (photo by Dr. J. Gery).

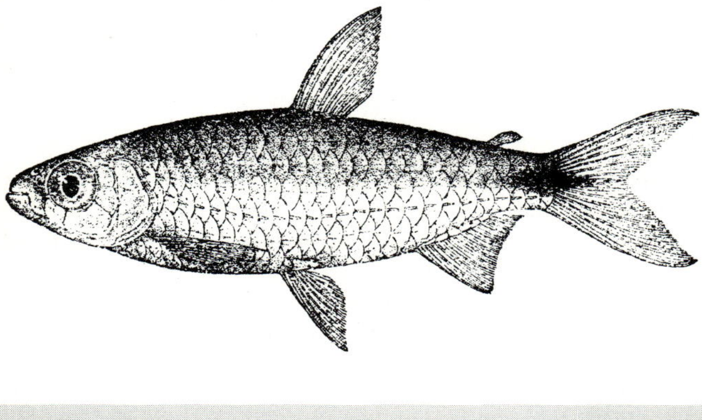

head of mature male barbs and species of *Labeo*) over the whole body in both sexes. Nothing is known about the significance of these tubercles (called nuptial tubercles or pearl organs in the barbs). Like *B. opisthotaenia, B. lateralis* has a black lateral band (after preservation) but no humeral spot. It inhabits Katanga and Angola up to the Zambezi River, and differs by having more scales (30-33 versus 24-27 lateral line scales in *B. opisthotaenia*). *B. humilis* from Angola and *B. abeli* from Ubanghi have a horizontally elongate caudal spot; they differ in the same way as the above, having respectively 30 and 24-26 lateral line scales.

Tribe Petersini

Small African Tetras

This group of about 60 species has recently been revised by Poll. He admitted 16 genera as valid, including some proposed by Hoedeman in 1956 which are questionable. It looks as if the group is oversplit, which seems evident when one compares the Petersini with their American counterpart the Hemibryconini. The latter group has 12 or 13 genera with about 80 species. However, the evolutionary conditions were not exactly the same, owing to the absence in South America, for example, of the barbs, which compete severely with the African tetras. Here only a part of the 16 genera (actually 18 including *Ladigesia* and *Virilia*, which were not described at the time of Poll's revision) will be accepted as diagnosed by the following key:

a. Teeth multicuspidate; two premaxillary series.
b. Predorsal line scaled; no outward-directed teeth.
c. A pair of conical or (rarely) cuspidate teeth behind the main mandibulary series.
d. Teeth usually with more than 3 cusps.
e. More than 4 outer premaxillary teeth total (lateral line usually complete) . . . (1) *Micralestes*
ee. 4 outer premaxillary teeth, 2 on each side (lateral line complete or incomplete) . . . (2) *Phenacogrammus* (This genus may be split into 3 or 4 groups)
dd. Teeth tricuspidate (lateral line incomplete; 4 outer premaxillary teeth) . . . (3) *Tricuspidalestes*
cc. Conical teeth behind the main mandibular series usually absent.
f. Scales of flank and abdomen of equal size; 4 or 6 outer premaxillary teeth (2-3 on each side); 8-11 mandibular teeth.
g. Posterior fontanel present; mandible not prominent (lateral line complete or incomplete) . . . (4) *Hemigrammopetersius* (This genus may be split into 4 or 5 groups of species)

gg. Posterior fontanel absent, at least in adults; mandible very prominent (lateral line complete) . . . (5) *Petersius*

ff. Scales of flank much larger than those of abdomen; 8 outer premaxillary teeth; 16-20 mandibular teeth . . . (6) *Arnoldichthys*

bb. Predorsal line naked; teeth partly outward-directed . . . (7) *Ladigesia*

aa. Teeth conical, in one series (lateral line incomplete).

h. Scales normal; mouth clupeoid . . . (8) *Clupeocharax*

hh. Body naked, without scales; mouth normal . . . (9) *Lepidarchus*

The Petersini are thus composed of 6 monotypic genera which are very well characterized (*Tricuspidalestes, Petersius, Arnoldichthys, Ladigesia, Clupeocharax* and *Lepidarchus*) and of three polytypic genera, two of which are quite complicated and perhaps not monophyletic.

(1) GENUS *MICRALESTES*

Micralestes is the largest genus of the Petersini as well as the most generalized. All the species are of small size (maximum length 100 or 110 mm for the type species, *Micralestes humilis*), are moderately elongate, usually have a longitudinal dark band, and are very similar to each other. According to Poll, they can be distinguished as follows:

Micralestes acutidens. Photo by Dr. Herbert R. Axelrod.

a. Pair of small teeth behind main mandibulary series conical, without cusps; no more than 7-9 cusps in the broadest teeth.

b. Usually less than 29 lateral line scales.

c. No black longitudinal band; body depth 2.6-2.9 in standard length . . . *M. holargyreus* (Congo basin)

cc. A black longitudinal band, more or less complete; body depth 2.9-4 in standard length.

d. Eye 2.6-3.3 in length of head (adult); more than 50 mm in standard length.

e. 22-24 lateral line scales.

f. Longitudinal band beginning at dorsal level; head 3.7-3.9 in standard length; 15-16 branched anal rays . . . *M. stormsi* (Congo basin; very difficult to separate from *M. humilis*)

ff. Longitudinal band complete; head 3.95-4.1 in standard length; 16-18 branched anal rays . . . *M. vittatus* (Lake Tanganyika)

ee. 25-29 lateral line scales (longitudinal band complete).

g. Lateral line complete.

h. Length of head 3.4-3.8 in standard length.

i. Scales of lateral line with black spot at base; 25-27 lateral line scales . . . *M. humilis* (Tchad, Congo and Zambezi Rivers)

ii. Scales of lateral line not marked; 24-25 lateral line scales . . . *M. voltae* (upper Volta basin; difficult to tell from *M. elongatus*)

hh. Length of head 3.75-4.1 in standard length.

j. 18-19 branched anal rays; tip of dorsal fin black (25-26 lateral line scales) . . . *M. occidentalis* (West Africa)

jj. 16-17 branched anal rays; tip of dorsal fin hyaline . . . *M. sardina* (upper Congo basin)

gg. Lateral line usually not quite complete, lacking a few pores on caudal peduncle; 16-19 branched anal rays; length of head 3.4-4 in standard length (species of West Africa).

k. Tip of dorsal fin black; body depth 3.45-3.95 in standard length . . . *M. elongatus* (upper Volta)

kk. Tip of dorsal fin hyaline; body depth 2.8-3.3 in standard length . . . *M. comoensis* (Ivory Coast)

dd. Eye 2.25-2.5 in length of head (adult); less than 50 mm in standard length . . . *M. fodori* (Congo)

bb. 29-33 lateral line scales (body depth 3.5-4.2 in standard length; longitudinal band complete).

l. 16-19 branched anal rays; scales of lateral line not marked.

m. Caudal peduncle 1.15-1.4 times longer than deep . . . *M. lualabae* (Congo basin)

mm. Caudal peduncle 1.5-1.75 times longer than deep . . . *M. argyrotaenia* (Angola)

ll. 19-21 branched anal rays; scales of lateral line with black spot at base (caudal peduncle as deep as long) . . . *M. congicus* (Congo basin)

aa. Pair of small teeth behind main mandibulary series with 2-4 cusps; up to 12 cusps in the broadest teeth (longitudinal band complete; tip of dorsal fin black; 14-16 branched anal rays and 26-29 lateral line scales) . . . *M. acutidens* (from Nigeria to Angola and Zambezi basin; West Africa?)

(2) GENUS *PHENACOGRAMMUS*

This composite genus includes the *Micralestes*-like species with an incomplete lateral line (however *Micralestes elongatus* and *comoensis* sometimes have an incomplete lateral line), plus certain forms which probably come from different evolutionary groups but are not separable at the generic level, their characters showing some overlap. These were considered as genera by Poll. It seems convenient to retain a similar division, but at a lower level, as follows:

a. Lateral line complete.

b. 4½ or 5½ transverse scales between predorsal line and lateral line . . . *Ph. altus*-group (or subgenus *Brachypetersius*)

bb. 6½-7½ transverse scales between predorsal line and lateral line . . . *Ph. ansorgii*-group (or subgenus *Nannopetersius*)

aa. Lateral line incomplete (or not quite complete in certain individuals) (4½-8½ transverse scales between predorsal line and lateral line).

c. 15-21 branched anal rays; unpaired fins often filamentous in adult male; body of moderate depth (2.6-4.3 in standard length) . . . *Ph. interruptus*-group (*Phenacogrammus* typical)

cc. 22-25 branched anal rays; unpaired fins unmodified in adult male; body usually deep (body depth 1.9-2.9 in standard length) . . . *Ph. caudomaculatus*-group (or subgenus *Bathyaethiops*, type species *greeni*)

The *Phenacogrammus altus*-group

The small species of this group (largest 100 mm) are rather well characterized by their deep body and relatively small number of transverse scales (four or five between dorsal fin and lateral line), which is unusual (the number of scale rows theoretically increases with the depth of the body), as well as by a peculiar spot on the lower part of the caudal peduncle, present in four species out of six (a fifth species has a band going onto the peduncle). These species can be identified as follows:

Phenacogrammus altus, juvenile. Photo by Dr. Herbert R. Axelrod.

a. 5 scales between dorsal fin and lateral line.
   b. A peduncular spot; 16-18 branched anal rays (28-30 lateral line scales) . . . *Ph. notospilus* (Ogooue, Gaboon)
   bb. No peduncular spot but a broad, diffuse longitudinal band present . . . *Ph. huloti* (Congo basin)

aa. 4 scales between dorsal fin and lateral line.
   c. A peduncular spot; 20-27 branched anal rays.
      d. 20-26 lateral line scales in a complete series; body very deep, 2.1-2.5 in standard length (adult).
         e. Peduncular spot round . . . *Ph. altus*, synonym *nummifer* (Congo basin)
         ee. Peduncular spot horizontally elongate . . . *Ph. pseudonummifer* (upper Congo basin)
      dd. 25-30 lateral line scales, the lateral line usually lacking one or two pores; body not very deep, 2.4-2.8 in standard length (adult) (peduncular spot horizontally elongate) . . . *Ph. cadwaladeri* (upper Congo)
   cc. No peduncular spot; 18-19 branched anal rays (a prominent humeral spot; species somewhat intermediary between this and the following group) . . . *Ph. gabonensis* (Gaboon)

The *Phenacogrammus ansorgii*-group

Ph. ansorgii and *Ph. lamberti* are small (75 mm) species inhabiting the coastal part of Gaboon, Congo and Angola. *Ph. ansorgii*, best known under the name *Petersius ubalo* (which was named for the female of *ansorgii*), is an elegant fish with strong sexual dimorphism, the males having a con-

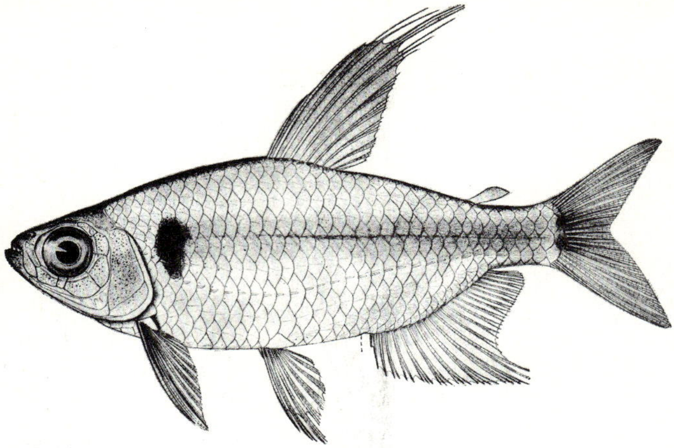

Type of *Phenacogrammus ansorgii* (after Poll, 1967).

spicuous humeral spot and a rather pale, diffuse lateral band up to the tip of the middle caudal rays, as well as filamentous fins, whereas the female has only the longitudinal color pattern. *Ph. lamberti* has the same filamentous fins in the males (chiefly the dorsal fin) and a much more conspicuous lateral band (from snout to tip of middle caudal rays), as well as a characteristic black mark (spot or band) on each caudal lobe. Apart from the coloration, the two species differ in the number of lower gill-rakers, 10-12 in *lamberti* against 14-17 in *ansorgii.*

The *Phenacogrammus interruptus*-group

This group is mostly known to aquarists by the type species of the genus, the Congo tetra, *Phenacogrammus interruptus,* which has a particularly brilliant coloration. Other species, such as *Ph. major, Ph. urotaenia* and *Ph. aurantiacus,* would be equally suitable for the aquarium. One species from the Congo, *Ph. polli,* is very different from all other species, having up to 35 or 37 longtudinal scales and 6 or 7 scales from dorsal fin to lateral line. It is very small (32 mm) with a more or less diffuse lateral band.

A second species, *Ph. pabrensis* from the upper Volta, is also quite different in having a peculiar modification of the anterior lobe of the anal fin in the adult male; the first rays are thickened and curved, the concavity toward the fore-part. This sexual dimorphism is also to be seen in *Micralestes elongatus, M. brevianalis* (a synonym of the West African populations of *Micralestes acutidens*?), and in *Hemigrammopetersius intermedius.* In other Petersini, mature males have the anterior lobe of the anal fin thickened, but the rays have their concavity toward the back. This is why a new genus, *Virilia,* had been proposed for *P. pabrensis.* This genus is not retained here.

Hepsetus odoe. Photo by Pierre Brichard.

Hepsetus odoe. Larger individual with spotted fins. Photo by Hilmar Hansen, Aquarium Berlin.

Phenacogrammus interruptus. Photo by G. Timmerman.

Another group includes a dubious species (*Ph. stigmatura* from Cameroon) and 5 very close forms. *Ph. stigmatura* has few longitudinal scales (about 20) and differs from *Ph. interruptus* in the presence of an unusual small spot at the base of the lower caudal lobe.

The last 5 species, which measure about 80 mm in standard length, have in common their sexual dimorphism, males of most of them having filamentous dorsal and caudal fins, a deeper body and a more brilliant coloration than the females, and many overlapping meristic characters. They fall nevertheless into two groups, depending upon color pattern: *Ph. interruptus* and *aurantiacus* on the one hand, have a diffuse, often not very conspicuous, broad lateral band; *Ph. major, Ph. urotaenia* and *Ph. deheyni* on the other hand, have a black band (after preservation) from the level of the dorsal fin, or farther back, to the end of the caudal rays, this band extending onto the lower part of the caudal peduncle only.

Ph. interruptus, from the Congo basin (called the Congo tetra by aquarists) differs from *Ph. aurantiacus,* also from the Congo basin (and Gaboon), mostly in the color of their fins, which are mostly pink to purple in *Ph. interruptus* and orange (ventral and anal fins) to carmine red (base of pectoral and dorsal fins and whole adipose and caudal fins) in *Ph. aurantiacus.* Both have a body shining with all the colors of a rainbow. Museum specimens can be told apart by the number of scales (20-23 in a longitudinal series, versus 23-26 respectively) and chiefly by the inner mandibular teeth which are very small or even lacking in 20% of the specimens of *Ph.*

aurantiacus. The latter species is evidently at the limits of the genus *Hemigrammopetersius* (where those teeth are always lacking).

The other three forms, *Ph. major, Ph. urotaenia* and *Ph. deheyni,* could very well be lumped in the future into a single polytypic species. They usually have a conspicuous humeral spot and the characteristic peduncle pattern described above. The coloration of only *Ph. urotaenia* is known with precision. In males the flanks are purple and the belly, the eye and all the fins are light carmine. The females are less colorful, more elongate, and they have two dark lines above their anal fin.

The *Phenacogrammus caudomaculatus*-group

The three Congolese species of the *Ph. caudomaculatus*-group, or subgenus *Bathyaethiops,* are deep bodied fishes, as deep as *Phenacogrammus altus,* which are rather well characterized by a large ovate black spot which extends (or almost extends) onto the entire depth of the pre-peduncular region, i.e. just above the last anal rays. They usually have a rather small humeral spot and a series of chevron-like stripes along the flanks. *Ph. caudomaculatus,* the African moon-tetra, is by far the most common of the three species. *Ph. greeni* (type species of *Bathyaethiops*) differs from it in having more transverse, longitudinal and peduncular scales ($7\frac{1}{2}$-$8\frac{1}{2}$/36-42/ $4\frac{1}{2}$-$5\frac{1}{2}$, 14-16 peduncular, in *greeni,* versus $5\frac{1}{2}$-$7\frac{1}{2}$/28-35/$3\frac{1}{2}$-$4\frac{1}{2}$, 10-12 ped., in *caudomaculatus*). *Ph. breuseghemi* has the same meristics as *Ph. caudomaculatus,* but it is somewhat less deep, with the peduncular spot more posterior and only on the lower part.

Phenacogrammus caudomaculatus (after Poll, 1959).

Bryconaethiops sp., apparently not *B. boulengeri* or *B. microstoma.* Photo by Dr. Herbert R. Axelrod.

Brycinus imberi from Chete Island cove, Lake Kariba. Photo by Dr. E. Balon.

Brycinus macrolepidotus juvenile. Photo by Dr. Herbert R. Axelrod.

Brycinus macrolepidotus adult from Gaboon. Photo by Dr. J. Gery.

Brycinus taeniurus. Photo by Dr. J. Gery.

Tricuspidalestes caeruleus (after Poll, 1967).

(3) GENUS *TRICUSPIDALESTES*

The single Congolese species, *T. caeruleus*, has the narrowest teeth of all Petersini (except *Clupeocharax* and *Lepidarchus*). It is a small fish (44 mm) having the form of a *Micralestes* but with long fin filaments, a large eye and a large maxillary bone, somewhat clupeoid in appearance. Its scales are numerous (36-41 in a longitudinal series, 3-6 of which are perforated, and 7 or 8 from dorsal fin to lateral line). The overall coloration is steel-blue, with the head and fins dark, and the back and flanks deep blue or mauve. It would certainly be a prized aquarium fish.

(4) GENUS *HEMIGRAMMOPETERSIUS*

Like *Phenacogrammus*, *Hemigrammopetersius* has been split into four subgroups which have generic status according to Poll. However, the characters alleged for the division show considerable overlap and the genera cannot be retained here. This overlap (with the exception of a specialized form, *H. barnardi*) is clearly apparent in the following key:

a. Preventral region not keeled; mouth terminal or not much superior; dorsal fin not much behind mid-body.

b. Lateral line complete or not quite complete, rarely incomplete.

c. Elongate species without filaments on the fins, derived from *Micralestes*, with usually more than 2 outer premaxillary teeth on each side; 2 subgroups, East African and West African . . . *H. tangensis*-group ("genus" *Rhabdalestes*)

cc. Deeper bodied species with filamentous fins on the males, resembling *Phenacogrammus* and perhaps derived from it, with 2 outer premaxillary teeth; all species Congolese except one from Nigeria.

d. Broadest premaxillary teeth usually with less than 9 cusps
. . . *H. caudalis*-group ("genus" *Alestopetersius*, type species *hilgendorfi*)

dd. Broadest premaxillary teeth usually with 9-14 cusps
. . . *H. tumbensis*-group (for which Poll has proposed the genus *Duboisialestes*)

bb. Lateral line incomplete (except sometimes in *H. intermedius*); species from the Cameroon and Tchad . . . *H. pulcher*-group (*Hemigrammopetersius* in the strict sense)

aa. Preventral region keeled; mouth superior; dorsal fin far behind mid-body (otherwise very similar to *Hemigrammopetersius intermedius*)
. . . *H. barnardi* (East Africa—probably a new generic type)

The *Hemigrammopetersius tangensis*-group

The *H. tangensis*-group (sometimes called *Rhabdalestes*) includes 6 or 7 small (maximum 75 mm) species mostly from East and South Africa (*H. tangensis, H. maunensis, H. rhodesiensis* and *H. leleupi*), with a few forms from West Africa (*H. septentrionalis, H. loennbergi* and *H. eburneensis*). They are usually elongate and resemble certain *Micralestes*, from which they differ only, or mainly, in the absence of the pair of small teeth behind the main mandibulary series.

The East and South African forms may be distinguished as follows:

a. 28-29 longitudinal scales, 10 around caudal peduncle (lateral line complete)
. . . *H. maunensis* (Bechuanaland)

aa. 35-37 longitudinal scales, 12-14 around caudal peduncle.

b. Lateral line complete.

c. An inconspicuous dark line on caudal peduncle
. . . *H. tangensis* (East Africa)

cc. Lateral band conspicuous, mostly posterior; anal fin base with a black band
. . . *H. rhodesiensis* (Rhodesia)

bb. Lateral line incomplete, without pores on caudal peduncle; coloration same as *rhodesiensis* . . . *H. leleupi* (East Africa, probably an isolated form derived from *H. tangensis*)

The West African forms are even closer; all have 14-17 branched anal rays, and the numbers of scales and gill-rakers overlap. They also have in common a black longitudinal band, more conspicuous posteriorly, and a black anal fin margin. The only differential characters are as follows:

Brycinus chaperi. Adult male. Photo by Dr. Herbert R. Axelrod.

Male (above) and female (below) *Brycinus lateralis* from Chikanka Island cove, Lake Kariba. Photo by Dr. E. Balon.

Young *Brycinus chaperi* from Lagos. Photo by Dr. Herbert R. Axelrod.

Brycinus longipinnis, male above and female below. Photo by J. Elias.

a. Tip of dorsal fin black.

b. Lateral line complete, with 22-26 tubes (sometimes with a pair of conical teeth behind the main mandibulary series; a connecting link with *Micralestes* ?) ... *H. eburneensis* (endemic in the Cavally in Ivory Coast)

bb. Lateral line perforating only 17-24 scales ... *H. loennbergi* (Gambia, Comoe in Ivory Coast, and Black Volta)

aa. Tip of dorsal fin hyaline ... *H. septentrionalis* (from Senegal to Cameroon)

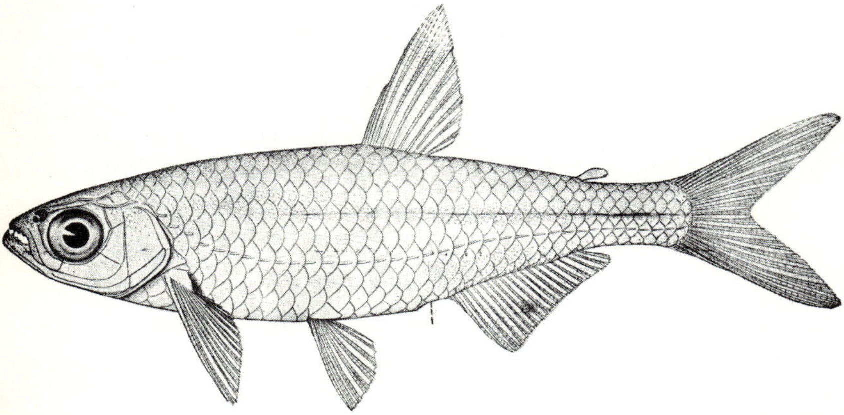

One of the types of *Hemigrammopetersius tangensis* (after Poll, 1967).

The *Hemigrammopetersius caudalis*-group

The seven species of the *H. caudalis*-group constitute a rather homogeneous group, with the exception of *H. hilgendorfi* (type of the genus *Alestopetersius*), which is quite aberrant in having 34-42 lateral line scales (instead of 27-33 for the bulk of the group) and up to 8 scales between the dorsal fin and lateral line. Six of the seven species are Congolese and the males of 6 out of 7 species have a filamentous dorsal fin, while 2 have the middle caudal rays prolonged as in certain species of *Phenacogrammus*.

The species are relatively easy to identify, as shown in the following key. The commonest species are *H. hilgendorfi* and *H. caudalis*, which is sometimes found in the aquarium trade under the name *Petersius xenurus*.

Key to the species of the *H. caudalis*-group:

a. 27-33 longitudinal scales (plus 1 or 2 on caudal fin); 5-7 transverse scales between dorsal fin and lateral line.
  b. 13-20 gill-rakers on lower part of first arch.
    c. Fins hyaline.
      d. No broad lateral band from eye to end of middle caudal rays (often a lateral streak which may get broader on caudal peduncle).
        e. Caudal fin of males without prolonged median rays.
          f. Body depth 3.0-3.15 in standard length; 27 + 2 longitudinal scales; 16 gill-rakers; a conspicuous triangular caudal spot; humeral spot scarcely visible ... *H. smykalai* (Lower Niger)
          ff. Body depth 2.4-3.0 in standard length; 30-33 + 1 or 2 longitudinal scales; 13-14 gill-rakers; no caudal spot; humeral spot large, vertically elongate ... *H. compressus* (Congo)
        ee. Caudal fin of males prolonged (body depth 2.5-2.8 in standard length; 27-32 + 1 or 2 longitudinal scales; 13-18 gill-rakers; humeral spot vertically elongate; a black lateral line extending onto caudal rays, where it is much broader) ... *H. caudalis*, synonym *H. xenurus* (Congo)
      dd. A broad lateral band, equally broad from eye to end of prolonged middle caudal rays (body depth 3.25-4.0 in standard length) ... *H. brichardi* (Congo)

Hemigrammopetersius caudalis. Photo by Milan Chvojka.

Micralestes stormsi. Photo by Dr. Herbert R. Axelrod.

Micralestes acutidens from Northern Rhodesia. Photo by Dr. Herbert R. Axelrod.

Young *Phenacogrammus interruptus.* Photo by Dr. Herbert R. Axelrod.

Male *Phenacogrammus interruptus,* the Congo tetra. Photo by Dr. Herbert R. Axelrod.

cc. Fins black, particularly the filamentous dorsal fin and the caudal fin (except at its base) (middle caudal rays not prolonged; a humeral spot; body depth 2.9-3.4 in standard length; a sibling species of *H. hilgendorfi* ?) ... *H. nigropterus* (Congo)

bb. 24-29 gill-rakers on lower part of first arch (dorsal and caudal fins not prolonged; a broad lateral band from eye to end of middle caudal rays; dorsal, pectoral and ventral fins black) ... *H. leopoldianus* (Congo)

aa. 34-42 longitudinal scales (plus 1 or 2 on caudal fin); 7-8 transverse scales between dorsal fin and lateral line (dorsal fin filamentous, middle caudal rays not prolonged; a vertically elongate humeral spot; fins black in adult male) .. *H. hilgendorfi* (Congo)

The *Hemigrammopetersius tumbensis*-group

This group, for which Poll proposed a new generic name *(Duboisialestes)* owing to its extremely broad teeth (6 to 14 cusps instead of usually less than 9 cusps for the *H. caudalis*-group), is composed of only 2 species, *H. tumbensis* from the Stanley Pool (Lower Congo), known as *H. duboisi* by aquarists, and a much rarer species, *H. bifasciatus*, from the Middle Congo basin. The latter has more scales ($7\frac{1}{2}/34\text{-}36/3\frac{1}{2}$, 13-15 ped., in *H. bifasciatus*, versus $5\frac{1}{2}\text{-}6\frac{1}{2}/30\text{-}33/3\frac{1}{2}$, 10-12 ped., in *H. tumbensis*) and conspicuous black marks on the base of the caudal lobes. *H. tumbensis* resembles *H. caudalis*, with which it has been confused having been described as a subspecies of *H. xenurus* (which is a synonym of *caudalis*).

Anal fin of a female (above) and male *Hemigrammopetersius intermedius* (after Gery, 1968).

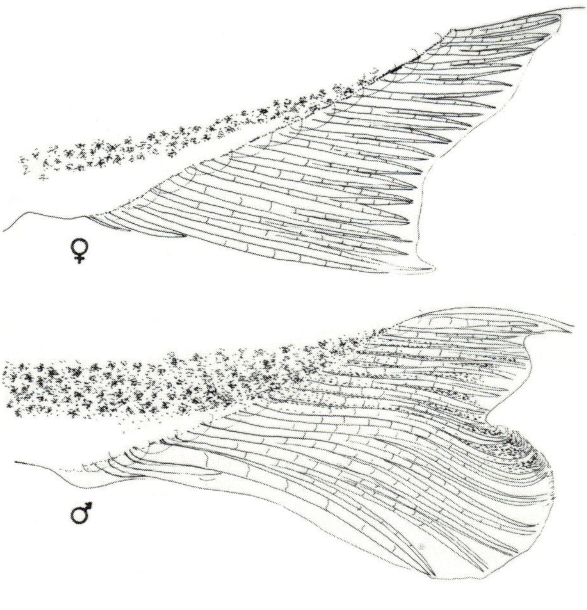

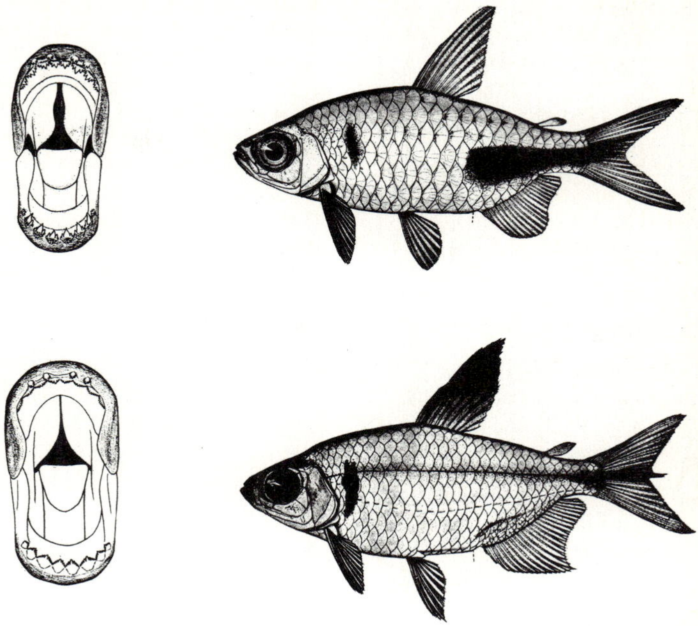

Hemigrammopetersius pulcher, type (after Poll, 1967) and *H. tumbensis,* type (after Poll, 1967).

The *Hemigrammopetersius pulcher*-group

One species surely belongs here, the poorly-known *H. pulcher* from the Ja River in Cameroon, because it is the type of the genus *Hemigrammopetersius.* Only two specimens are known, which look similar to the *Phenacogrammus major-urotaenia-deheyni* complex, but without the pair of small conical teeth behind the main mandibulary series.

Three additional species are provisionally referred to the group by Poll, *H. brevidorsalis* and *H. intermedius* from the Tchad and Chari River, and *H. barnardi* from East Africa. The first two species are said to differ mostly in the number of gill-rakers. However, *H. brevidorsalis* is known only from damaged adult specimens and some juveniles. Males of *H. intermedius* are rather similar to those of *H. septentrionalis,* and certain specimens even have the lateral line complete or almost complete. They have a thickened, rounded anal fin, surrounded by a broad black band.

H. barnardi is different from all other species of the group, being adapted to a life near the surface. It may be a new generic type, but for the moment is retained in *Hemigrammopetersius.*

Phenacogrammus aurantiacus. A male individual without very much color in its fins. Photo by Dr. Herbert R. Axelrod.

Phenacogrammus huloti. Photo by Pierre Brichard.

Male *Phenacogrammus urotaenia.* Photo by Dr. J. Gery.

Phenacogrammus ansorgii. Photo by Dr. Herbert R. Axelrod.

(5) GENUS *PETERSIUS*

The single species *Petersius conserialis*, from around Dar-es-Salaam in Tanganyika (coast of East Africa), is quite unusual among this group of genera in attaining a relatively large size (145 mm), the mouth upturned with the mandible prominent, and in lacking a posterior fontanel. It has a caudal spot extending onto the middle caudal rays. This is about all that is known of this species.

(6) GENUS *ARNOLDICHTHYS*

The single (?) species *Arnoldichthys spilopterus*, the Niger tetra or Arnold's characin, is well known to aquarists. It is unusual not only in having different sizes of scales, but also in having a larger number of teeth than other Petersini. The "rule" of 4 inner premaxillary teeth, for example, which has few exceptions, is not valid here. The Niger tetra has up to 7 inner teeth on a side. The same is true for the lower row of teeth and, in fact, lacking the pair of conical teeth on the mandible, this African tetra has almost the same dentition as that of its American cousins. However, the shape of its body, and chiefly of its snout, is distinctly alestid, indeed, much more *Brycinus*-like than *Micralestes*-like.

Its rainbow coloration is well known and need not be described here. The sexual dimorphism of its anal fin is less well known. In the female, the anal fin, with the border almost straight like that of other alestins, has a black spot in the last rays; in the male the anal fin, with a convex edge, is tricolor, red and yellow with a black border. According to certain aquarists, there is a polymorphism in this color pattern, some males (or males of a sibling species?) having an irregularly striated anal fin.

(7) GENUS *LADIGESIA*

A single species from Sierra Leone, has recently been described, *Ladigesia roloffi*, but slightly different forms have been collected since in other parts of West Africa.

The jelly bean tetra is another peculiarity of the African tetras. It has a good part of the predorsal region entirely naked, and its upper teeth point outward. Its body shape and black color pattern resemble that of an "ordinary" *Hemigrammopetersius* such as *H. septentrionalis* or *H. intermedius*, but the adult males (largest size 35-40 mm) are entirely bright vermilion.

(8) GENUS *CLUPEOCHARAX* (synonym *CLUPEOPETERSIUS)*

The single species *Clupeocharax schoutedeni* (do not confuse it with *Clupeacharax*, which is a South American genus) is again aberrant: it has

Petersius conserialis and *Clupeocharax schoutedeni* (after Poll, 1967).

retained the body shape of a Petersini, but has acquired a clupeoid, upturned mouth with a blade-like maxillary in a remarkable convergency with some Aphyoditeini (American), *Brittanichthys* for example. This small fish (47 mm) is a species from the region of Lake Tumba (near the confluence of the Ubanghi with the Congo), where it performs seasonal migrations, according to Matthes, as well as from another lake of the central Congo. It is a pelagic plankton-eater and, as such, is not very colorful, being without marks except for a narrow lateral line.

(9) GENUS *LEPIDARCHUS*

The single species *Lepidarchus adonis* is one of the most striking discoveries of recent years. This minute (less than 20 mm standard length) West African fish is entirely devoid of scales and lateral line. The teeth, like those of *Clupeocharax*, are in a single series on both jaws, minute, conical, but the maxillary is not clupeoid.

The male has numerous purple spots on the distal half of its body as well as on the caudal fin. The ventral and anal fins are likewise spotted or even barred, whereas the front part of the dorsal fin is brown in both sexes.

Hemigrammopetersius caudalis, the yellow-tailed Congo tetra. Photo by Dr. Herbert R. Axelrod.

Phenacogrammus caudomaculatus, the African moon-tetra. Photo by Dr. Herbert R. Axelrod.

Hemigrammopetersius intermedius. Photo by H.J. Franke.

Hemigrammopetersius rhodesiensis. Photo by Warren E. Burgess.

SUBFAMILY HYDROCYNINAE

African Tiger Fishes

Hydrocynus (not *Hydrocyon* as it is frequently spelled) species are *Alestes* modified toward predation. They have acquired strong, acute, uniserial canine teeth, those of the upper jaw alternating with those of the lower and acting as a trap. This makes the tiger fishes some of the most formidable fish-killers of African waters.

Six species are known, differing in the number of scales and in certain proportions. *H. forskalii* and *H. vittatus*, very close species, are widely distributed in Africa. They are found from the Nile, Senegal and Niger to the Congo, with *H. vittatus* even extending to South Africa. *H. brevis* and *H. somonorum*, siblings in Niger and Tchad, are restricted to West Africa, whereas *H. vittiger* and the giant tiger fish, *H. goliath*, inhabit the Congo. Most of the species reach 40 or 50 cm, or even 68 cm for *H. brevis*. The record size is held by *H. goliath*, which attains a length of 132 cm and a weight of 38 kg, and which is said to reach 150 cm and 50 kg on occasion.

Certain species are brilliantly colored during the breeding season. *H. vittatus* for example, the most ubiquitous, has a series of longitudinal lines formed by spots at the base of each scale, has the eye and ventral fins red, and has the other fins at least partially orange.

Family Citharinidae

Citharinids

The citharinids, strictly African, are sometimes split into 2 families (Distichodidae and Citharinidae) or even 3 (with the Ichthyboridae). These groups will be treated here as subfamilies which have the following characters in common:

—Scales ctenoid (with the exception of the genus *Citharinus*).

—Ventral rays usually more numerous than in the other characoids (except hemiodids).

—Only two epurals.

—Teeth small, often movable, usually cuspidate (again a convergence with the hemiodids).

—Presence of pharyngeal "bags".

—According to Daget, intercalarium very small and incorporated into the neurocranium (whereas in a non-citharinid such as *Brycon*, it is a separate small osseous plate covering the prootic and other bones of the cranium).

—Joint between the two sides of the mandible simple, without interlocking devices, often ankylosed.

Most of these characters are quite specialized. The three subfamilies can be told apart by means of the following key:

Key to subfamilies of Citharinidae:

a. Premaxillary bones short and not movable; joint between two sides of mandible movable.
 b. Scales cycloid (except in one species); body short and compressed, bream-like ... CITHARININAE
 bb. Scales always ctenoid; body short and compressed only in a few instances (*Xenocharax* and *Distichodus*) ... DISTICHODINAE

aa. Premaxillary bones long and movable; joint between two sides of mandible not movable ... ICHTHYBORINAE

SUBFAMILY CITHARININAE

Citharins

This small group of large, usually deep-bodied, limnivorous species is exactly homologous to the American prochilodins and, like the latter, are of economic importance in certain regions of tropical Africa. The largest species reaches a size of 80 cm and a weight of 18 kg. The scales are cycloid in *Citharinus* and ctenoid in *Citharidium*, and are numerous (46 to 90 in the lateral line, which is complete). The dorsal and anal fins are relatively long (16-24 rays for the former and 19-31 for the latter) in relation to the great depth of the body. The mouth structures are well adapted for feeding on the mud at the bottom of rivers: it is very broad, with very weak teeth inserted into the lips; the maxillary bones are not toothed; the gill-rakers are numerous; and the intestinal tract is very long.

GENUS *CITHARINUS*

Seven species belonging to the genus *Citharinus* (which is split by some into three subgenera) are known. The following key will help in the identification of these species (after J. Daget):

a. Relatively generalized forms with body not very deep and adipose fin base not very long (shorter than its distance to the dorsal fin).
 b. 7 or 8 transverse scale rows between lateral line and ventral fin; 46-51 lateral line scales; body depth 2.2-2.7 in standard length ... *C. distichodoides* (Tchad and Niger basin)
 bb. 10-20 transverse scale rows between lateral line and ventral fin; 55-90 lateral line scales; body depth usually less than 2.3 in standard length.
  c. 10-12 transverse scale rows between lateral line and ventral fin; 55-60 lateral line scales ... *C. macrolepis* (Congo basin)
  cc. More than 12 transverse scale rows; more than 60 lateral line scales.
   d. 13 or 14 transverse scale rows between lateral line and ventral fin.
    e. 60-66 lateral line scales ... *C. congicus* (Congo basin)
    ee. 68-73 lateral line scales ... *C. eburneensis* (Ivory Coast)

Unidentified African tetra. Photo by Klaus Paysan.

Arnoldichthys spilopterus, the Niger tetra. Photo by Dr. Herbert R. Axelrod.

Ladigesia roloffi, the jelly bean tetra. Photo by E. Roloff.

Lepidarchus adonis. Photo by E. Roloff.

dd. 17-20 transverse scale rows between lateral line and ventral fin (77-90 lateral line scales) ... *C. citharus*, type species of the genus *Citharinus* (from Senegal to Nile basin)

aa. Specialized, deep-bodied forms with an elongate adipose fin base (longer than its distance from dorsal fin).

f. 10-12 transverse scale rows between lateral line and ventral fin; 53-60 lateral line scales (same scalation as that of *C. macrolepis*) ... *C. gibbosus* (Congo basin)

ff. 13-15 transverse scale rows between lateral line and ventral fin; 59-71 lateral line scales ... *C. latus* (from Senegal to Nile basins)

GENUS *CITHARIDIUM*

The second genus, *Citharidium*, with the single species *ansorgii*, from the Niger basin, is exactly like the deep-bodied *Citharinus* with a long adipose fin, but its scales (46-50 in lateral line and 11 or 12 between lateral line and ventral fin) bear longitudinal (along axis of body), parallel ridges terminating in spines. According to Daget, these scales are different from the typical ctenoid scales of the next subfamily.

SUBFAMILY DISTICHODINAE

Distichodins

The Distichodinae is a relatively large group of about 50 or 55 micro-predator and herbivorous species. They are distributed among fewer than 10 genera, as follows:

a. Generalized forms with gill membranes not forming a continuous bridge under throat; body deep, head broad and snout obtuse (two or three rows of teeth on both jaws) ... (1) *Xenocharax*

aa. More specialized forms with gill membranes forming a continuous bridge under throat; body shape variable (two series of teeth on both jaws).

b. Gill membrane not fused with throat or isthmus; maxillary teeth often present (small rather elongate species with head narrow and snout more or less pointed).

c. 31-48 lateral line (or longitudinal) scales.

d. Lateral line complete; adipose fin present ... (2) *Nannaethiops*

dd. Lateral line incomplete; adipose fin present or absent.

e. 10-14 dorsal fin rays; maxillary teeth present ... (3) *Neolebias*

ee. 15-16 dorsal fin rays; maxillary teeth present or absent.

f. A series of pit-lines on head; maxilla toothless; adipose fin present ... (4) *Congocharax*

ff. No pit-lines on head (?); maxilla toothed; adipose fin absent ... (5) *Dundocharax*

cc. 59-73 lateral line scales; a genus forming the transition with the Ichthyborinae ... (6) *Microstomatichthyoborus*

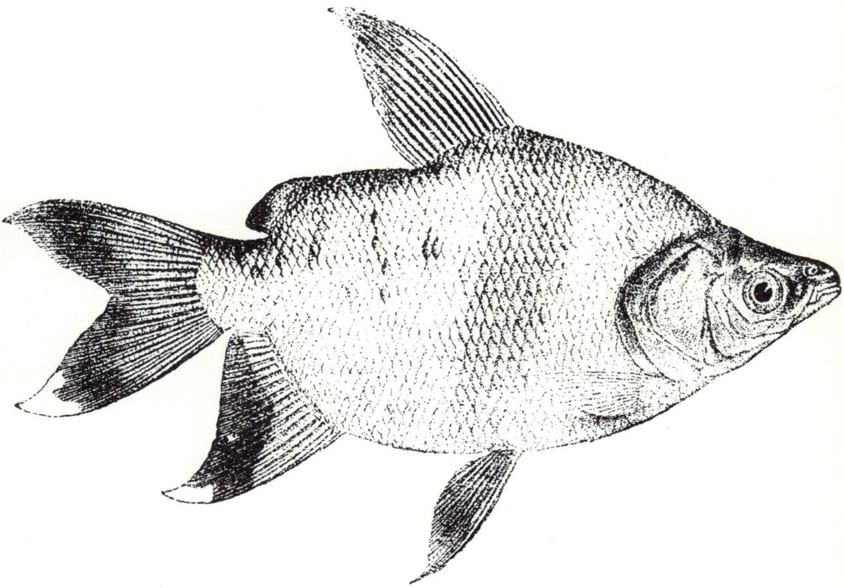

Type of *Citharinus gibbosus* (after Boulenger, 1899) and type of *Citharidium ansorgii* (after Boulenger, 1909).

Hydrocynus sp., possibly *H. goliath* or may be a giant specimen of *H. vittatus*. Photo by Dr. J.P. Gosse.

Hydrocynus vittatus juveniles from Lake Kariba. Photo by Dr. E. Balon.

Citharinus citharus. Photo by Klaus Paysan.

bb. Gill membrane fused with isthmus; maxilla toothless.

g. Two series of bifid teeth on both jaws; dorsal fin long, with 16-27 rays.

h. Large deep bodied species, quite comparable to the Citharininae; adipose and caudal fins scaled ... (7) *Distichodus*

hh. Small, elongate species; adipose fin not scaled ... (8) *Paradistichodus*

gg. A single series of teeth on both jaws; dorsal fin relatively short, with 12-16 rays.

i. Lateral line complete ... (9) *Nannocharax*

ii. Lateral line incomplete ... (10) *Hemigrammocharax*

(1) GENUS *XENOCHARAX*

Xenocharax spilurus, called Ihajo by the Kota fishermen in Gaboon, is a deep bodied citharinid which differs from the other large species (of *Citharinus, Citharidium* and *Distichodus*) only by a few details such as the structure of the gill membranes (which are separated in the middle as in most characoids). It reaches 25 cm in size and, chiefly when young, is quite nicely colored with the fins orange or red. It inhabits Cameroon, Gaboon and the Congo.

A second species, *Xenocharax crassus,* found in the Upper Congo basin, is said to differ from the preceding mostly in some proportions and in the presence of a black spot on the dorsal fin.

(2-5) *NANNAETHIOPS* AND ALLIED GENERA

This group of 10 or 11 pigmy distichodins comprises the genera *Nannaethiops* and *Neolebias,* very popular among aquarists, as well as two other more recent (and much less known) genera from the Congo basin and Angola, *Congocharax* and *Dundocharax*. *Dundocharax* was previously classified near *Hemigrammocharax,* with which it does not seem to have much affinity.

The first two genera have the following characteristics: small size; maxillary bone always toothed; two rows of teeth on both jaws, the outer ones bifid, the inner ones also bifid (*N. unitaeniatus, N. trilineatus, N. trewavasae* and maybe others) or conical; 11–15 dorsal rays; 31–36 longitudinal scales. The following key is inspired by the rather recent revision of Poll & Gosse (1963). The subgenera proposed by Daget the following year are not used.

Key to the genera *Nannaethiops* and *Neolebias*:

a. Lateral line complete (adipose fin present; inner teeth bifid on both jaws; usually 2 maxillary teeth; a longitudinal dark band extending onto middle caudal fin rays;

relatively large size: 62 mm) ... genus *Nannaethiops* (a single species, *N. unitaeniatus*, from Niger to the Congo basin)

aa. Lateral line incomplete (adipose fin present or absent; inner teeth bifid or conical on both jaws; 2-7 maxillary teeth) ... genus *Neolebias*

b. 16 scales around caudal peduncle; a single longitudinal band.

c. About 7 maxillary teeth; a broad dark band along the side and a dark bar at base of caudal fin; adipose fin usually absent ... *Neolebias ansorgii* (from Cameroon—called *N. landgrafi*—to lower Congo basin)

cc. Always less than 5 maxillary teeth; a rather narrow band along the side and no dark bar at base of caudal fin.

d. A longitudinal black band extending to caudal fin.

e. Longitudinal black band extending along middle of caudal peduncle; usually a caudal spot; maxillary bone not reaching level of eye.

f. 33-35 longitudinal scales; body depth 3.0-3.7 in standard-length; dorsal fin behind mid-body; usually no adipose fin ... *Neolebias unifasciatus* (type of the genus, Portuguese Guinea to Gaboon)

ff. Usually 36-39 longitudinal scales; body depth 3.9-4.15 in standard length; dorsal fin in front of mid-body; adipose fin present ... *Neolebias gracilis* (central Congo basin)

ee. Longitudinal black band extending along lower part of caudal peduncle; no caudal spot; maxillary bone almost reaching to level of pupil; (adipose fin present; 33-36 longitudinal scales; body depth 2.5-3.3 in standard length; dorsal fin in front of mid-body) ... *Neolebias trewavasae* (Nile to lower Congo)

dd. A longitudinal band interrupted above anal fin, not continued on caudal peduncle (no adipose fin; maxillary bone not reaching level of eye; dorsal fin at mid-body; 31-33 longitudinal scales) ... *Neolebias axelrodi* (a species described from 2 small specimens less than 19 mm in standard length collected near Lagos, Nigeria, by Dr. H. R. Axelrod)

bb. 12 or 13 scales around caudal peduncle; 2 or 3 longitudinal bands.

g. Two longitudinal bands; no caudal spot; no adipose fin ... *Neolebias philippei* (Central Congo basin)

gg. Three longitudinal bands; a small caudal spot; adipose fin usually present ... *Neolebias trilineatus* (Congo basin); often called *Nannaethiops tritaeniatus*

Xenocharax spilurus juvenile. Photo by Dr. J. Gery.

Neolebias ansorgii. Photo by G.J.M. Timmerman.

Neolebias unifasciatus, a race from Gaboon. Photo by Dr. J. Gery.

Neolebias unifasciatus. Photo by E. Roloff.

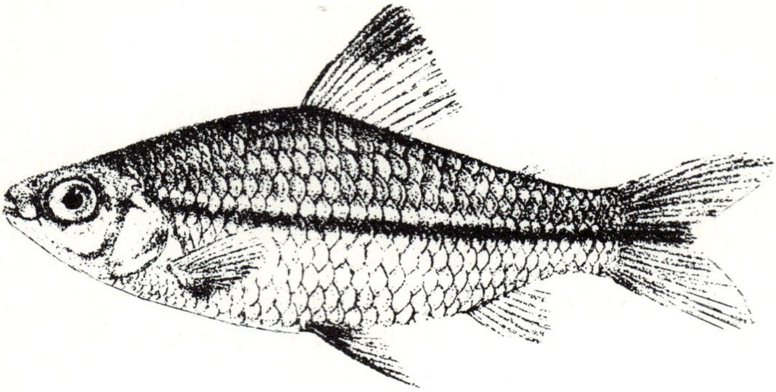

Type of *Nannaethiops unitaeniatus* (after Guenther, 1871).

All species are micropredators, living on small animals found on the bottom or among aquatic plants. The following species are easy to acclimate to the aquarium and are often seen. (1) *N. ansorgii* and a paler form called *N. landgrafi*, which probably belong to the same species. The former variety is green-blue (a quite unusual tint among fishes) with red fins; the latter form is pale blue. (2) *N. trilineatus*, which has its median longitudinal band surrounded by a golden metallic stripe like that of *Rasbora pauciperforata* or *Hemigrammus erythrozonus*, as well as reddish finnage. (3) *N. unifasciatus*, which is said to have the same color pattern but without the upper and lower dark bands. (4) *N. trewavasae*, which was often confused with the last, though it is somewhat less elongate and with the fins not strongly colored.

GENERA (4) *CONGOCHARAX* AND (5) *DUNDOCHARAX*

The genus *Congocharax* was erected for a species previously classified within *Hemigrammocharax (olbrechtsi)* as well as for *Neolebias spilotaenia*. The type species has the maxillary bone not toothed, it has slightly more dorsal rays and more scales than the *Neolebias* species, and has some pit-lines on the head. The two species are distinguished as follows:

a. 40-43 longitudinal scales; a lateral band, which may be broken into a series of spots . . . *Congocharax spilotaenia* (Chiloango basin)

aa. 44-48 longitudinal scales; a humeral spot and a large caudal spot . . . *Congocharax olbrechtsi* (Central Congo basin)

With regard to *Dundocharax*, the single species *D. bidentatus* (from Angola) differs from the preceding species by having 30–32 longitudinal scales, some maxillary teeth and no adipose fin. Judging from its description, the genus is scarcely separable from *Neolebias*.

(6) GENUS *MICROSTOMATICHTHYOBORUS*

This long, difficult name designates quite interesting Congolese species somewhat intermediate between the distichodins and the ichthyborins. The body and head are elongate, the teeth much like those of a *Neolebias*, the scales numerous (59–73 in the lateral line, which is complete), and the caudal fin is obliquely marked like that of a scissortail rasbora. The two known species are micropredators like *Neolebias*, and not ichthyophagous or fin-eaters like most ichthyborins. They can be distinguished as follows:

a. 59-61 lateral line scales . . . *M. katangae*
aa. 63-73 lateral line scales . . . *M. bashfordeani*

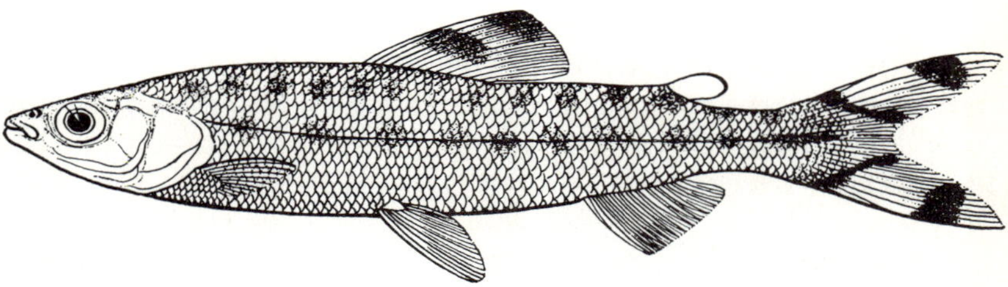

Microstomatichthyoborus bashfordeani (after Fowler, 1936).

(7) GENUS *DISTICHODUS*

Here we have a series of large (maximum size 70 cm) detritivorous and herbivorous species which differ from a preceding genus, *Citharinus*, mostly in having the scales ctenoid, the teeth biserial and the adipose fin as well as the caudal fin lobes covered with small scales; the caudal lobes are frequently rounded. Boulenger, in his classical *Catalogue of the Fresh-water Fishes of Africa*, listed 17 species of *Distichodus*. Later, a few more species were described, and the following key will help to characterize and identify the species:

Neolebias trewavasae. Photo by Pierre Brichard.

Neolebias trilineatus. Photo by Dr. Herbert R. Axelrod.

Nannaethiops unitaeniatus. Photo by Klaus Paysan.

Young *Distichodus fasciolatus* (left) and *Distichodus maculatus* (right). Photos by B. Kahl.

a. Less than 60 lateral line scales (16-20 dorsal rays; snout broad or feebly compressed).

b. Anal fin with 19-22 rays, at least as long as dorsal fin.

c. 7/37-39/10 scales ... *D. affinis* (Congo basin)

cc. 9/40-42/11-12 scales ... *D. altus* (upper Congo)

bb. Anal fin with 12-17 rays, shorter than dorsal fin.

d. Mouth terminal; dorsal fin with a black blotch or bar; 38-46 lateral line scales.

e. 12 or 13 anal rays; very small size; a series of spots on body (6-7/38-41/6-8 scales) ... *D. decemmaculatus* (a pigmy species from the Central Congo basin)

ee. 14-16 anal rays; moderate size; body plain.

f. 6 or 7 scales from lateral line to base of ventral fin; caudal lobes pointed ... *D. notospilus* (from South Cameroun to Angola)

ff. 8 scales from lateral line to base of ventral fin; caudal lobes rounded ... *D. noboli* (upper Congo)

dd. Mouth strongly inferior; dorsal fin plain; 53-57 lateral line scales (12 or 13 anal rays; 6 scales from lateral line to base of ventral fin; caudal lobes obtusely pointed) ... *D. hypostomatus* (Gaboon)

aa. 60-110 lateral line scales.

g. Snout deeper than long.

h. Snout broad or feebly compressed (mouth terminal to strongly inferior).

i. Mouth terminal or only slightly inferior.

j. 60-78 lateral line scales.

k. Snout shorter than interorbital space, 1.33-1.66 in its width (dorsal fin spotted)

l. 9 or 10 scales between lateral line and base of ventral fin; 60-66 lateral line scales; about 14 narrow transverse bars on body ... *D. antonii* (Congo basin)

ll. 13 or 14 scales between lateral line and base of ventral fin; 68-75 lateral line scales; a humeral spot ... *D. petersii* (East Africa)

kk. Snout about equal to or slightly shorter than interorbital space (11 or 12 scales between lateral line and base of ventral fin; 67-70 lateral line scales) ... *D. mossambicus* (Zambezi River★)

jj. 80-110 lateral line scales (snout much shorter than interorbital space; mouth sub-inferior; large species from the Nile and Niger, very close to each other, the dorsal fin always more or less spotted, the body spotted in the young, usually plain in adult (except *brevipinnis*).

m. Body plain (adults).

★ The individuals mentioned in the literature as having the mouth inferior may belong to *D. langi*.

n. 15-16 scales from mid-dorsal line to lateral line (13-15 scales from lateral line to base of ventral fin) ... *D. rostratus* (Nile, Senegal and Niger)

nn. 18-20 scales from mid-dorsal line to lateral line (15-18 scales from lateral line to base of ventral fin) ... *D. niloticus* (Nile only)

mm. Body spotted, even in adults (17-20 scales from mid-dorsal line to lateral line; 18-20 from lateral line to base of ventral fin) ... *D. brevipinnis* (Nile, Senegal and Niger)

ii. Mouth strongly inferior.

o. 16-19 dorsal rays; 11 anal rays (series of spots on flank; 9-10/68-75/10-12 scales) ... *D. maculatus* (Congo basin)

oo. 22-27 dorsal rays; 13-16 anal rays.

p. Snout shorter than interorbital space, 1.33-1.66 in its width; (6 transverse bars on body; 13-14/68-77/14-15 scales) ... *D. atroventralis* (Congo basin)

pp. Snout about equal to or slightly shorter than interorbital width.

q. Body plain (71 lateral line scales; 12 between lateral line and base of ventral fin; very close to *D. mossambicus* with the exception of the position of the mouth) ... *D. langi* (Congo basin)

qq. Body patterned.

r. About 18 transverse bars on body (13-15/66-78/14-16 scales) ... *D. fasciolatus* (Congo basin and Angola)

rr. Series of spots on body (15-16/75-85/16-17 scales) ... *D. engycephalus* (Nile and Niger basins)

hh. Snout strongly compressed (mouth sub-inferior; 24-25 dorsal rays, 12-14 anal rays; 14-16/60-68/14-16 scales; 6 or 7 broad transverse bars on body) ... *D. sexfasciatus* (Congo basin and Angola)

gg. Snout longer than deep (mouth more or less terminal; 25-26 dorsal rays, 13-14 anal rays; 15-17/70-85/15-17 scales; 6 to 8 broad transverse bars on body) ... *D. lusosso* (Congo basin and Angola)

The author does not possess enough data concerning *Distichodus albini* from East Africa (apparently near *D. petersii*), *D. kolleri* (synonym *vexillifer*)

Young *Distichodus notospilus.* Photo by Dr. J. Gery.

Adult *Distichodus notospilus.* Photo by Dr. J. Gery.

A juvenile *Distichodus*, possibly *D. noboli.* Photo by Dr. Herbert R. Axelrod.

Distichodus noboli. Photo by Dr. Herbert R. Axelroa.

from Cameroon (close to *D. notospilus*), and *D. schenga* (close to *mossambicus?*) to fit them into the key.

(8) GENUS *PARADISTICHODUS*

Paradistichodus differs from *Distichodus* by the more elongate body and the adipose fin not being covered with small scales. The single species, *Paradistichodus dimidiatus*, rather strongly resembles a *Nannocharax*. It differs from *Nannocharax* by its biserial teeth, as well as by having a longer dorsal fin than in most *Nannocharax* species (formula iii or iv, 13–15). It has a complete lateral line, with 51–63 perforated scales, and can at once be recognized due to its characteristic dorsal spot (which may act in nature as a recognition signal). Some ichthyologists acknowledge two subspecies, the nominal one from Casamance to the upper Niger basin, and a second one from Tchad.

Paradistichodus dimidiatus. Photo by Dr. J. Gery.

GENERA (9) *NANNOCHARAX* AND (10) *HEMIGRAMMOCHARAX*
AFRICAN DARTERS

Nannocharax and *Hemigrammocharax* are the exact counterpart in Africa of *Characidium* (and allied genera) of South America. Both strongly convergent groups are bottom fishes with large pectoral fins and sometimes other adaptations to life in fast-running waters, even torrents. A specimen of *Nannocharax* is usually impossible to distinguish from *Characidium* with the naked eye (provided that the collecting locality is not given), except for a certain rugosity of the body surface due to the ctenoid scales, which contrasts with the smoothness of the cycloid scales of the *Characidium*.

As contrasted with the other distichodins, *Nannocharax* and *Hemigrammocharax* have a single series of bifid teeth on both jaws. Like their Neotropical cousins, they are rather poorly known, as far as their taxonomy is concerned, primarily because most of the original descriptions were incomplete. This is why the following key will indicate only the outlines of their classification:

a. Lateral line complete; mouth usually inferior; body usually somewhat compressed ... genus *Nannocharax**.

b. Dorsal fin inserted in front of level of ventral fins (usually less than 50 lateral line scales except in *N. luapulae*).

c. 14 branched dorsal rays; body depth 3.5 in standard length ... *N. ogoensis* (Gaboon)

cc. 9-12 branched dorsal rays; body depth more than 3.5 in standard length.

d. Dorsal fin inserted at mid-body (even more posteriorly in young) ... *N. parvus* (Niger to Ogowe; possible synonyms are *ansorgei* and *micros*)

dd. Dorsal fin inserted in front of mid-body.

e. Less than 46 lateral line scales; body depth 4.5 in standard length.

f. About 35-38 lateral line scales ... *N. pteron* (Ubanghi)

ff. About 40-44 lateral line scales ... *N. macropterus* (Congo, Angola)

ee. 48-50 lateral line scales; body depth 4.5-6.5 in standard length.

g. 6 or 6½ transverse scales below lateral line; 6 or 7 round spots along body ... *N. gracilis* (Congo)

gg. 7½ transverse scales below lateral line; about 10 vertically elongate spots along body ... *N. taenia* and *N. luapulae* (Ubanghi to South Katanga)

bb. Dorsal fin inserted at level of ventral fins or behind it, the ventral fins usually being very anterior (37-57 lateral line scales).

h. 37-42 lateral line scales.

i. 6 branched anal rays (broad spots on the dorsum) ... *N. brevis* (upper Congo basin)

ii. 7 or 8 branched anal rays.

j. Depth usually less than 4.33 ... *N. altus* and *N. hollyi* (Ubanghi and Gaboon)

jj. Depth 4.25-5.0 ... *N. procatopus* and *N. schoutedeni* (Congo to Angola)

hh. 42-57 lateral line scales.

k. Body depth usually less than 5.5 in standard length; several vertical bars on body.

l. 4½ scales above lateral line; 42-49 lateral line scales ... *N. fasciatus*, type of the genus (with a subspecies in the Liberian-Guinean forest and the nominate form from Niger to Gaboon)

ll. 5½ scales above lateral line; 47-55 lateral line scales ... *N. intermedius* (South Cameroon)

* The author lacks data concerning *N. gobioides* Roman from the Upper Volta.

A young *Distichodus sexfasciatus.* Photo by B. Kahl.

Distichodus fasciolatus. Photo by Dr. Herbert R. Axelrod.

Distichodus sexfasciatus adult. Photo by Dr. Herbert R. Axelrod.

kk. Body depth usually more than 5.5 in standard length; body marked with roundish spots.

m. 1 unbranched ray at beginning of ventral fin . . . *N. lineomaculatus* (Niger and Tchad)

mm. 2 unbranched rays at beginning of ventral fin . . . *N. niloticus* (with 4 subspecies from the Nile, nominal form, to the Congo basin)

aa. Lateral line usually incomplete (said to be complete in certain individuals of *H. multifasciatus*); mouth usually terminal or subinferior; body usually quite cylindriform (an ocellus at root of caudal fin) . . . genus *Hemigrammocharax*

n. 34-44 longitudinal scales.

o. 12 scales around caudal peduncle.

p. Adipose fin always (?) present.

q. 17 or 18 dorsal rays and 15 or 16 anal rays (35 longitudinal scales) . . . *H. uniocellatus* (type species of the dubious subgenus *Microdistichodus*; Congo basin)

qq. 12-16 dorsal rays and 10-12 anal rays (34-44 longitudinal scales)

r. Dorsal fin inserted in front of level of ventral fins, or just at its level.

s. 21-23 lateral line scales; length of pectoral fin 1.15 in length of head; 7-10 irregular transverse bands . . . *H. angolensis* (Angola)

ss. 10 lateral line scales; length of pectoral fin 1.8 in length of head; 14 transverse bars . . . *H. monardi* (Zambezi River)

rr. Dorsal fin inserted behind level of ventral fins (a longitudinal band, crossed by 9-12 short vertical bars) . . . *H. lineostriatus* (Angola)

pp. Adipose fin always (?) absent (13 or 14 dorsal rays and 12 anal rays; dorsal fin behind ventral level; 34-36 longitudinal scales, of which only 6-7 are perforated) . . . *H. machadoi* (Angola)

oo. 15-16 scales around caudal peduncle (adipose fin always (?) present; dorsal fin inserted in front of level of ventral fins; 13-15 dorsal rays and 10-12 anal rays; body with 13-20 more or less regular vertical bars).

t. Body depth 3.7-4.7; length of head 3.4-3.8 in standard length; usually 12-32 lateral line scales.

u. Rarely more than 18 lateral line scales
... *H. wittei* (Katanga; Angola)

uu. Usually 27-32 lateral line scales
... *H. stigmaturus*-complex with *H. stigmaturus* (Zambezi River), *H. minutus* (South Katanga) and *H. polli* (upper Volta basin)

tt. Body depth up to 5.5; length of head 4.3 in standard length; lateral line very long or even complete (?) in some individuals
... *H. multifasciatus* (a species from the Zambezi River, which is often synonymized with *H. stigmaturus*)

nn. About 50 longitudinal scales
... *H. ocellicauda* (type of the genus; Cameroon)

Hemigrammocharax wittei (after Poll, 1967) and *Nannocharax cf. fasciatus* (photo by Dr. J. Gery).

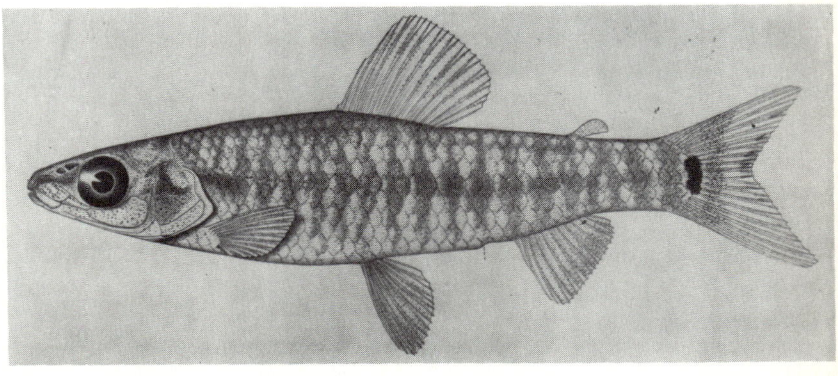

Distichodus sexfasciatus, young fish. Photo by Dr. Herbert R. Axelrod.

Distichodus lusosso, young fish. Photo by Dr. Herbert R. Axelrod.

Distichodus sexfasciatus, adult. Photo by G. Meola, African Fish Imports.

Distichodus lusosso, adult. Photo by Dr. Herbert R. Axelrod.

SUBFAMILY ICHTHYBORINAE

Fin-eaters

This small group of highly specialized predators is very interesting from a biological point of view, some of its members having a semi-parasitic life, living, at least in part, by eating the fins of other species. Indeed, with their scissor-like jaws (the upper one movable upwards), their cutting teeth and pike-like body, species of the genera *Eugnathichthys, Belonophago, Ichthyborus* and *Phago* (and possibly others) are well adapted to this specialized predation, whereas other genera such as *Phagoborus* and *Mesoborus* eat whole fishes, as their mouth, armed with some canine-like teeth, indicates. Finally, the least specialized of all in some respects, *Hemistichodus*, forming a sort of transition with the preceding subfamily (for example with *Microstomatichthyoborus*), can only capture small insects with its bifid lateral teeth. It should be noted that most of the species have an extremely small maxillary bone, in contrast with other predators.

The genera can be distinguished as follows:

a. No teeth on the front part of the upper jaw (a single series of teeth on the sides) . . . (1) *Hemistichodus*

aa. Front teeth present, lobed, conical or canine-like.

- b. A single series of teeth on both jaws.
  - c. 2 canines above and 3 below in front; snout not longer than postocular part of head . . . (2) *Phagoborus*
  - cc. 8 or 9 canines above and below; snout more than 2 times longer than postocular part of head . . . (3) *Gavialocharax*
- bb. Two rows of teeth on both jaws.
  - d. Inner teeth consisting of a pad of brush-like, very small teeth; outer teeth caniniform in front, bifid on sides . . . (4) *Ichthyborus*
  - dd. Inner teeth in a row, not brush-like.
    - e. 2 canines above and below . . . (5) *Mesoborus*
    - ee. No canines (fin-eaters for the most part).
      - f. Snout not longer than postocular part of head.
        - g. Snout rounded, jaws massive, outer bicuspid teeth all of the same size; cheek partly naked; gill membranes free from isthmus . . . (6) *Eugnathichthys*
        - gg. Snout narrow, beak-like, outer upper bicuspid teeth larger in front than on sides; cheek almost entirely covered by suborbitals; gill membranes united to isthmus.
          - h. 87-90 lateral line scales, 10 between dorsal fin and lateral line . . . (7) *Paraphago*
          - hh. Less than 50 lateral line scales, 1 or 2 between dorsal fin and lateral line . . . (8) *Phago*
      - ff. Snout longer than postocular part of head . . . (9) *Belonophago*

Hemistichodus lootensi and *H. mesmaekersi* (after Poll and Daget, 1968).

(1) GENUS *HEMISTICHODUS*

Two or, more probably, three small (50-100 mm) species belong to this genus. *H. vaillanti*, the type species from Gaboon, has a characteristic color pattern: the dorsal and caudal fins are marked with conspicuous black spots surrounded with orange-yellow; the spot on the lower caudal lobe is the largest, extending onto the distal half of the lobe. It is interesting to note that almost all Ichthyborinae have some caudal pattern, usually one or two bands crossing the lobes obliquely. It has been hypothesized that these markings act as a recognition signal, telling the other individuals not to attack their own species.

The species, with two of them sympatric in the central Congo, can be distinguished by means of the following key:

a. No marks on caudal lobes; last unbranched dorsal ray longer than head (64-74 longitudinal scales; lateral line interrupted near middle)
. . . *H. lootensi* (Central Congo)

aa. Caudal lobes with a pattern; last unbranched dorsal ray not longer than head.

b. About 65 longitudinal scales, lateral line interrupted near middle; two horizontal bars on dorsal fin, two oblique bars on each caudal lobe
. . . *H. mesmaekersi* (Central Congo)

bb. About 75-80 longitudinal scales, lateral line complete; a single spot on dorsal fin and a single spot on each caudal lobe (lower one very large)
. . . *H. vaillanti* (Gaboon)

Juvenile *Distichodus schenga* (above) and *D. mossambicus* from Namazambwe. Photo by Dr. E. Balon.

Adult *Distichodus schenga.* Photo by Dr. E. Balon.

Distichodus hypostomatus. Photo by Dr. J. Gery.

Adult *Distichodus mossambicus.* Photo by Dr. E. Balon.

(2) GENUS *PHAGOBORUS*

The type species, *Phagoborus ornatus* (formerly *Neoborus ornatus*), inhabits the Central Congo basin as do most Ichthyborinae. It attains a length of up to 200 mm and is a predator of small fishes. It has three conspicuous black lateral bands on its body, and its orange caudal lobes are barred with three longitudinal bands, more regular on the upper lobe.

A second, smaller (?) species, *P. quadrilineatus*, with the same caudal pattern, has been described from a very remote river basin, that of the Casamanza in Portuguese Guinea. It has fewer lateral line scales (90 instead of 98-110), fewer anal rays (15 instead of 17-18), and 4 longitudinal bands on the body.

(3) GENUS *GAVIALOCHARAX*

The single species *Gavialocharax monodi* has been found in Cameroon. It has the same size and body shape as that of *Phagoborus*, but it is more elongate with the snout strongly resembling that of an American ctenolucid. It has only 2 longitudinal bands on its body, and a very similar caudal fin pattern.

Ichthyborus besse; dentition and caudal fin pattern (after Poll 1957).

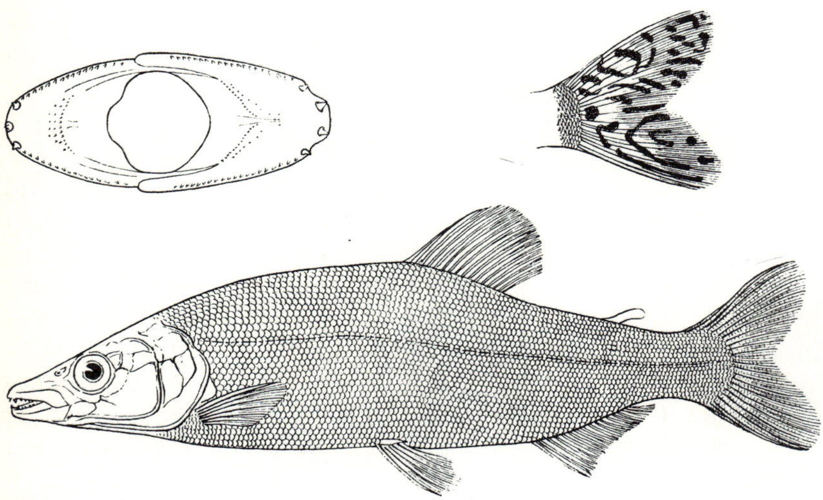

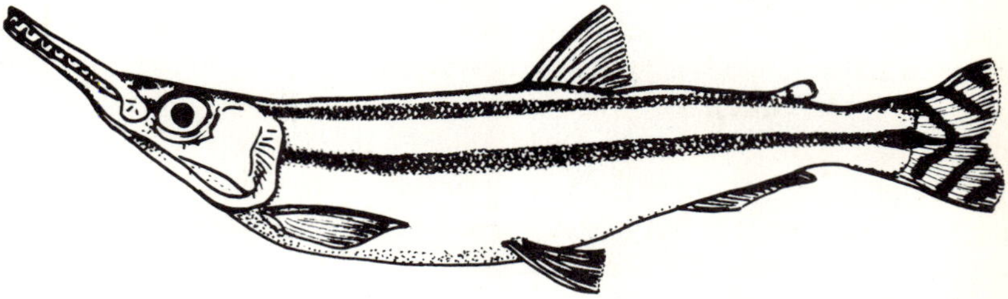

Type of *Gavialocharax monodi* (after Pellegrin, 1928).

(4) GENUS *ICHTHYBORUS**

The single species, *Ichthyborus besse*, is remarkable in having evolved (according to Daget) differently in the Nile, Tchad and Benoué basins on the one hand, and in the Congo basin on the other. In the Tchad basin, for example, it is a fin-eater, at least partly because there is no other fish to occupy this very peculiar ecological niche. In the Congo basin, where the fin-eaters are well established and well adapted, the subspecies *I. besse congolensis* contents itself with small whole fishes and some crustaceans.

The species reaches 200 mm in size and the two forms can be distinguished by their caudal fin color pattern. In the nominal form *(I. besse besse)* the yellow caudal fin is spotted. In the form from the Katanga *(I. besse congolensis)* these spots are united in vermicular, irregular lines and there is an ocellus at the fin base.

(5) GENUS *MESOBORUS*

The two *Mesoborus* species are apparently purely ichthyophagous and are not well known, being rare. The type species, *Mesoborus crocodilus*, from the Upper Congo, is a rather large fish (250 mm long) decorated with 2 series of round spots on the body and plain fins (or perhaps faded). The second species, *M. pellegrini*, also from the Congo basin but from different rivers (in South Cameroon and Kasai), has more branched dorsal rays (15 or 16 instead of 14) and the dorsal fin and caudal lobes are decorated with 3 oblique dark bands. The body patterns are the same, and in its general shape and coloration a *Mesoborus* looks much like an *Eugnathichthys* or a *Paraphago*.

* Not *Ichthyoborus*, which is the generic name of certain birds.

Distichodus decemmaculatus. Photo by B. Kahl.

Distichodus decemmaculatus. Photo by A. Norman.

Nannocharax fasciatus. Photo by Hilmar Hansen, Aquarium Berlin.

Hemigrammocharax sp., probably *H. multifasciatus.* Photo by Warren E. Burgess.

Mesoborus pellegrini (after Steindachner, 1913, as *M. crocodilus*); *Eugnathichthys eetveldii* (after Boulenger, 1898).

(6) GENUS *EUGNATHICHTHYS*

The two sympatric *Eugnathichthys* species from the Congo basin, both fin-eaters, are the ecological counterparts of the above-described *Mesoborus* species (ichthyophagous), in the same way as are the two subspecies of *Ichthyborus besse* (*I. besse besse* is a fin-eater, *I. besse congolensis* an ichthyophage). In contrast with *Mesoborus, Eugnathichthys* has no caniniform teeth but has bicuspid, compressed teeth forming a continuous, jaw-like cutting tool adapted to the fin-eating habit, as was shown by Matthes in the English journal *Nature* in 1961. Both species have the dorsal fin and caudal lobes barred with 3 (4?) black bands, and a spotted body.

Eugnathichthys eetveldii, the type species, attains a rather large size (30 cm) and has more scales (95-107 lateral line scales) than the smaller *E. macroterolepis* (68-80 lateral line scales), and the spots on the sides are roundish instead of being vertically elongate.

(7) GENUS *PARAPHAGO*

Technically, the single species *Paraphago rostratus*, from the Congo, differs from *Phago* only in its numerous scales, which is a rather poor generic character. However, a look at its body shape and coloration (the so-called habitus of the species) shows that it is much more like an elongate *Mesoborus* (body depth 5.5-6 in the standard length) that has lost its canine teeth and has become a fin-eater, with the same teeth as those of *Phago*.

?Phago boulengeri. Photo by Timmerman.

(8) GENUS *PHAGO*

The three or four species of *Phago* are relatively small (110-160 mm), and have the teeth compressed and tricuspidate, very trenchant and set close together (about 17-22 outer teeth on each side of each jaw), which enables them to cut the fins of other, sometimes larger, fishes. The resemblance of their "beak" to scissors is increased by the peculiar upper jaw of the Ichthyborinae, which moves upward when the mandible goes down. Another peculiarity is the scales, which are rugose, with a median spine forming a sort of longitudinal crest, giving somewhat the aspect of the osseous plates of a seahorse.

The species have a fusiform body shape, their fins are barred as in most other genera, and their elongate body is usually spotted.

The following key gives sufficient data for their identification:

a. "Beak" shorter than postocular part of head; 42-45 lateral line scales, 8 around caudal peduncle (body depth 6-7 in standard length)

. . . *P. boulengeri* (Central and upper Congo basin; Tchad ?)

Hemistichodus vaillanti. Photo by Dr. J. Gery.

Paraphago rostratus. Photo by Dr. Herbert R. Axelrod.

Phago maculatus. Photo by Klaus Paysan.

Two characteristic habitats of the African characoids. *Above:* Pool in Sierra Leone. Photo by E. Roloff. *Below:* Tributary of the Congo River. Photo by Dr. Herbert R. Axelrod.

aa. "Beak" longer than postocular part of head; 47-48 lateral line scales, 6 around caudal peduncle.

b. 2 scales between dorsal fin and lateral line; body depth about 6 in standard length . . . *P. intermedius* (Congo basin)

bb. 1 scale between dorsal fin and lateral line; body depth 8-9 in standard length.

c. 19 outer teeth on each side . . . *P. loricatus*, type species (Niger)

cc. 22 outer teeth on each side . . . *P. maculatus* (a dubious species, also from the Niger)

Belonophago tinanti. Photo by Timmerman.

(9) GENUS *BELONOPHAGO*

The species of this genus are very unusual in a group where odd forms are numerous. They resemble a needlefish (family Belonidae) as the generic name implies, with their very elongate body (body depth 10-11 times in the standard length) and snout (which is 1.5 to 2 times longer than the postocular part of the head). The scales, with their median spine, are similar to those of *Phago*, as is the scale formula (30-35 in the lateral line, one between dorsal fin and lateral line, and 6 around the caudal peduncle), and the fish is something like a very elongate *Phago*, with the same fin-eating habits. As concerns the coloration, the usual black marks on the caudal fin are replaced by red spots (one on each caudal lobe base).

The two species, both from the Congo basin, can be separated as follows:

a. "Beak" twice as long as postocular part of head . . . *B. hutsebouti* (Central Congo basin)

aa. "Beak" only 1.5 to 1.75 times longer than postocular part of head . . . *B. tinanti* (Stanley Pool)

Part II
The American Characoids other than the Characidae

KEY TO THE AMERICAN FAMILIES

a. Teeth, when present, in a single series at least on upper jaw; anal fin short, usually with less than 3 unbranched rays (except in Chilodinae) and less than 10 branched rays (except in *Abramites* and a few species in the Curimatidae); scales usually of the alestid-type, i.e. with their circuli of the caudal (apical) part toward the median line or even concentric (except in the Characidiidae).

b. *Amia*- or pike-like predators; teeth usually conical; occipital crest truncate; fontanels lacking.

c. *Amia*-like predators; dorsal long in *Hoplias*; caudal fin rounded; maxilla long; scales large, cycloid . . . ERYTHRINIDAE

cc. Pike-like predators; dorsal short, inserted far back (together with anal fin); caudal fin with well formed lobes; maxilla short; scales small, ctenoid . . . CTENOLUCIIDAE

bb. Usually non-predaceous fishes of medium to small size; teeth conical only in small or pigmy species (Characidiidae for example); occipital crest and fontanels variable.

d. Poeciloid body shape; dorsal fin long; sexual dimorphism present; a paired frontal organ (anterior fontanel lacking) . . . CRENUCHIDAE

dd. Body-shape variable; dorsal fin short (except in *Elachocharax*); usually no external differences between sexes (except in *Curimatopsis*); no paired frontal organ.

e. Teeth numerous, often in two rows on mandible; pterygoid teeth often present; fontanels usually absent (body usually elongate).

f. Scales of the *Hemigrammus*-type; dorsal fin with up to 19 rays; usually strongly adapted to life on river bottoms, with a large pectoral fin, flattened ventral surface, inferior mouth, etc.; maxilla usually not toothed . . . CHARACIDIIDAE (3 subfamilies)

ff. Scales of the *Alestes*-type; dorsal fin short; species adapted to life at the surface or in mid-water, with small pectoral fins, rounded ventral area, terminal or upturned mouth, etc. . . . LEBIASINIDAE (2 or 3 subfamilies)

Erythrinus erythrinus. Photo by Harald Schultz.

The trahira, *Hoplias malabaricus,* from the Rio Aguaro. Photo by Dr. Herbert R. Axelrod.

Hoplerythrinus unitaeniatus. Photo by Harald Schultz.

Young *Hoplias malabaricus.* Photo by Dr. Herbert R. Axelrod.

ee. Teeth, when present, few in number, with a single series on mandible; no pterygoid teeth; fontanels usually present; maxilla not toothed (except in Parodontinae); body shape variable.

g. Teeth always present in adult, at least on upper jaw, and more or less firmly set into the bone (except in *Bivibranchia* et al.).

h. Not more than 4 teeth, strongly fixed into the bone on each jaw (on each side); maxilla never toothed; mouth usually terminal or upturned, rarely inferior . . . ANOSTOMIDAE (2 subfamilies)

hh. Usually more than 4 compressed teeth, not firmly fixed into premaxilla; maxilla and mandible toothed only in the Parodontinae; mouth usually subinferior or inferior . . . HEMIODIDAE (2-3 subfamilies)

gg. Teeth usually absent, or, when present, set in the lips . . . CURIMATIDAE 4 subfamilies)

aa. Teeth variable, usually in more than one row on upper jaw (exceptions occur in the Characinae, Cheirodontinae and Serrasalmidae); anal fin moderate or long, with at least 3 unbranched rays and 10 branched ones (with a few exceptions in certain regressed Tetragonopterinae and Cheirodontinae); scales usually cycloid (except in certain Characinae) with circuli of caudal (apical) zone parallel or even divergent with axis of body (called *Hemigrammus*-type, with exceptions in specialized groups such as the Serrasalmidae and certain Characidae).

i. Coracoids forming a large, rounded keel; pectoral fins well developed, permitting flight (body very compressed; dorsal fin with up to 17 rays) . . . GASTEROPELECIDAE (2 subfamilies)

ii. Coracoids and pectoral fins usually not developed; when developed, not being adapted to real flight.

j. Maxilla reduced, not toothed; dorsal fin long, with at least 16 rays; usually a predorsal spine and a series of ventral spines (serrae) (body very compressed, usually disciform; teeth variable according to diet; scales small, the circuli concentric) . . . SERRASALMIDAE (3 subfamilies)

jj. Maxilla usually toothed and not reduced; dorsal fin short; rarely a predorsal spine (in the Stethaprioninae); no real serrae (a strong keel with pointed scales in *Stichonodon*) . . . CHARACIDAE (12 subfamilies)

Family Erythrinidae

Trahiras

The trahiras, as the Brazilians call them, are quite similar to the bowfin *(Amia calva)* of sluggish waters of the eastern United States, although they come from a very different evolutionary line. It is a small group, without economic importance (except for the giant trahira, a most delicious fish,

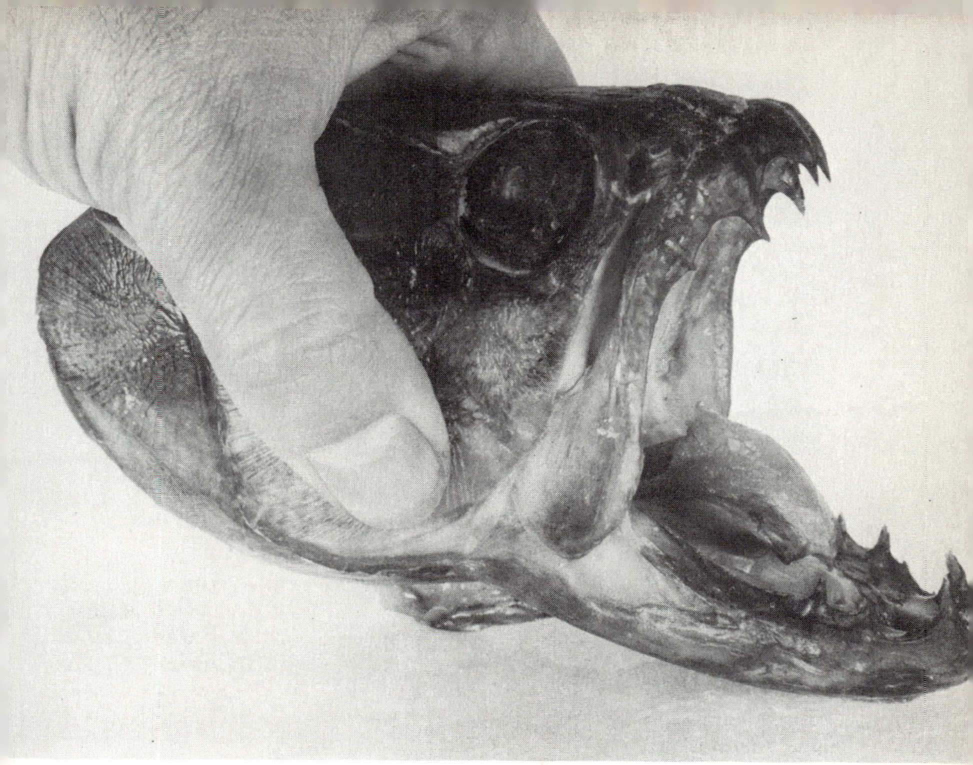

Head of a giant trahira (*Hoplias macrophthalmus*) about 50 cm long. Photo by Dr. Herbert R. Axelrod.

which is becoming rare),* but interesting from the zoologist's point of view. Some are predators which have become adapted to life in small brooks or ponds, where they can exist without much oxygen, thanks to accessory systems which can absorb it directly from the air. At least one species is capable of retaining moisture and air for a long time and walking on the ground at night in the search for a more suitable pond than the one just abandoned.

As regards their anatomy, particularly that of the bones of the head, certain experts believe that they are exceptional among the characoids. They are probably the most archaic in many respects (though specialized for predation), having 5 branchiostegal rays, a cylindrical body form with a short anal fin and large scales, numerous teeth on the palate and (apparently) a primitive pectoral girdle like that of the bowfin.

The trahiras may be divided into two groups at the generic level, as follows:

a. Dorsal rays iii, 11 to 15; at least 37 lateral line scales; maxillary bone with 2 or 3 small canines plus a series of conical teeth . . . Genus *Hoplias*

aa. Dorsal rays iii, 8 or 9; 32-37 lateral line scales; maxillary bone without canines . . . Genera *Erythrinus* and *Hoplerythrinus*

*At least in the Guianas.

Head of *Erythrinus erythrinus.* Upper Juruena, Mato Grosso. Photo by Harald Schultz.

Boulengerella lateristriga. Photo by Dr. J. Gery.

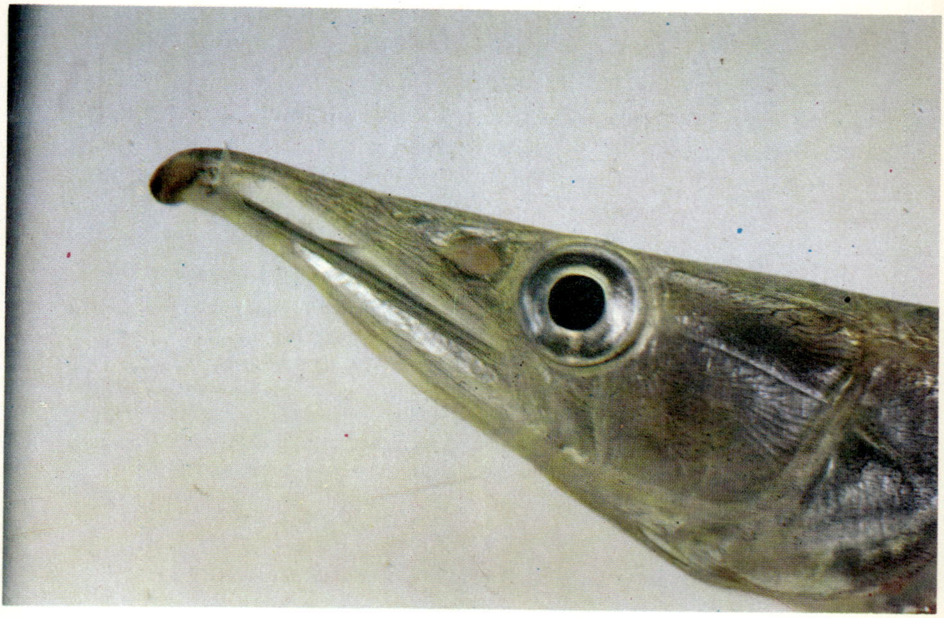

Head of *Ctenolucius hujeta.* Photo by Dr. Herbert R. Axelrod.

Ctenolucius hujeta. Photo by Dr. Herbert R. Axelrod.

GENUS *HOPLIAS*

There are probably only 3 species in this genus, including the true trahira *(Hoplias malabaricus)*, which is distributed in most parts of South America and is somewhat confused as concerns its nomenclature. It is one of the most often cited of the characoids (more than 80 references), but often under quite different names (see article by Dr. Viggo Schultz in *Tropical Fish Hobbyist*, September, 1971).

The three species can be distinguished by means of the following key:

a. Large size (up to one meter); eye relatively large, 14 or 15 times in standard length of adults, partly visible from above; abdomen flat and not colored . . . *H. macrophthalmus* (Guianas, where it is called Aimara)

aa. Rather small size (up to 25 cm or so); eye small, about 18 to 20 times in standard length of adults, lateral, not visible from above; abdomen round and colored

b. 43-47 lateral line scales; ii, 13-15 dorsal rays . . . *H. microlepis* (Pacific slope of southern Ecuador, Panama and Costa Rica)

bb. Usually 37-43 lateral line scales; iii, 11 or 12 dorsal rays, rarely 13 . . . *H. malabaricus* (South America, including the countries between the localities of *H. microlepis*)*

**H. lacertae*, classically considered as a synonym of the trahira *(H. malabaricus)*, may be a distinct species.

Erythrinus erythrinus **in typical surroundings. Upper Juruena River. Photo by Harald Schultz.**

GENERA *ERYTHRINUS* AND *HOPLERYTHRINUS*

The two genera, each containing a single species, are very close to each other. The following key summarizes their most important features:

a. Maxillary bone not very elongate, not reaching posterior margin of eye in adults; eye very small, 22-25 times in standard length; dorsal fin pointed and elongate in males; caudal fin spotted; color pattern of flanks rather variable, usually marbled . . . *Erythrinus erythrinus* (South America and Trinidad)

aa. Maxillary bone elongate, its tip well beyond posterior margin of eye in adults; eye small, about 19 times in standard length; dorsal fin low in both sexes; caudal fin plain; dark longitudinal band, caudal spot and opercular ocellus present, which are characteristic of the species . . . *Hoplerythrinus unitaeniatus* (South America)

Family Ctenoluciidae

South American Pike-characoids

The Ctenoluciidae are medium to large size predators (20-100 cm), having acquired a very long and tapered pike-like body, with a gavial-like snout and short dorsal and anal fins inserted posteriorly. The lower jaw, slightly less elongate than the upper one, is included; the numerous teeth are small, usually acute, close together and directed backward, and occur on both jaws as well as on the ectopterygoid bone; either the snout or the chin has fleshy appendages. According to T. Roberts, who studied their anatomy extensively, they bear the closest resemblance to the African Hepsetidae. Both families are rather remote from the Acestrorhynchini, the most elongate members of which have acquired the same pike-like aspect by evolutionary convergence.

The four known species can be distinguished by means of the following key:

a. Premaxillae joining in front in a semicircle; fleshy flaps present on chin; a short, internal second row of teeth on the mandible; ectopterygoid bone (in roof of mouth) with conical teeth only; lateral line incomplete; 42-50 strongly ctenoid scales in a longitudinal row; anal fin before middle of dorsal fin, its last rays never filamentous (an ocellus on caudal fin base) . . . Genus *Ctenolucius* (Panama and N.W. South America) (a single species, *C. hujeta*, with 3 subspecies: *C. hujeta beani* in Panama and Colombia, *C. hujeta insculptus* in Colombia, and *C. hujeta hujeta* in Venezuela; synonym *Luciocharax insculptus*)

Head of *Boulengerella maculata.* Photo by Harald Schultz.

Boulengerella maculata. Photo by Harald Schultz.

Crenuchus spilurus male. Photo by Harald Schultz.

A pair of *Poecilocharax weitzmani*. Photo by E. Roloff.

aa. Premaxillae joining in front at an acute angle; no mental barbels, but a fleshy appendage on snout; both jaws with a single series of teeth; ectopterygoid bone finely granulated, otherwise having a few weak, conical teeth anteriorly; lateral line complete (subgenus *Spixostoma*) or incomplete (subgenus *Boulengerella* typical); 78-100 or more finely denticulated scales in a longitudinal row; last anal rays filamentous, at least in young ... Genus *Boulengerella*

b. Lateral line complete; 97-110 scales along a longitudinal series above or below lateral line; dorsal fin situated at a level between ventral and anal fins, its longest rays not reaching adipose fin (sometimes one or more series of small spots on dorsal part of sides; ocellus on caudal fin base; a longitudinal band (in young only)) ... Subgenus *Spixostoma* (a single species *B. (Spixostoma) lucia,* from Guyana and, rarely, Amazon basin; synonyms *B. ocellatus, B. cuvieri* and *?B. longipinne*)

bb. Lateral line incomplete; 78-91 scales along a longitudinal series; dorsal fin partly above anal fin, as in *Ctenolucius* ... Subgenus *Boulengerella*

c. 78-82 longitudinal, 12-14 transverse and 60-64 predorsal scales; longest dorsal rays not reaching adipose fin; a brown longitudinal band, no spots on sides ... *B. (Boulengerella) lateristriga* (Rio Negro)

cc. 82-92 longitudinal, 15-17 transverse and about 75 predorsal scales; longest dorsal rays reaching adipose fin; series of spots on flanks ... *B. (Boulengerella) maculata,* synonym *B. taedo* (Amazon basin)

Family Crenuchidae

The Crenuchidae, like the Geisleriinae of the Characidiidae, have a most unusual appearance and a number of atypical features when compared to a generalized characoid like *Astyanax*. They have a cichlid- or cyprinodont-like body, a long dorsal fin (more than the usual ii,10 or ii,11 number of rays), and one of the two genera lacks an adipose fin. The teeth, in *Poecilocharax weitzmani* at least, are tricuspidate in females and young and more numerous and acutely conical in the adult male. At the same time the maxillary bone becomes very elongate and vertical and the mandible becomes prominent, resembling (very remotely) the modification of the male salmon when it migrates upstream. Moreover, this change is accompanied by a sexual dimorphism. The male *Crenuchus* has light spots on the anal fin like those of most mouth-brooding cichlids (genus *Haplochromis* etc.), and male *Poecilocharax weitzmani* have the same spots on the dorsal fin, together with a velvet-black maxillary bone and gill membrane (a system for intimidation during fighting periods of the males?).

Boulengerella (Spixostoma) lucia. Photo by Harald Schultz.

A pair of *Crenuchus spilurus,* male above. Photo by Timmerman.

Photo by Dr. Herbert R. Axelrod.

Photo by Dr. Herbert R. Axelrod.

Photo by Dr. Fernandez-Yepez.

These photos represent five different species of the genus *Characidium.* For accurate identifications at the species level an anatomical study of the specimens would be necessary. Top photo on this page by Dr. Herbert R. Axelrod, bottom photo by Stanislav Frank.

Poecilocharax bovallii, male above, female below (after Eigenmann, 1912).

Finally, the most peculiar feature of the Crenuchidae is a mysterious organ on top of their head, which is housed in a pair of recesses in the frontal bone at eye level and which is not possessed, as far as known, by any other teleost group. The organ is formed by rod-shaped cells (similar to those of the retina) encircled by a net of capillary blood vessels. The whole structure suggests a receptor of radiations (infrared??) which are compared with some internal parameter (perhaps the internal temperature?), thanks to the blood vessels. As the physiological study of this phenomena has not even begun, the above is pure speculation and a fascinating problem remains to be solved.

GENUS *CRENUCHUS*

The single known species, *Crenuchus spilurus*, from the Guianas and a large part of the Amazon basin (including the Upper Solimoes and Rio Negro), differs from the two species of *Poecilocharax* mostly in having an adipose fin. The adult male reaches less than 50 mm in standard length, and resembles certain dwarf cichlids, due to its body shape as well as to its brilliant coloration.

It has long been considered an aggressive fish and has never been popular in aquaria (hence not collected). Therefore, little is known about its reproductive behavior. Some have said that it spawns on large leaves or stones (again like certain cichlids). With the great popularity of cichlids, it is hoped that this very unusual characoid will become a favorite of aquarists.

GENUS *POECILOCHARAX*

The type of the genus, *Poecilocharax bovallii*, is a small, nicely colored, killy-like fish inhabiting, exclusively, the surroundings of the famous Kaieteur Falls in Guyana on the Potaro River. According to its first collector and describer, the famous Carl Eigenmann, the male has a geranium-red caudal fin decorated with pearl-white spots. As early as 1954 Eigenmann's pupil and follower, G. S. Myers, recommended it as a potential ornamental aquarium fish.

The second species, *P. weitzmani*, more recently discovered, is even more striking. It inhabits the Upper Solimoes region (habitat of the well-known neon tetra), as well as the Upper Negro-Orinoco regions. Adult males of less than 30 mm standard length have a yellow body, darkening above and tending toward red on its posterior part, with a longitudinal black band on which shines (like a liner in the night with all portholes illuminated) the neon-blue center of each scale; superimposed on it is a blood-red band; the large fins are yellow with brown and red, as well as pearl-white spots; the eye is green and red, and the head is marked with velvet-black bands. According to the well-known German ichthyologist Dr. Hermann Meinken, its behavior seems to be nearer that of an ordinary tetra than that of *Crenuchus spilurus*.

Family Characidiidae
South American "Darters"

Most members of the Characidiidae, all of small to extremely small size, have evolved toward a life suited to the peculiar medium of river bottoms and, sometimes, to the immersed roots of trees. The original members of the family (of unknown line) have probably recent representatives in the Characidiinae from southeastern Brazil, which still have a middle-water life. The family, which attracted attention only rather recently, has been split into subunits as follows:

Lebiasina astrigata. Photo by Dr. Herbert R. Axelrod.

Lebiasina sp., possibly *L. panamensis.* Photo by Dr. Herbert R. Axelrod of a specimen from Rio Calima, Colombia.

Nannobrycon unifasciatus, form lacking caudal ocellus. Photo by Dr. Herbert R. Axelrod.

a. Dorsal fin short, usually with 10-13 rays (sometimes 14 in *Jobertina rachovi*); small species mostly adapted to life on river bottoms, the extreme reached in the elongate body, flat breast, paddle-like pectoral fins, inferior mouth and superior eyes of species of the torrents; lateral line complete or incomplete . . . subfamily CHARACIDIINAE

aa. Dorsal fin long, usually with 17 or more rays; very small species, mostly adapted to life around submerged roots or falling leaves on the edge of small streams, with a short body and the pectoral fins not greatly enlarged; lateral line incomplete or absent.

b. Teeth tricuspidate, with only one row on premaxilla; no maxillary teeth; pterygoid teeth present; suborbital bones reduced; posterior fontanel present . . . subfamily ELACHOCHARACINAE

bb. Teeth conical, with two rows on premaxilla; maxillary teeth present; pterygoid teeth absent; suborbital bones absent; fontanels absent . . . subfamily GEISLERIINAE

SUBFAMILY CHARACIDIINAE

South American Darters

This group has long been considered as comprised of only a few species, with most of the specimens collected given the passe-partout name *Characidium fasciatum* (the oldest named characidiid). Thanks to recent studies, chiefly those of the Brazilian expert Haroldo Travassos, the complexity of the group has been revealed and the number of described and still undescribed species may be estimated at 50 or more. Within the Characidiinae, three genera may be recognized:

a. No maxillary teeth (pectoral fin with 3 to 5 unbranched rays and 6 to 12 branched ones).

b. First dorsal ray not rudimentary; lateral line incomplete; species from southeastern South America, living in mid-water (with exceptions) . . . *Jobertina*

bb. First dorsal ray rudimentary; lateral line complete; species not restricted to southeastern South America, usually living on stream bottoms . . . *Characidium*

aa. 4-8 maxillary teeth (pectoral fin with 4 or 5 unbranched rays and only 4 to 8 branched ones) . . . *Klausewitzia*

GENUS *JOBERTINA*

The genus *Jobertina* is still in a confused state, being not well defined. It includes a well-delimited group of four or five southern species (*J. interrupta*, *J. lateralis*, *J. bahiensis*, *J. rachovi*, and *J. theageri*, respectively from Rio de Janeiro, upper Paraguay, Bahia, Paranagua, and Uruguay) and another, composite Amazonian-Guianan group (*J. eleotroides*, from Guiana plus a few other, still undescribed, species).

Characidium species. Photo by Timmerman.

The true *Jobertina* live in mid-water and look quite generalized. The other ones look like tiny, regressed *Characidium*, and probably live on the bottom in small rivers and brooks. Certain species have lost not only their lateral line, but even their adipose fin, and they would probably not be classified in the future within the genus *Jobertina*.

GENUS *CHARACIDIUM*
SOUTH AMERICAN DARTERS

Unlike most characoid groups, *Characidium* and its allies have plenty of characters on which to base a convenient and stable classification. These characters, such as the pectoral and ventral rays, presence or absence of scales on the isthmus, position of the eyes and form of the teeth, correspond with the way of life of the fish, depending on whether it lives on sand or rocks, in fast running waters or slow moving ones, etc. However, adaptations are rarely very restrictive in the teleostean fishes, and certain species may tolerate quite a few different biotopes.

Nannostomus espei. Photo by Dr. Herbert R. Axelrod.

Nannostomus bifasciatus. Photo by Dr. Herbert R. Axelrod.

Nannostomus espei spawning. Photo by H.J. Richter.

Male (above) and female (below) *Nannostomus espei* approaching spawning site. Photo by H.J. Richter.

The complex genus *Characidium* will be treated here very briefly, by means of the following artificial key:

a. First three pectoral rays unbranched, sometimes thickened.
  b. First ventral ray unbranched.
    c. Isthmus (pectoral region just behind the gill membrane) scaled . . . *C. fasciatum*-group in the broadest sense. (This group includes at least 20 Guianan and Amazonian spp., which are mostly identified by the structure of their teeth and their proportions)
    cc. Isthmus naked, or with 3 to 7 or more scales lacking (sometimes only one scale lacking) . . . *C. purpuratum*-group in the broad sense. (This group includes 15 to 20 spp., mostly from northwest and southeast South America, very few from the Amazon basin)
  bb. First two ventral rays unbranched . . . undescribed species from Rio de Janeiro
aa. First four pectoral rays unbranched and sometimes thickened.
    d. First ventral ray unbranched . . . *C. crandelli*-group (2 spp. from Rio Branco and Ecuador)
    dd. First two ventral rays unbranched, at least on one side, and sometimes thickened (isthmus usually naked) . . . *C. blennioides*-group (5 very specialized Guianan spp., for the most part undescribed)

Type of *Elachocharax pulcher.* Original drawing by Mr. Pablo Bravo in 1930, published for the first time by permission of Dr. G.S. Myers.

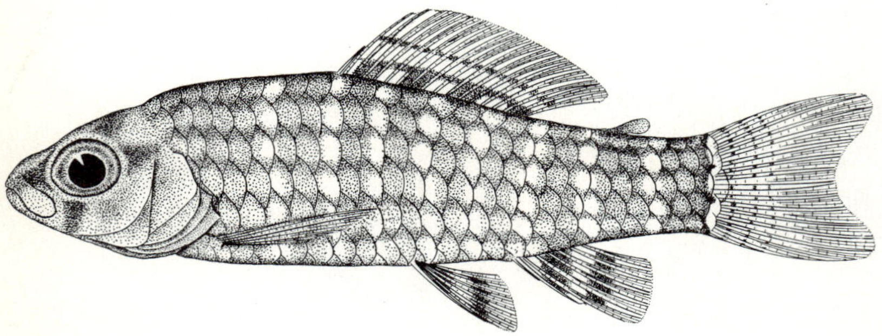

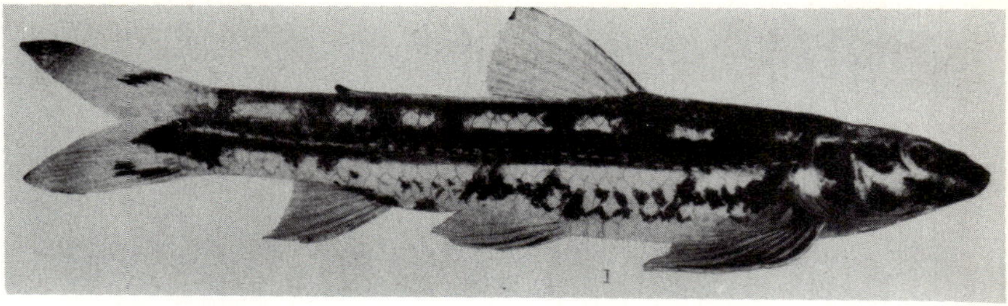

Type of *Klausewitzia vintoni* (after Eigenmann, 1912).

GENUS *KLAUSEWITZIA*
SOUTH AMERICAN DARTERS

This interesting genus, which differs from *Characidium* in the presence of maxillary teeth as well as in the small number of branched pectoral rays, has been recognized only recently. To the already known species *K. vintoni* (synonym *K. laterale*) and *K. ritae* (the type species), one must add a curious 50 mm-long species which has been observed in nature by Dr. Rolf Geisler. It is elongate, green, and camouflages itself with the algae-like plants undulating in the current. To add to this mimicry, it always stays near the top of the narrow leaves, head upstream. *Klausewitzia vintoni* is quite similar in meristic characters and it is not impossible that it has the same ecology.

SUBFAMILY ELACHOCHARACINAE

This curious group includes one genus with at least three extremely small species living in the fine roots of plants growing in the banks of brooks. They are shorter than most species of *Characidium*, and have a striped brown pattern which may be homochromic with the surroundings.

Possibly a new species of *Elachocharax* from Madeira basin. Photo by Dr. Herbert R. Axelrod.

Nannostomus beckfordi. Photo by Harald Schultz.

Male *Nannostomus beckfordi* in full color. Photo by H.J. Richter.

Xanthochromic variety of *Nannostomus beckfordi;* usually it is called "gold anomalus." Photo by Dr. Herbert R. Axelrod.

Another form of the *Nannostomus beckfordi* "complex." Photo by Stanislav Frank.

Elachocharax pulcher, the first known species, lives in Venezuela. *E. georgiae* lives in small tributaries of the Rio Negro near Manaus, in the Upper Amazon, as well as in the Rio Madeira basin, where a "giant" specimen has been found (25 mm in standard length). A third species, lacking an adipose fin, is still undescribed.

SUBFAMILY GEISLERIINAE

The single known species, *Geisleria junki*, is even more aberrant than the *Elachocharax* spp. just mentioned. If not for the structure of the first and last vertebrae, which are characteristic of a characoid, it could easily be thought to be a killy. Its ecology is likely to be the same as that of *Elachocharax*.

Holotype of *Geisleria junki,* from the Rio Madeira basin (after Gery, 1973, in Vie et Milieu, with permission), and radiograph of the specimen, made by Dr. W. Born, Freiburg, Germany.

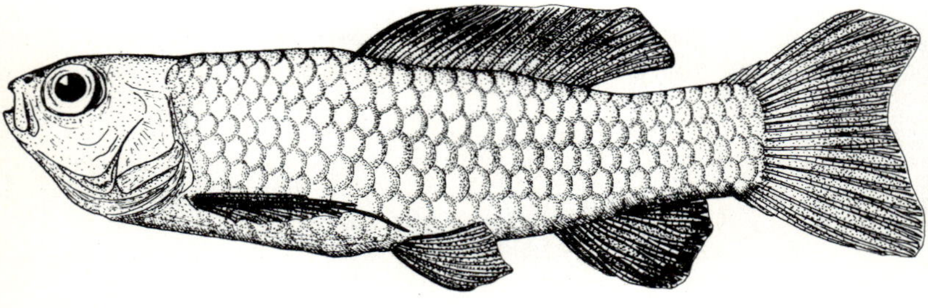

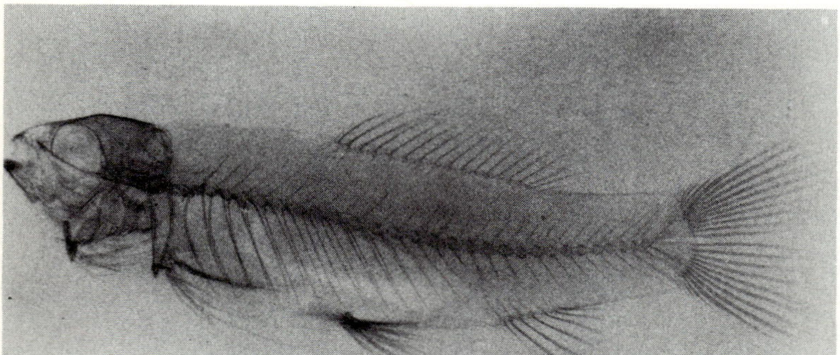

Family Lebiasinidae

Pencil Fishes and Pyrrhulinins

The lebiasinids have many characteristics in common with the trahiras (erythrinids). For example, almost nothing differentiates a *Lebiasina* externally from an *Erythrinus*. Both have the same body shape and head form, both lack the adipose fin, and both are adapted to a life in poorly oxygenated waters. However, certain features of their internal anatomy, as well as different dentition (the lebiasinids usually have a short maxillary bone, 2 rows of teeth, at least on the lower jaw and never the strong conical or canine-like teeth of the trahiras), seem to be in contradiction: the lebiasinids are presently considered as coming from a different stem than the erythrinids.

Within the group, two subgroups are easily recognizable: (1) Lebiasininae proper, and (2) Pyrrhulininae, which look like pigmy lebiasinids and include the well-known pencilfishes and pyrrhulinins.

Technically, they can be separated as follows:

a. Median-sized fishes with a relatively long maxillary bone and tricuspidate teeth; a patch of teeth on palate; four branchiostegal rays (upper caudal lobe often longer than lower one; lateral line incomplete; adipose fin irregularly present) . . . Subfamily LEBIASININAE

aa. Small fishes with a short maxillary, usually not functional, teeth conical (*Pyrrhulina* et al.) or multicuspidate incisiform (*Nannostomus* et al.), never tricuspidate; no teeth on palate; three branchiostegal rays . . . Subfamily PYRRHULININAE

b. Mouth terminal; teeth multicuspidate (except the inner mandibular row, which is composed of a few conical teeth); lateral line incomplete; caudal lobes equal or the lower one longer (in species which swim head up); adipose fin irregularly present . . . Tribe Nannostomini

bb. Mouth upturned; teeth conical on both jaws; lateral line completely lacking; upper caudal lobe longer than lower one; adipose fin never present . . . Tribe Pyrrhulinini

SUBFAMILY LEBIASININAE

This group classically includes two genera, *Lebiasina* and *Piabucina*. The characters which separate them, however, are variable and the genus *Piabucina* is treated here as a synonym of the older genus *Lebiasina*.

All species are from the northern part of South America, with the exception of *Lebiasina intermedia*, said to come from Santarem on the Lower Amazon. Some, like *L. bimaculata*, have been introduced into other countries in the fight against mosquitoes. The following key gives the main points distinguishing the species:

Nannostomus trifasciatus. Photo by Dr. Herbert R. Axelrod.

Nannostomus marginatus. Photo by H.J. Richter.

Above and below: A pair of *Nannostomus marginatus* spawning. Photos by H.J. Richter.

Lebiasina sp. from the Cauca Valley, 15 miles from Cali, Colombia. Photo by Dr. Herbert R. Axelrod.

a. 37 scales in a longitudinal series and 9 in a transverse series; dorsal fin at mid-body or close to it (a vertical caudal spot; adipose fin usually present)
. . . *L. erythrinoides*, type of the genus *Piabucina* (Venezuela)

aa. 25-32 scales in a longitudinal series and 6-8 in a transverse series; dorsal fin behind mid-body.

b. Depth of body 3.5-4.0 in standard length (usually 28 longitudinal scales; branches of mandibles, as seen from below, forming a U; humeral and dorsal spots present).

c. Length of head 4.2-4.5 in standard length; adipose fin present only in 5% of the samples; caudal spot and median longitudinal band present
. . . *L. bimaculata*, type of the genus *Lebiasina* (originally from Peru and Ecuador)

cc. Length of head 3.7-4.25 in standard length; adipose fin usually present; no caudal spot, a longitudinal band below median line
. . . *L. festae* (Panama, N.W. Colombia)

bb. Depth of body usually more than 4 in standard length.

d. A series of spots along mid-axis (depth of body 4.5-5.0 and length of head 4.0-4.75 in standard length).

e. Adipose fin present in only 10% of the samples; spots small and numerous in front of dorsal level; no dorsal spot
. . . *L. multimaculata* (N.W. Colombia)

ee. Adipose fin usually present; spots large and roundish; usually a dorsal spot
. . . *L. astrigata* (Ecuador and maybe Rio Meta in Colombia)

dd. A longitudinal band (sometimes two) (adipose fin usually present).
f. Depth of body 4-4.6 in standard length; 10 or 11 anal rays.
g. 7 or 8 scales between dorsal and ventral fins.
h. 27 longitudinal scales (branches of mandibles forming a V) . . . *L. unitaeniata* (Guyana)
hh. 28-32 longitudinal scales.
i. Length of head 3-4 in standard length.
j. Longitudinal band often split up into irregular blotches . . . *L. panamensis* (Panama)
jj. Longitudinal band continuous, though not well delimited . . . *L. boruca* (Costa Rica)
ii. Length of head 4-4.33 in standard length.
k. No dorsal spot; two lateral bands and a small caudal spot. . . . *L. aureoguttata* (Ecuador)
kk. A dorsal spot; one lateral band and a roundish caudal spot (branches of mandibles forming a V) . . . *L. pleurotaenia* (Venezuela)
gg. 6½ scales between dorsal and ventral fins (32 longitudinal scales; depth of body 4.5 and length of head 4.25 in standard length; lateral line going up to dorsal level; humeral, dorsal and caudal spots present, plus a lateral band) . . . *L. intermedia* (Santarem)
ff. Depth of body 4.7–5 in standard length; 12 anal rays (branches of mandibles forming a V; humeral and dorsal spots present, plus a lateral band) . . . *L. elongata* (Upper Amazon in the northern part of Peru and south of Ecuador)*

SUBFAMILY PYRRHULININAE

Tribe Nannostomini

Pencil Fishes

The pencilfishes are among the best known characoids. Although somewhat delicate in aquaria and difficult to induce to breed, they are very popular among aquarists. Their fusiform shape, their brilliant colors and the curious "head up" swimming of certain species amply justify their popularity. All pencilfishes have a very small, always open mouth with the anterior part of the mandibular bone movable over the posterior part, which enables them to pick up very small prey on various substrata (mostly plants, rocks and fallen branches).

According to a recent revision, only two genera are recognized within the Nannostomini: *Nannostomus* and *Poecilobrycon*.† They are separable on the basis of a character which is difficult to observe by the layman, i.e. the presence or absence of a canal in the second circumorbital bone of

*The author lacks data concerning *Lebiasina uruyensis* from Venezuela.

† A still more recent revision, that of Weitzman and Cobb (March, 1975), recognizes only one genus, *Nannostomus*.

Nannobrycon unifasciatus. Photo by Harald Schultz.

Nannostomus harrisoni. Photo by H.J. Richter.

Nannobrycon eques. Photo by Harald Schultz.

Nannobrycon eques. Photo by H.J. Richter.

the "cheek," which is associated with the length of the snout (distinctly longer in *Poecilobrycon*). Actually, two species of the genus *Poecilobrycon* (those belonging to the subgenus *Nannobrycon*) swim in a head-up position and the lower lobe of their caudal fin is distinctly larger than the upper one. These characters seem more convincing and the genus *Poecilobrycon* (type-species *P. harrisoni*) is here abandoned in favor of *Nannobrycon* (type-species *N. eques*), *Poecilobrycon* becoming a subgenus of *Nannostomus*.

GENUS *NANNOSTOMUS*
Subgenus *Nannostomus*
PENCILFISHES

In contrast to several other characoid groups, the seven or so species of *Nannostomus* have distinctive diurnal color patterns which make them rather easy to identify, at least if one does not bother with details. The term diurnal is used on purpose, because all nannostomins have a so-called nocturnal color pattern, which is usually composed of two or three blotches, very different from the stripes of most species during the day.

The following key is based mostly on coloration:

a. 5 oblique blotches on flanks, as if the fish were in a sort of permanent nocturnal pattern (adipose fin present) . . . *N. espei* (Guyana), the barred pencilfish

aa. Diurnal color pattern composed of stripes along body.

b. Only one stripe somewhat below mid-axis of body (adipose fin absent) . . . *N. beckfordi* (Guianas, lower Rio Negro, lower and middle Amazon basin)†

bb. Two or three stripes present.

c. 2 stripes, one similar to that of *N. beckfordi* and the other above mid-axis of body.

d. Adipose fin absent; a few scales with lateral line pores; anal rays iii, 9 . . . *N. bifasciatus* (Surinam and French Guiana)

dd. Adipose fin present; no lateral line; anal rays iii, 8 . . . *N. digrammus* (Rio Madeira, lower and middle Amazon, and Guyana), the two-striped pencilfish

cc. Three stripes, the two first like in the preceding species and the last one low on abdomen.

e. Adipose fin usually present; about 26 or 27 longitudinal scales.

† A few forms, which may be sibling species or subspecies, can be recognized, for example *N. minimus* and *N. aripirangensis; N. anomalus* could be a synonym of the nominate species; it is still a popular name of the Beckford's pencilfish, chiefly for the xanthochromic strain, an artificial variety.

f. Median stripe ending on caudal peduncle
. . . *N. trifasciatus* (upper Amazon), the three-lined pencilfish

ff. Median stripe extending posteriorly onto caudal fin
. . . *N. erythrurus* (Guyana; middle and lower Amazon, including lower Rio Negro), also a three-lined pencilfish

ee. Adipose fin absent; 21-23 longitudinal scales
. . . *N. marginatus* (Lower Amazon, Guyana and Surinam, Colombia), the dwarf pencilfish

Despite the apparent simplicity of the above classification, the genus *Nannostomus* is probably quite complex, actually actively evolving, somewhat in the manner of the African cyprinodonts.* For example, there are indications that the species exist by "pairs" which are almost indistinguishable externally. Two cases are reasonably demonstrated, the *N. digrammus-bifasciatus* pair and the *N. trifasciatus-erythrurus* pair. The same situation is likely to occur in *N. beckfordi* and in *N. marginatus*, which are quite variable. Members of the pairs live in different localities (allopatric forms), and their status as species can only be ascertained on the basis of their different chromosomes. This does not eliminate the possibility of sibling (sympatric) species of each member of the pair, and the variable chromosome formulae (karyotypes) give some indications of it (J. J. Scheel, personal communication).†

Subgenus *Poecilobrycon*

The single species *Nannostomus harrisoni*, rarely seen in the aquarium, looks much like an elongate *Nannostomus beckfordi*, with the following coloration: a single stripe below mid-axis of body extending onto caudal fin; anal and ventral fins blue with a red spot in the male, which also has red spots on the caudal base above and below the extension of the stripe. It swims horizontally and differs from a typical *Nannostomus* only in the longer snout and the presence of the second infraorbital canal; it possesses an adipose fin. This species is found in Guyana and also in the Upper Amazon and (probably) the Middle Amazon.

* For the evolution of *Aphyosemion* see J. Scheel, 1968. *Rivulins of the Old World*, TFH Publications.

† In the revision cited in an earlier footnote, Weitzman and Cobb have just described a new species, *Nannostomus marilynae*, that looks somewhat like a small *N. trifasciatus* with slightly different diurnal and nocturnal coloration.

Pyrrhulina of the *P. brevis* group. Photo by Dr. Herbert R. Axelrod.

Pyrrhulina sp., also of the *P. brevis* group. Photo by Dr. Herbert R. Axelrod.

Pyrrhulina of the *brevis* group, the form generally called *P. "lugubris."* Photo by Harald Schultz.

Pyrrhulina of the *P. brevis* group from Rio Aguaro. Photo by Dr. Herbert R. Axelrod.

Pyrrhulina laeta includes *P. semifasciata* as a synonym. Photo by Harald Schultz.

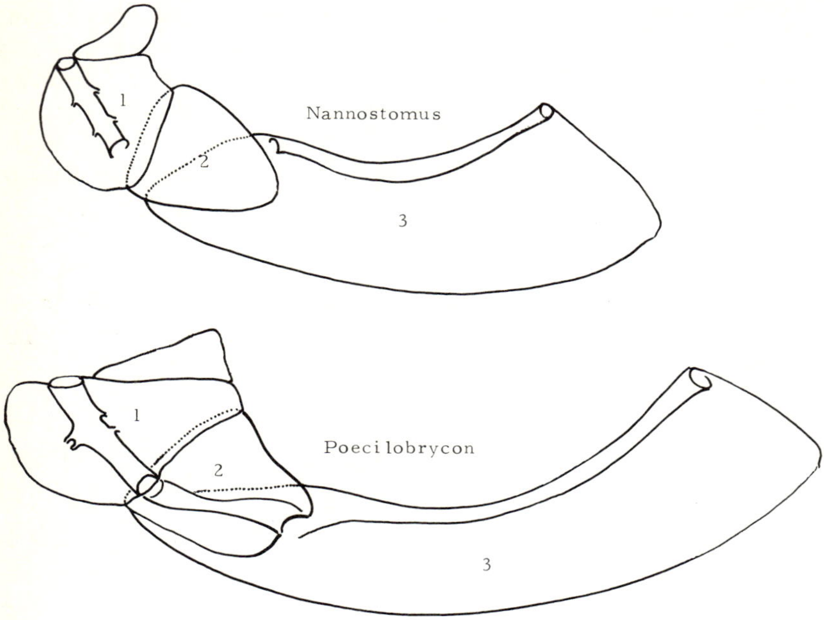

Distinctions between circumocular bones in *Nannostomus* and *Poecilobrycon.* Original drawing by the author.

GENUS *NANNOBRYCON*

In this genus, whose species swim head up at a 45-75° angle, the lower caudal lobe is distinctly larger and is formed by 11 principal rays (instead of 9 in *Nannostomus* as well as, actually, in almost all characoids), and the snout and second infraorbital are *Poecilobrycon*-like.

The two species have often been confused. *Nannobrycon unifasciatus*, the one-lined pencilfish, from the Middle and Upper Amazon basins, the Rio Negro, the Upper Rio Orinoco and Guyana, has 28-30 longitudinal scales with the first 2-5 scales perforated; the adipose fin is apparently always present. It has a longitudinal band similar to that of *N. harrisoni* and *N. beckfordi*, which extends onto the enlarged lower caudal lobe in various ways depending on the population. In specimens from Guyana the center of the lobe is light (red in life) and there is an ocellus above it, at the inferior part of the upper lobe base (this population has been described as a distinct species, *N. ocellatus*); in specimens from Brazil, among others, there is no such ocellus, or an indistinct one. As in the genus *Nannostomus*, a pair of semi-cryptic species may be suspected.

Nannobrycon eques, usually sympatric with the preceding species (same localities) and resembling it, has only 24 or 25 longitudinal scales with not a single one perforated, and the adipose fin is sometimes absent. The brown longitudinal stripe, well below the mid-axis of the body, also extends onto the enlarged lower caudal lobe; it is underlined by a series of spots on the center of the scales of the belly and bordered above by a silvery band, above which, on the upper part of body, there are two narrow lines. The two oblique transverse bands of the nocturnal pattern of the nannostomins are often superimposed on this longitudinal pattern, even during the day. The anal fin is brown with a red spot in the middle, and the brown lower caudal lobe is spotted with orange. This species is much less variable than its cousins.

Tribe Pyrrhulinini
Pyrrhulinins

The pyrrhulinins are much less well known than the pencilfishes, and they have never been revised (the author is informed that an important work was in progress at the time this was written). They are surface fishes, living mostly in small ponds together with the rivulins, but they are not able to travel on moist soil from pond to rivulet and so on, as do *Rivulus* species. They have a relatively large mouth with acute teeth, as contrasted with the pencilfishes, and some species (at least) are capable of ingesting ants that have fallen into the water. They have no adipose fin, no lateral line and no fontanels. Almost all have a plain colored body or are with a stripe and a dark spot on the dorsal fin. At least some have a very elaborate reproductive behavior, which is found mostly in more advanced groups, protecting their eggs until hatching.

The 30 or so described species fall into three rather well defined groups, at least anatomically, as follows:

a. Opercular membrane united to isthmus far forward; two complete rows of teeth in upper jaw; maxillary bone straight or slightly curved, never "S"-shaped, usually toothed; nostrils close to each other . . . genus *Pyrrhulina*

aa. Opercular membrane united to isthmus far backward; one row of teeth on upper jaw, plus usually one or two teeth behind it near the middle, not constituting a real row of teeth; maxilla always toothed.

b. Body short, the body depth not more than 4 in standard length; fins without filaments; outer edge of maxillary bone not "S"-shaped; nostrils close to each other . . . genus *Copeina*

bb. Body elongate, the body depth 4.25 to 6.0 in standard length; fins filamentous in adults; outer edge of maxillary bone forming a double curve in the shape of an "S", with a rudimentary apophysis; nostrils separated from each other by a cutaneous bridge . . . genus *Copella*

Pyrrhulina vittata. Photo by Dr. Herbert R. Axelrod.

Pyrrhulina spilota. A form with a short body approaching that of *P. vittata.* Photo by Dr. Herbert R. Axelrod.

Pyrrhulina sp. Photo by Klaus Paysan.

Pyrrhulina spilota. Photo by Dr. Herbert R. Axelrod.

GENUS *PYRRHULINA*
PYRRHULINAS

The charming dwarf fishes of the genus *Pyrrhulina* (the name means "small bullfinch"), which are surface fishes with the mouth upturned, the top of the head flat and without fontanels, and the upper caudal lobe enlarged, are among the lesser known characoids. Not one species, except the most recently named such as *P. stoli* and *P. spilota*, has been properly described or even figured. It is hoped that the prospected revision referred to above will put some order into the genus. Meanwhile, it is possible to recognize several groups: (1) *P. rachoviana*-group; (2) *P. brevis*-group; (3) *P. laeta*-group; (4) *P. filamentosa*-group and (5) *P. vittata*-group. These groups, perhaps quite artificial, are mostly based on external characters such as color pattern (except for the first group).

(1) *Pyrrhulina rachoviana:*

Rachow's pyrrhulina from La Plata basin is the only *Pyrrhulina* with a straight, un-toothed maxillary bone. In the other species, the outer maxillary edge is always more or less curved and always bears some teeth (usually 7 or 8 even in species considered as having a toothless maxilla, such as *Pyrrhulina vittata*). It is a deep bodied species with a longitudinal dark stripe from snout to caudal peduncle; its most colorful parts are the eye and the base of the ventral and anal fins, which are red in life. This species lays its eggs on a broad submerged leaf or, less commonly, on the bottom.

(2) *Pyrrhulina brevis*-group:

There is much confusion concerning this group, which is characterized by a relatively deep body (body depth 3.4 to 4.25 in the standard length), few scales in a longitudinal row (20-23) and a very short stripe which extends onto the eye or usually to the opercle.

In the author's opinion, there may be only one species, with slightly divergent characters depending on the locality. In the Rio Negro around Manaus the typical species has a short and well curved maxilla; in the south, the subspecies *P. brevis australe*, from La Plata-Parana-Paraguay basin and perhaps the Rio Guapore basin (with possible synonyms *P. macrolepis* and *P. stigmasemion*), is slightly more elongate (body depth usually 4.0-4.25 in the standard length versus 3.4-4.0 in *P. brevis brevis*), with the stripe variable. *P. lugubris* is an intriguing form with the dorsal, ventral and anal fins black edged and the maxilla more elongate and less curved than that of *P. brevis brevis*. It may be a variety from the upper Amazon basin (and also Rio Meta in Colombia), but it may also be a sibling species.

Pyrrhulina brevis is said to spawn on a flat stone.

(3) *Pyrrhulina laeta*-group:

P. laeta and associated forms are very close to *P. brevis*. However, they always have more longitudinal scales (23-28), the stripe extends onto the

Pyrrhulina species, from the Brazilian Amazon. Photo by Harald Schultz.

second or third scale or more and, especially important, the species have one or two elongate spots on the predorsal region, easily seen from above.

Pyrrhulina eleanorae, from the Rio Ucayali, and *Pyrrhulina stoli* from Surinam, Guyana and Upper Rio Meta (?), have the stripe extending up to the second or third longitudinal scale, not more. They are extremely close to each other. However, a close examination shows that *P. eleanorae* has slightly fewer predorsal scales (11) than *P. stoli* (12-13). They may be two geographic varieties of the same species.

Pyrrhulina laeta, from the Amazon basin, is a pain in the neck of an ichthyologist. Its type, dating from 1871, lacks the anterior part of the body, including the head! One may believe that, in such a case, it would be better to destroy the specimen, and start off again with a new name and a new description. The Rules of Zoological Nomenclature forbid such an act: the half-specimen may be useful in the future, if a new technique to characterize the species, a chemical one for example, is discovered. We are presently left with the oldest valid name without a really good way to attach it to such-or-such a species! By elimination, the most "probable" species is the one which has been named a few years later by Steindachner, *P. semifasciata* (and, later on, *P. maxima* by Eigenmann). *P. laeta* has a stripe extending onto the flank to the level of the dorsal fin. Its reproductive behavior is not known.

Two photos showing a spawning sequence of *Copeina guttata.* Photos by R. Zukal.

Copella of the *nattereri*-group (possibly *C. eigenmanni*). Photo by Harald Schultz.

Copeina guttata. Photo by Dr. Herbert R. Axelrod.

(4) *Pyrrhulina filamentosa*-group:

Pyrrhulina filamentosa, the type species of the genus from the Guianas, is a relatively large-sized species which resembles the genus *Copella* by its elongate shape, the body depth of the adult males attaining 4.25-4.75 in the standard length (versus usually less than 4.25 in all other pyrrhulinas except *beni*). It has 26-28 longitudinal scales and the usual stripe extends to the eye (as in *Pyrrhulina brevis*).

Pyrrhulina beni, from Bolivia, may be its closest relative, also being elongate (body depth 4.45 in the standard length), with 25 longitudinal scales. The longitudinal band, like that of *Pyrrhulina laeta*, extends up to the dorsal level.

Nothing is known concerning the breeding of either species.

(5) *Pyrrhulina vittata*-group:

The two species forming this group are very characteristic, having a spotted color pattern instead of a striped one as in the other species. They are in a somewhat similar position as *Nannostomus espei* is in comparison with the other nannostomins.

Pyrrhulina vittata, the banded pyrrhulina, from the Amazon basin including Rio Madeira, has a stripe extending to the opercle, 2 conspicuous spots on the body and a transverse precaudal band. It is a small, deep bodied species with few scales (20-22) and 3 rows of teeth in the lower jaw (which is unusual). Like *P. rachoviana*, it spawns on a broad submerged leaf.

Pyrrhulina spilota, of unknown origin, is an interesting novelty. It has about the same pattern as *P. vittata*, but usually has one more spot and a slightly more elongate body. It also has more scales (25 or 26). The method of reproduction is not known.

GENUS *COPEINA*
COPEINAS

In a certain sense, the one or two *Copeina* species (if two species, they are very close to each other) are the least specialized of the pyrrhulinins, and are rarely found at the surface. They resemble lebiasinins the most with their large size, their not very elongate body and their dorsal fin near the middle of the body. Their maxilla is short, rounded, with none to five teeth.

The type species of the genus is *Copeina guttata* (synonym *Copeina argyrops*), from the Amazon basin, which has 4 or 5 maxillary teeth. In the Peruvian Amazon it is replaced by a very similar form, *Copeina osgoodi*, which can nevertheless be distinguished by the absence of maxillary teeth (at least according to a cursory examination of the type specimen). *Copeina* species have a typical color pattern without a stripe but with a bright red spot on most scales of the middle of the flanks, a red border on the ventral and anal fins, as well as on the lower caudal lobe, all standing out in contrast to an overall shiny blue tint.

The male *Copeina guttata* fertilizes the eggs as they are being laid in a depression in the sand. Photo by Ruda Zukal.

C. guttata (and probably also *C. osgoodi*, if the latter is a distinct species) spawns in a hole in the sand, and the eggs are guarded by the male.

GENUS *COPELLA*
COPELLAS

The genus *Copella* was erected by Dr. George Myers to include *Copeina* species which have a double curved maxillary bone forming a S (at least in the adult male). At the time of the description of *Copella* (1956), only three species were included in the genus. The examination of several species has shown that all forms previously known as *Copeina* are indeed *Copella* (except evidently the type species *guttata* and its close associate *osgoodi*). Actually both sexes, when adult, have the maxilla S-shaped, which is more pronounced in the male, as well as other characteristic features such as nostrils separated by a dermal flap, elongate body and filamentous fins.

(1) *Copella arnoldi*-group:

This group is characterized by a black stripe extending to the eye or opercle. The species resemble *Pyrrhulina filamentosa* to such a point that the aid of a binocular dissecting scope is necessary to classify a composite sample. As far as known, *Copella arnoldi*, the well known splashing tetra, is known only from aquarium specimens, said to come from the lower Amazon. It is moderately elongate (body depth about 4.25-4.75 in the standard length)

Unidentified *Pyrrhulina.* Photo by Dr. Herbert R. Axelrod.

Unidentified *Pyrrhulina.* Photo by Dr. Herbert R. Axelrod.

Copella nattereri. Photo by Dr. Herbert R. Axelrod.

Copella nattereri. Photo by Dr. Herbert R. Axelrod.

and very graceful with its red, elongate fins and shiny scales. The pair spawns on a broad leaf above the surface. The male remains under the leaf and moistens the eggs and embryos with "splashes" of its tail.

Copella callolepis is considered as a synonym of *C. arnoldi* or *C. nattereri* by some authors.

Copella carsevennensis, from the Guianas, can be distinguished from *C. arnoldi* as follows: it has more scales (26 versus 23 or 24) and the dorsal fin is slightly nearer the head than caudal fin (not so in *C. arnoldi*).

(2) *Copella nattereri*-group:

This group is characterized by a more or less distinct longitudinal band to the caudal fin and a slightly more elongate body than the preceding copellas.

Comparison between the heads of a *Pyrrhulina* (*P. filamentosa*, left) and a *Copella* (*C. carsevennensis*, right). Original drawing by the author.

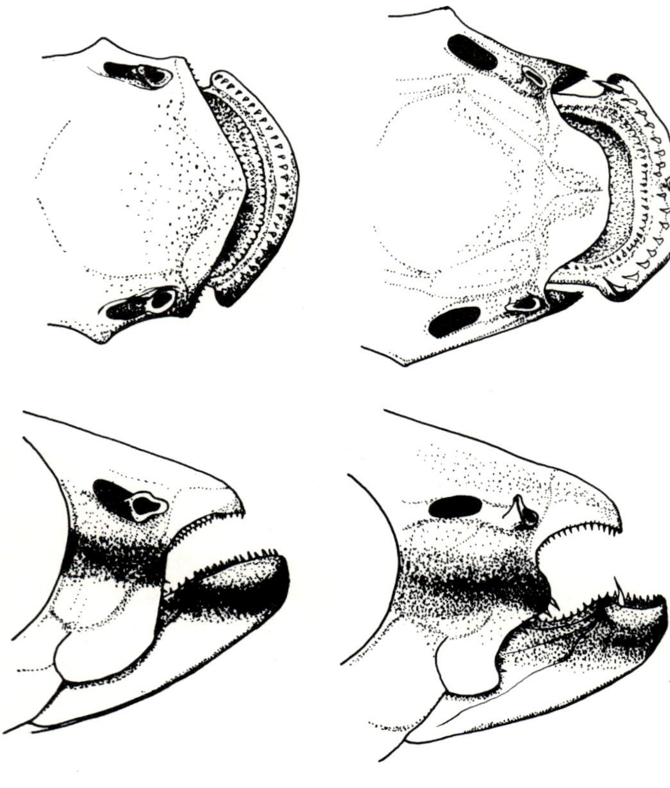

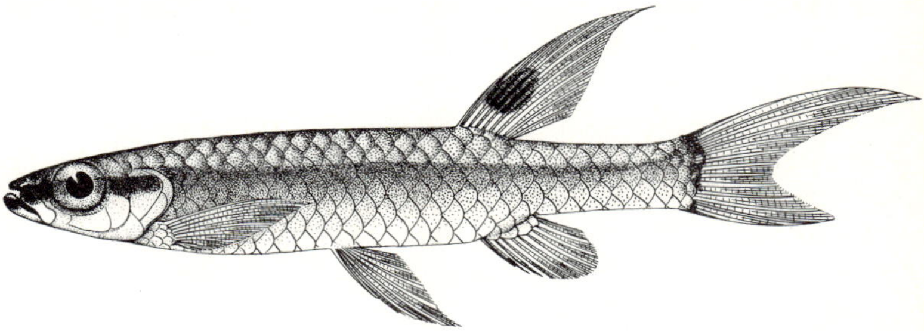

Male paratype of *Copella compta.* Original drawing by Mr. Pablo Bravo in 1930, published for the first time with permission of Dr. G.S. Myers.

Copella nattereri, from the Rio Negro and the Middle Amazon, has a pale spot on each scale, a dusky band along the body and a black triangular spot at the base of the lower caudal lobe, which seems characteristic of this species.

Copella metae, from Rio Meta, and *Copella eigenmanni*, from Guyana and Para, are to be distinguished from the preceding by the more distinct longitudinal band and the somewhat more numerous scales (23 or 24 in a longitudinal series and 14 in predorsal series, versus 20-22 and 12, respectively). The two forms (distinct species ?) are extremely close to each other and differ in the dorsal fin position, which is more posterior in *C. metae. Pyrrhulina nigrofasciata* is probably synonymous with the latter species.

Nothing is known about the reproductive behavior of these species, which are quite difficult to identify in the aquarium.

(3) *Copella compta*-group:

The two species of this group, again extremely close to each other, "represent the acme of gracefulness reached by characins of the Pyrrhulininae" (G. S. Myers). The body depth attains 5.5 times in the standard length, and the fins in the adult males are very elongate.

Copella compta, from the upper Rio Negro, has a color pattern similar to that of *C. nattereri*, at least judging from preserved specimens. Its coloration in life is not known. The rainbow copella, *Copella vilmae*, from the upper Amazon not very far from Tabatinga, is of a magnificent wine-red color with some metallic blue spots on the flanks. Technically it differs from *C. compta* in having fewer longitudinal scales (24 or 25 versus 26) and fewer predorsal scales (14-16 versus 17), about in the same manner as *C. nattereri* differs from *C. metae*, proportionately speaking.

The reproductive behavior of neither species is known.

Copella of the *arnoldi*-group. Photo by Hilmar Hansen, Aquarium Berlin.

Another male *Copella* of the *arnoldi*-group. Photo by Dr. Herbert R. Axelrod.

Copella sp. from Rio Negro, possibly *Copella compta.* Photo by Dr. Herbert R. Axelrod.

Copella vilmae, the rainbow *Copella.* Wild male in full color. Photo by Harald Schultz.

Family Anostomidae

Headstanders and Leporins

The Anostomidae, including among others the well-known leporins and certain headstanders (not the chilodins), are a very satisfying group from the taxonomists' point of view: they are well classified, with a reasonable number of fairly well known genera, and the limits of the higher groups are rather sharply defined. Only at the species level do the difficulties begin: no one knows, for example, how many species of *Leporinus* really do exist.

The principal characters of the Anostomidae are:

—body usually elongate (except in *Abramites*)

—anal fin short, with less than 10 branched rays (except in *Abramites*)

—gill membrane united with the isthmus

—anterior nostril forming a tube

—teeth in one series of only 3 or 4 on each jaw, usually incisiform, firmly set into the bone; no teeth on maxilla, none on palate.

Two groups are recognizable within the Anostomidae, the subfamily Leporellinae with the single genus *Leporellus* which has the nostrils close to each other and the caudal fin scaly, and the subfamily Anostominae, with several genera which usually have the nostrils set apart from each other and the caudal fin naked.

Head and teeth of a generalized anastomid, *Leporinus fasciatus*, type of the genus *Leporinus*. Original drawing by the author.

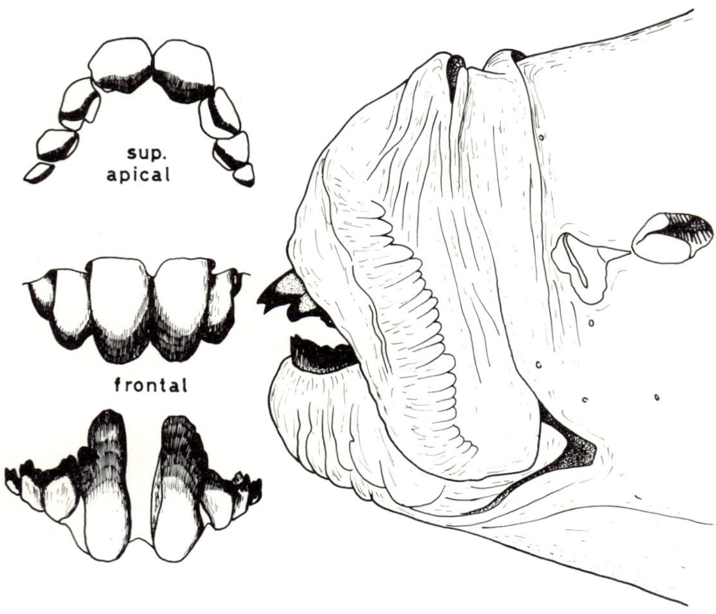

SUBFAMILY LEPORELLINAE

GENUS *LEPORELLUS*

Apart from the characters given above, the species of the genus *Leporellus* are like those of *Leporinus*, with a dentition somewhat similar to that of subgenus *Hypomasticus* of the latter genus. Seven forms have been described: *L. vittatus, L. pictus, L. maculifrons, L. timbore, L. cartledgei, L. retropinnis,* and *L. sexdentatus. L. maculifrons, L. timbore,* and *L. cartledgei* are probably synonyms of *L. pictus*. They are very close to each other and, apart from *Leporellus retropinnis* from the Rio Piracicabia in southeast Brazil (which has the dorsal fin at midbody instead of well in advance of midbody), they may belong to the same species. They are recognizable on first sight owing to the peculiar oblique bands of their unpaired fins, strongly reminiscent of those of certain *Prochilodus* (*insignis*-group). *Leporellus*, although everywhere rare, inhabits all the principal South American basins, i.e. Orinoco, Amazon and Parana-Paraguay.

SUBFAMILY ANOSTOMINAE

Leporins and Headstanders

The genera of this important group are well defined by a few anatomical characters, chiefly of their mouth and dentition, as follows:

a. Mouth never upturned.
  b. Teeth incisiform with 2 to 4 or 5 cusps (body elongate) . . . *Schizodon*
  bb. Teeth without cusps or simply notched.
    c. Body not very elongate; teeth usually rather thick, normally truncate.
      d. Body not very high, not compressed; anal fin with less than 10 branched rays; no post-ventral keel; most species swim in normal (horizontal) position . . . *Leporinus*
      dd. Body high and compressed; anal fin with more than 10 branched rays; a post-ventral keel; species swim in head-down position . . . *Abramites*
    cc. Body very elongate, cylindriform; teeth thin, with irregular, finely denticulated, cutting edges (convex above and concave below) . . . *Rhytiodus*

aa. Mouth more or less upturned.
  e. 3 or 4 teeth on each side of lower jaw.
    f. Body rather deep; upper teeth sometimes tricuspid, lower teeth rounded, as in some *Leporinus* (mouth upturned, but the opening not quite vertical) . . . *Anostomoides*
    ff. Body elongate.
      g. Teeth tri-to multicuspid, at least in upper jaw *(Anostomus)*.
        h. Mouth with opening not quite vertical, as in *Anostomoides*; rather large size; upper teeth with 3 or 5 cusps, lower teeth truncate when adult . . . *Anostomus (Laemolyta)*

Dr. Martin Brittan preserving a collection of fishes, mostly anostomids, from Tefe, Brazil. Photo by Dr. Herbert R. Axelrod.

Schizodon fasciatum adult collected at Solimoes, between Coari and Jutica. Photo by Dr. Herbert R. Axelrod.

Leporellus vittatus. Photo by Aaron Norman.

hh. Mouth strictly superior; small size (except a few forms); teeth usually bi-or tricuspid on both jaws

... *Anostomus (Anostomus)*

gg. Teeth usually rounded at the tip or at most bifid.

i. 4 teeth on each side of both jaws, usually unicuspid, the lower symphyseal pair not greatly enlarged

... *Synaptolaemus*

ii. 4 teeth on the upper jaw, bifid; only 3 teeth on lower jaw, the symphyseal pair very long, protruding vertically out of mouth

... *Sartor*

ee. Only one tooth on each side of lower jaw near symphysis, very long, spoon-shaped at its end and extending above the snout as in *Sartor*; lower jaw extremely narrow and slender

... *Gnathodolus*

GENUS *SCHIZODON*

The species of this genus do not differ externally from most *Leporinus*. However, once the terminal mouth is examined, the pluricuspidate teeth, forming a continuous crenulate cutting border, are quite characteristic. It is much more difficult to distinguish the species, and the genus is certainly in need of a revision.

The best known species, and the one which is commonly seen, is *Schizodon fasciatum* from the Amazon basin; it has the median teeth with 4 cusps, about 8+1+13 gill-rakers and 4 vertical black bands on the body, plus a caudal spot which is well separated from the last band located above the anal fin.

In *Schizodon vittatum*, also from the Amazon basin (Rio Araguaia), the last transverse band is joined with the caudal spot and the center of each scale is paler than its border, giving the impression of longitudinal dark stripes between scale rows. It is likely to be only a color variation of *S. fasciatum*. Another form, which also may belong to *S. fasciatum sensu lato*, is *Schizodon dissimile* from the basin of the Rio Paranahyba. It is best identified by its elongate body and lack of a caudal spot.

In a second group of species we may include *Schizodon corti*, *S. isognathum*, *S. kneri*, and *S. platae*, which differ from the *S. fasciatum*-group in having 5 or 6 transverse scales above and below the lateral line instead of 4. *Schizodon corti*, described as a subspecies of *S. fasciatum*, seems indeed to be the representative of it in Venezuela, having exactly the same color pattern; however, apart from the transverse scales, it differs in tooth structure, never having as many as 4 cusps on either jaw, but only 2 or 3.

Schizodon isognathum from the Rio Paraguay Basin, *S. kneri* from the Rio Sao Francisco basin, and *S. platae* from Rio de la Plata (the latter two species perhaps synonymous with the former) have a different coloration, having lost the lateral marks but retaining the caudal spot. These forms are not very elongate (depth about 3.66 times in the standard length).

A third group is best characterized by the slightly inferior mouth together with the coloration as in *S. isognathum*: it is composed of only two very similar species, *Schizodon nasutum* from Matto Grosso (and maybe from Rio Sao Francisco) and *Schizodon rostratum* from Rio Paranahyba.

Finally, the systematic postion of *Nanognathus borellii*, which has a dentition similar to that of a *Schizodon*, is not clear. Some authors have identified specimens of a *Schizodon* from Paraguay as *Schizodon borellii*. However, in the original description of *N. borellii* it is expressly stated that the nostrils are close to each other and the gill membrane is free from the isthmus, which seems to exclude *Nanognathus* from the Anostomidae. On the other hand, *Leporinus platycephalus* Meinken may be a *Schizodon*. All *Schizodon* species attain a rather large size (maximum 260 mm in standard length). In certain places such as Manaus, where they are rather abundant, they are found in the fish market together with members of the next genus.

GENUS *LEPORINUS*
LEPORINS

This is one of the largest genera of characoid fishes (together with *Curimata, Prochilodus, Hyphessobrycon, Hemigrammus, Astyanax* and *Moenkhausia*), and one of the least known. In typical *Leporinus* (as pointed out by G. S. Myers) "the teeth are heavy and thick at the base, those of upper jaw arranged in a step or stair-like pattern, the median pair set far forward and largest; each tooth which follows is set as much behind as to the side of the preceding tooth . . . and the cutting edge is approximately transverse to the long axis of the fish; the lower teeth are slanted forward to the median pair which is largest . . . with an oblique, usually concave cutting edge."

The genus can be split into four subgroups as follows:

a. Mouth terminal or subinferior.
  b. Lower median pair of teeth not developed into laterally compressed tusks.
    c. Small size, body rather compressed with ventral region keeled; mouth terminal; 3 upper teeth only, bifid, not arranged in a step-like pattern (or very feebly so); gill membrane united far forward of isthmus, thus forming an acute angle . . . subgenus *Leporinops* (2 species ?)
    cc. Usually rather large size; body not very compressed, with ventral region rounded or flattened; mouth terminal or subinferior; 3 or 4 upper teeth arranged in a step-like pattern (see above definition); gill membrane united back to isthmus, thus forming an obtuse angle . . . subgenus typical *Leporinus* (several species)
  bb. Lower median pair of teeth developed in adults into laterally compressed tusks (rather large size; ventral region somewhat flattened; 3 upper incisiform teeth,

Schizodon fasciatum juvenile. Photo by Harald Schultz.

Leporinus of the *pellegrini-alternus*-group (Upper Amazon form). Photo by Dr. Herbert R. Axelrod.

Leporinus of the *pellegrini-alternus*-group. Photo by Klaus Paysan.

Young *Leporinus* of the *fasciatus*-group. Photo by Dr. Herbert R. Axelrod.

not arranged in a step-like pattern; gill membrane united far back to isthmus)
... Subgenus *Myocharax*
(one species)

aa. Mouth inferior, visible only from below, the snout almost vertical (small to medium size; body and gill membrane like that of *Leporinus* typical; 3 or 4 upper teeth vertically set, not arranged in a step-like pattern)
... Subgenus *Hypomasticus*
(at least 5 described species)

Subgenus *Leporinops*

This very small group of leporins with a slight ventral keel has *Leporinus moralesi* as the type species (a name given by Fowler to replace *Leporinus muelleri*, which was preoccupied). It is a small fish (less than 100 mm in standard length), an inhabitant of the upper Amazon basin, with a color pattern formed by three elongate spots ending in a longitudinal band extending to the top of the caudal rays, resembling that of *Leporinus agassizi*, etc.

Another species probably belonging to the same group is *Leporinus jamesi* from the middle Amazon basin. It has no longitudinal pattern but only a series of bars across the dorsum like in most *Leporinus* (chiefly when young).

Subgenus typical *Leporinus*

Leporins

It is beyond the scope of the present book to describe all the known species (50 or 60) of this large group and to give their taxonomic characters (which are not even precisely known for each species). For want of a classification based on well-studied features, we may artificially split the genus *Leporinus* into a few types of color patterns: (1) species with a transverse pattern; (2) spotted species; (3) species with a longitudinal pattern; and (4) plain species without a conspicuous pattern.

(1) Species with transverse bands on body:

This group, which includes the type species of the genus, *Leporinus fasciatus*, may in turn be divided into subunits:

(a) Species with 4 bands (3 bands on body, plus a precaudal band):

This group includes *Leporinus trifasciatus*, common in the middle Amazon basin (sold on the Manaus market), and *Leporinus wolfei*, a close form from the Upper Amazon, which is deeper bodied (depth 3.1-3.33 versus 3.3-3.7 in the standard length) than *L. trifasciatus* with fewer longitudinal scales (40 versus 42 or 43).

(b) Species with 7 or 8 transverse bands:

This group includes *Leporinus conirostris* from Rio Parahyba, *Leporinus octofasciatus* from Santa Catarina State in Brazil and *Leporinus latofasciatus* from Rio Orinoco with 8 bands. These 3 species are poorly

Leporinus jamesi (above) and *Leporinus moralesi* (below) (both after Borodin, 1929).

known. *Leporinus pellegrini, L. alternus* and *L. paralternus*, from the Guianas and the upper Amazon, have 4 broad bands alternating with 4 narrower ones. *Leporinus pearsoni* and *L. yophorus*, from Colombia and Bolivia respectively, strongly resemble each other, having 8 transverse bars, but 2 or 3 of these are united (at the lateral line level) to form a Y.

(c) Species with 9 or 10 transverse bars in the adult (*fasciatus*-group):

Several names, *L. fasciatus, L. multifasciatus, L. fasciatus altipinnis, L. fasciatus tigrinus* and *L. holostictus*, probably apply to the same rather variable form, whose transverse bars are quite numerous when young, but some uniting during growth (about fourteen becoming ten). *Leporinus affinis*, from the state of Para, is perhaps different, having only 9 bars and the anal fin rounded (instead of pointed). The forms of the *L. fasciatus*-group are quite abundant in the Guianas (type locality of *Leporinus fasciatus*) as well as in the Amazon basin.

(2) Species with spots on the body:

This group, which includes no less than 30 described species, is probably the least well known of all leporin groups. Like the preceding group, it can be split quite unnaturally into subunits depending on the number of spots, which are usually large, round, and situated astride on the lateral line. Smaller supplementary spots or dorsal lines may be present in

Leporinus fasciatus. Photo by Harald Schultz.

Leporinus of the *fasciatus*-group. Photo by Harald Schultz.

Adult *Leporinus fasciatus* from Tefe. Photo by Dr. Herbert R. Axelrod.

Unidentified anostomid from Venezuela, possibly of the genus *Laemolyta.* Photo by Dr. Herbert R. Axelrod.

some species, at least when young, this being a usual pattern in the spotted *Leporinus*. It cannot be taken into consideration except for the so-called "*maculatus*-group" which seems to have retained at the adult stage its juvenile coloration. Certain forms may be distinguished thanks to their very elongate body (this is associated with horizontally oval spots).

(a) Species with one or two spots on body when adult (no caudal spot):

There is one species with only one large spot at the dorsal fin level, i.e. *Leporinus lacustris* from Sao Paulo State in Brazil. It is one of the deepest, largest scaled species. There is also a pair of Amazonian species (*Leporinus bimaculatus*, *L. nattereri*) and one from Ceara (*L. piau*), all with 2 spots, which are in need of revision. It is possible that the 3 names cover a single entry. On the other hand, the close nostrils described in *L. nattereri* would characterize a very distinct species, even a distinct genus, whereas *Leporinus piau* may eventually be found to be the young of *L. reinhardti*.

(b) Species with three or four spots on body when adult (including a caudal spot):

The classification of the *L. friderici*-group is one of the worst Chinese puzzles of characoid systematics. For brevity, this group will be treated here by means of a tentative key to their identification. It is more than probable that this key will soon be outdated.

Unidentified *Leporinus* of the *friderici*-group. Photo by Harald Schultz.

Key to the *friderici*-group of *Leporinus* spp.

a. Body depth usually less than 3.66 times in standard length (*friderici*-group *sensu stricto*).

b. Less than 40 lateral line scales.

c. 4/3½ or 4 transverse scales.

d. Depth about 3 in standard length.

e. 35 lateral line scales; head 3.5 in standard length; anal fin with 10 rays, rounded, reaching caudal fin when depressed ... *L. bahiensis* (Bahia)

ee. 37 lateral line scales; head 4.2 in standard length; anal fin with 12 rays, concave, not reaching caudal fin when depressed ... *L. steindachneri* (Rio Jequitinhonha)

dd. Depth 3.4-3.66 in standard length (36-39 lateral line scales; head 4-4.2 in standard length; anal with 10-11 rays, concave, not reaching caudal fin when depressed; teeth pointed, 3-4/4) ... *L. copelandi* (Rio Parahyba)

cc. 5-6/5-6 transverse scales.

e. 4 lateral spots ... *L. leschenaulti*, synonym *L. parae* (Oyapock and Para)

ee. 3 lateral spots (including caudal spot).

f. Head 3.33-3.4 in standard length (depth 2.85-3.15 in standard length; 35-38 lateral line scales; 12 predorsal scales; teeth slightly notched, 4/3 ?) ... *L. subniger* (Colombia)

ff. Head 3.8 in standard length.

g. Teeth slightly notched, 3/3; 35 lateral line scales; 10-11 predorsal scales; 14 peduncular scales ... *L. reinhardti* (Rio S. Francisco)

gg. Teeth truncate, 4/4; 36-38 lateral line scales; 10 predorsal scales; 16 peduncular scales ... *L. friderici*, syn. *L. acutidens* (Guianas, Amazon, probably other basins; possible subspecies: *L. spilopleura*, *L. ecuadoriensis*)

bb. 40-43 lateral line scales.

h. 5/5 transverse scales; body depth 3.33-3.5 in standard length (sometimes more, or sibling species ?) ... *L. muyscorum* (Rio Magdalena and Rio Atrato in Colombia)

hh. 6½-7½/5½-6 transverse scales.

i. Body depth 2.9 in standard length; 6/5½ transverse scales; 3/3 teeth ... *L. obtusidens*, synonyms *L. pachyurus*, *L. brachyurus* and ? *L. aguapeiensis* (La Plata and S.E. Brazil to S. Paulo)

ii. Body depth 3.5-3.66 in standard length; 6½-7½/6 transverse scales; 4/4 teeth ? ... *L. silvestrii* (Coxipo, Mato Grosso)

Leporinus sp. (*nattereri*-group?). Photo by Harald Schultz.

Leporinus steyermarki. Photo by Harald Schultz.

Leporinus of the so-called "*maculatus*-group" (*Leporinus granti* auct.). Photo by Harald Schultz.

Leporinus agassizi. Photo by Harald Schultz.

Leporinus melanopleura? Photo by Dr. Herbert R. Axelrod.

aa. Body depth more than 3.66 in standard length (*cylindriformis*-group).

j. 37-41 lateral line scales (see also couplet h., *L. muyscorum*).

k. Body depth 3.7-4 in standard length
. . . *L. klausewitzi* (Amazon basin)

k k. Body depth 4.5-4.66 in standard length
. . . *L. niceforoi* (Colombia)

j j. 44 lateral line scales (body depth 4.75 in standard length)
. . . *L. cylindriformis* (Lower Amazon)

(c) Species with several spots on body at all ages (so-called *maculatus*-group):

The description of this group is hampered by a delicate problem of nomenclature: the species called *L. maculatus* by ichthyologists and aquarists (see for example page F.377.00 in the Looseleaf Edition of *Exotic Tropical Fishes* by Axelrod et al.) is certainly not the one described by Müller and Troschel under this name (it is not the proper place here for a nomenclatorial change; it suffices to say that *L. maculatus* is in reality what is called

Color pattern of three forms of spotted *Leporinus*. Top to bottom: *L.* (*Hypomasticus*) *"maculatus"* auct. (note inferior mouth), *"granti"* auct., and *steyermarki* (schematic).

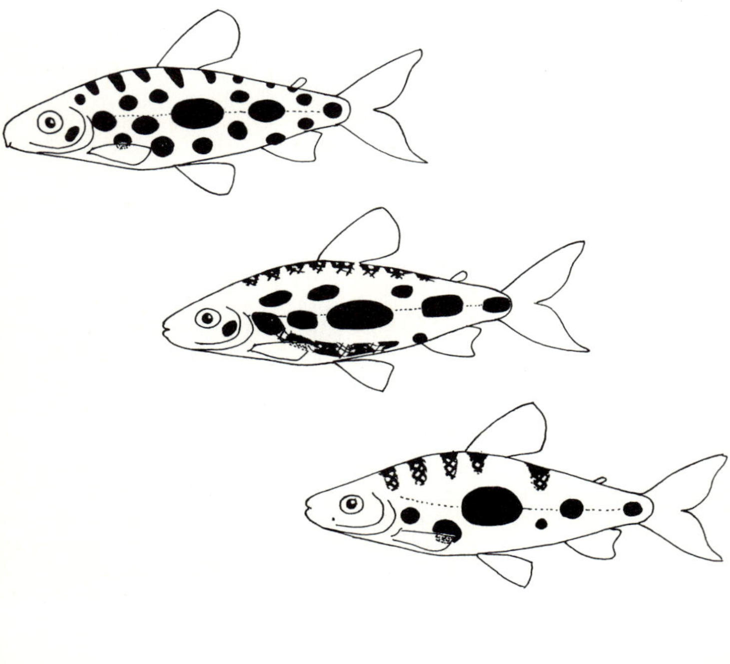

at present *L. pellegrini*). The situation is even worse than one may believe because *L. maculatus* was applied to at least two different species belonging to 2 subgenera; one with an inferior mouth and 3/3 teeth (see description of the subgenus *Hypomasticus* below), and one with a terminal or scarcely sub-inferior mouth. For the last one, which occurs in the Guianas and the Amazon, four names are available: *L. megalepis* (at least part of the types), *L. marcgravii*, *L. granti* and *L. badueli*. A re-examination of all *megalepis*-material is necessary to determine which name is valid.

A final species with several spots on its body, *L. steyermarki* from Venezuela, differs from the preceding (which will be called here provisionally *Leporinus granti*) as noted in the following key:

a. 3 oval or almost rectangular spots, first under dorsal fin, second midway to caudal fin, preceded by two smaller, less well delimited spots, and the third, smallest, at end of peduncle; a series of 3 smaller spots above main series, beginning well in front of first large spot; an oblique series of small spots below main series, usually more or less obscure; a series of blotches or bars on dorsum; usually 16 circumpeduncular scales . . . *Leporinus "granti"*

aa. 3 spots of main series round; spots of series above confused with bars across dorsum; spots of series below only 3, the last one very small (color pattern seemingly intermediate between *friderici*-type and "*maculatus*"-type); 12 circumpeduncular scales . . . *Leporinus steyermarki*

(3) Group of species with a longitudinal pattern:

The seven or so species with one or more longitudinal dark bands are slightly better known than the preceding ones. They can be distinguished as follows:

a. Only one longitudinal band along axis of body (see also *Leporinus moralesi*, of subgenus *Leporinops*).
  b. Longitudinal band beginning at dorsal fin level.
    c. Depth 3.2-3.45 and head 3.7-4 in standard length; 38-40 lateral line scales, 5/5 transverse scales . . . *L. agassizi* (synonym *L. semivittatus*) (Amazon basin)
    cc. Depth 4.5-4.55 and head 4-5 in standard length; 40-42 lateral line scales, 5 or 6/5 transverse scales . . . *L. nigrotaeniatus* (synonym *L. margaritaceus*) (mostly Guianas; occurence in the Amazon not proved)
  bb. Longitudinal band complete, beginning behind opercle (depth 3.5-4.0 in standard length; 35-38 lateral line scales; teeth 3/4).
    d. 4/4 tranverse scales . . . *L. melanopleura* (Bahia) (squamation said to be 3½/35/2 in *L. macrolepidotus*, an uncertain species from Rio de Janeiro)
    dd. 5/5 transverse scales . . . *L. taeniatus* (Rio das Velhas)

Leporinus friderici. Photo by Dr. Herbert R. Axelrod.

Unidentified *Leporinus*, probably close to *L. jamesi.* Photo by Dr. Herbert R. Axelrod.

Leporinus sp., probably of the *nattereri*-group. Photo by Dr. Herbert R. Axelrod.

Leporinus "maculatus" auct. Photo by Dr. Herbert R. Axelrod.

aa. Three longitudinal bands on flanks plus one dorsal band.

e. Medium sized fish, with largest size recorded 175 mm standard length; depth 3.5-4.0 in standard length; teeth 3/4; profile of head straight over eyes; dorsal fin midway between snout and adipose fin; no black spot at pectoral fin base . . . *L. striatus* (Paraguay, Uruguay, Mato Grosso, Sao Paulo, ? Bolivia, Colombia and Ecuador)

ee. Large sized fish, with maximum size recorded 400 mm standard length; depth a little more than 3 in standard length; teeth 4/4; profile of head slightly convex over eyes; dorsal fin midway between snout and a point between adipose fin and beginning of caudal fin; a black spot at pectoral fin base . . . *L. arcus* (Guianas, Venezuela, Upper Amazon and ? Para)

(4) Group of species without conspicuous color pattern when adult, or with only a caudal spot (certain species being known only from old collections in museums may have some coloration when alive):

Unidentified species of *Leporinus*. Photo by Timmerman.

a. Mouth strictly terminal.

b. Body depth 3.9 and head-length 4.55 in standard length; 6 or 6½/42/5-5½ scales ... *L. elongatus* (Buenos Aires)

bb. Body depth 3 and head-length 3.5-4.0 in standard length; 4½ or 5/35/3½ or 4 scales; a peduncular spot ... *L. melanostictus* (Oyapok in French Guiana)

aa. Mouth subinferior (species being at the limit of the subgenus *Hypomasticus*).

c. Body depth 4.15 in standard length; 4/4 teeth (5/38-41/5 scales; lips thin; scales edged with dark) ... *L. brunneus* (Rio Negro)

cc. Body depth 3.0-3.75 in standard length; 3/3 teeth (4½/37/3½ scales; lips thick; no caudal spot when adult) ... *L. crassilabris* (Rio Jequitinhonha)

Subgenus *Myocharax*

The single species, *Leporinus desmotes* from the Rupununi River in Guyana, has an elongate body, a long tubular anterior nostril, a subinferior mouth with thick lips, and 3/3 peculiar teeth: the upper ones directed forward, somewhat obliquely, the lower ones also directed forward, almost horizontally, the median pair forming long tusks abutting against the superior ones. In contrast with the lower teeth of *Leporinus fasciatus*, a species which also develops "tusks" when old, these teeth are laterally compressed. The color pattern is strikingly similar to that of *Abramites* and *Synaptolaemus*.

Subgenus *Hypomasticus*

The species of this peculiar group have a strictly inferior mouth, the upper teeth clearly vertical or even pointing backward. Certain species eat fresh-water sponges, but so do other leporins with a terminal mouth ("*L. pellegrini*" auct. for example), as well as anostomids with a superior mouth (*Sartor respectus*, for example). Although only 4 or 5 species have been described, it is the author's feeling that a number of forms are to be discovered in the future (at least 5 new species are already in collections). Curiously, representatives of the subgenus *Hypomasticus* are usually mimics of a form of the nominal subgenus *Leporinus*: for example *L. (H.) despaxi* copying *L. arcus*, "*L. (H.) maculatus*" (auct.) copying *L. granti*, and so on.

As mentioned above, *Leporinus* "*maculatus*" (in the sense of most authors) presently has no official name. It is found in the Guianas as well as in the southern Amazon basin and can be distinguished from its "model," *Leporinus granti* (also a tentative name), not only by its small size, inferior mouth and some meristic characters, but also in details of color pattern, notably the lower row of small spots which are well separated and not fused

Newly laid eggs (below) and developing embryo (above) of *Leporinus* *"maculatus."* Photo by H. Azuma.

Leporinus "maculatus." Above: one month old juveniles of 15-16 mm. *Below:* four months old and 30 mm in length. Photos by H. Azuma.

Leporinus thayeri (above) and *L. garmani* (below) (both after Borodin, 1929).

as in *L. granti*. *L.* "*maculatus*," one of the most beautiful of the leporins, is unfortunately quite rare. Otherwise it would have become very popular in the aquarium hobby. It has rather recently reproduced in an aquarium, however, and the present situation may change rapidly.

Another striking, but very rare, *Hypomasticus*, *L. despaxi*, also a small form, cannot be told apart from young *L. arcus* or *L. striatus* by color pattern alone. It is restricted, as far as we know, to Guianan rivers: the Maroni, Mana, and Oyapok.

The other already described *Hypomasticus* are plain or with a faint caudal spot. They are also of small size (135 mm maximum standard length), extremely rare, and restricted to rivers of southeastern Brazil. They may be identified as follows:

a. 36 or 37 lateral line scales; depth 4 in standard length; 4/4 teeth; no spot on body . . . *L. mormyrops*, type of the subgenus, synonym *L. thayeri* ? (Rio Parahyba, etc.)

aa. 34 lateral line scales; depth 4.75 in standard length; 3/4 teeth; a faint spot at end of caudal peduncle . . . *L. garmani* (Rio Jequitinhonha) (see also two close species with subinferior mouth from the same region: *L. conirostris* and *L. crassilabris*).

GENUS *ABRAMITES*
HEADSTANDERS

This genus has the same characteristics as the genus *Leporinus* (chiefly its subgenus *Leporinops*) with the exception of a higher body, a post-ventral keel and a longer anal fin. The most common forms have the well-known color pattern consisting of about 8 oblique bands on the body, the 3rd extending onto the ventral fin, the 4th on the dorsal fin and the 6th on the anal fin. An almost identical color pattern is seen in *Leporinus fasciatus*, *Leporinus desmotes* and *Synaptolaemus cingulatus*.

There was a revision of the genus in the twenties, but it is now out of date. The following tentative key will serve to distinguish the few forms recognized by the author:

a. Anal fin with 13 or 14 branched rays; 6½/41-42/5 scales; 4 transverse bands, their direction perpendicular to the long axis of the body ... *A. eques* (species restricted to the Rio Magdalena in Colombia)

aa. Anal fin with 11 or 12 branched rays; usually 6/39-41/6 scales; 8 transverse oblique bands ... *A. hypselonotus*, a superspecies or polytypic species. Two forms may eventually be distinguished:
—*Abramites hypselonotus ternetzi*, restricted to the Rio Paraguay basin. Possible synonyms are *Leporinus solarii* and *A. nigripinnis*. It is said to have the dorsal fin farther back and the caudal peduncle deeper than the Amazonian form.
—*Abramites hypselonotus hypselonotus*, the common form from the Amazon and Orinoco basins. A probable synonym is *Abramites microcephalus*, which cannot be told apart from *A. hypselonotus* (by usual taxonomic methods at least). The young fish are distinctly more elongate than the adults.

Leporinus arcus. Photo by Dr. Herbert R. Axelrod.

Leporinus striatus. Photo by Dr. Herbert R. Axelrod.

Leporinus sp., possibly *L. nigrotaeniatus.* Photo by Klaus Paysan.

Abramites hypselonotus. Photo by Harald Schultz.

A young *Abramites hypselonotus.* Photo by Dr. Herbert R. Axelrod.

GENUS *RHYTIODUS*

This peculiar genus includes the most elongate anostomids (depth up to 7 in the standard length); they have finely denticulated teeth, usually 4 above and 3 or 4 below on each side, their cutting edges respectively convex and concave, strongly compressed from front to behind and relatively thin, as compared to the robust teeth of most leporins. The species are of rather large size (300 mm in standard length is not rare), and live in large rivers (pelagic forms); they are of variable color, usually yellowish with dark mottlings, or the reverse. They may be identified as follows (*Rhytiodus elongatus*, which is curiously the least elongate, was not seen by the author):

a. 48-56 lateral line scales, 7-8/4-5 transverse scales.
  b. Depth 4.8 in standard length; 48 lateral line scales; teeth said to be 3/3 . . . *R. elongatus* (Rio Purus)
  bb. Depth 5.2-7.0 in standard length; 52-56 lateral line scales; teeth 4/4 . . . *R. argenteofuscus* (Amazon basin) (including "variety" *unifasciatus* ?)
aa. 76-90 lateral line scales, 11-12/8-9 transverse scales (depth 5.5-6.2 in standard length; teeth 4/3 or 4) . . . *R. microlepis* (Amazon basin)

GENUS *ANOSTOMOIDES*

This genus includes two deep bodied species which have teeth like certain leporins (somewhat trifid above, nevertheless) and the mouth upturned like in the subgenus *Laemolyta*. These two species, *Anostomoides atrianalis* from Venezuela and *Anostomoides laticeps* (first described as a *Schizodontopsis*, which is synonymous with *Laemolyta*) from the Essequibo River in Guyana, are very close to each other. They differ in the following way:

a. No marks actually visible on body; anal fin with 10 branched rays . . . *A. atrianalis*
aa. 3 transverse bands plus some darkening of caudal base; anal fin with 8 branched rays . . . *A. laticeps*

GENUS *ANOSTOMUS*
LISAS AND HEADSTANDERS

This important genus, the second (in number of species) after *Leporinus*, has the body elongate and the teeth multicuspid (at least above) like *Schizodon*, but the mouth clearly upturned. It includes two groups of species which can be considered as distinct genera or subgenera, a matter of opinion. These two groups, which will be treated here as subgenera because of their common characters, are *Laemolyta* and *Anostomus* (typical).

Anostomoides laticeps (after Eigenmann, 1912).

Laemolyta species are usually of a rather large size, but this is also the case of one species of *Anostomus*; they usually have the mouth slightly less upturned (not quite on the top of the head), but this character is somewhat subjective; their teeth are classically multicuspid above and truncate below, but the lower cusps disappear only at a certain age (corresponding to 60 or 70 mm in standard length), and the distinction between a young *Laemolyta* and a young *Anostomus* is quite difficult and often one must guess; finally some *Laemolyta*, at least, frequently adopt a head-down position as do most species of *Anostomus*. As concerns the geographical distribution, the more generalized *Laemolyta* species inhabit, as a rule, the center of the Amazon basin as well as the middle Orinoco, with some exceptions, whereas the more specialized *Anostomus* species (together with *Synaptolaemus, Sartor* and *Gnathodolus*) are more peripheral, inhabiting mostly the Guianas and the south of the basin, again with exceptions.

As with *Leporinus*, the *Laemolyta-Anostomus* species will be characterized in the shortest possible way, i.e. by means of identification keys.

Key to the subgenus *Laemolyta* (lisas):

a. A well-delimited longitudinal band; back with longitudinal lines; less than 6 scales between dorsal fin and lateral line (body compressed, the thickness 1.5-1.8 in the depth).

b. Transverse scales 4/4 . . . *A. (L.) garmani* (Central and Upper Amazon basin; also Upper Rio Xingu)

bb. Transverse scales 5/4 or 5 . . . *A. (L.) taeniatus* (Central Amazon basin; common in the Rio Negro)

Rhytiodus microlepis (above) and *Rhytiodus argenteofuscus* (below) from Rio Solimoes. Photo by Dr. Herbert R. Axelrod.

Unidentified young anostomid. Photo by Klaus Paysan.

A group of anostomid fishes. Photo by Klaus Paysan.

Adult *Anostomus* (*Laemolyta*) *proximus*. Photo by Dr. Herbert R. Axelrod.

aa. Longitudinal band usually diffuse or broken; back with transverse bands, at least in young specimens; more than 5 scales between dorsal fin and lateral line.

c. Transverse scales 6/5 or 6 (*proximus-varius-nitens* complex).

d. Longitudinal band not well delimited but continuous; depth 4.25-4.7 in standard length ... *A. (L.) proximus*, possible synonym *A. varius* (sympatric with *A. taeniatus*)

dd. Longitudinal band broken; depth 5 in standard length ... *A. (L.) nitens* (sympatric with *A. taeniatus*, also present in Lower Amazon basin)

cc. Transverse scales 7 to 9/6 to 7 (50 to 55 lateral line scales).

e. Longitudinal band extends to end of caudal fin; depth 4 and head 4.66 in standard length; thickness of body less than 1.8 in depth (?) ... *A. (L.) orinocensis* (Orinoco)

ee. Longitudinal band not extending onto caudal fin; depth 4.2 and head 4.25 in standard length; thickness of body more than 1.8 in depth.

f. 9/50/7 scales ... *A. (L.) fernandezi* (Upper Orinoco)

ff. 7 or 7½/54-55/6 or 6½ scales ... *A. (L.) petiti* (Rio Araguaia)

Key to the subgenus *Anostomus* (headstanders):

In this group of species, the teeth are bi- or tricuspid in both jaws and the mouth is exactly superior, that is the lower jaw is vertical when the mouth is closed. As with the other genera, the species will be defined by a key:

a. 42-46 lateral line scales (dorsal fin far in advance of midbody; color pattern of lateral spots and longitudinal lines through the center of the scales).

b. Usually a large spot on flank under dorsal fin, and a small peduncular spot; several bars astride back; depth 3.2-3.9 in standard length in adult; dorsal fin usually midway between snout and adipose fin; scales 5 or 6/42-44/4½ or 5, 11½ to 13 in predorsal series; about 120-140 mm ... *A. (A.) trimaculatus* (the typical form, from Mato Grosso, is perhaps different from the specimens collected in the Amazon, Rio Negro, Guyana and Venezuela. It has the depth 3.2-3.4 in standard length, as opposed to 3.6-3.9 for the Amazonian-Guianan form)

bb. Usually 4 rounded spots on flank; no dorsal bars; depth 4.15-4.20 in standard length in adult; dorsal fin between snout and point behind adipose fin; 5 or 5½ /44-46/4½ or 5 scales, 13-14 in predorsal series; about 160-180 mm ... *A. (A.) gracilis* (Rio Guapore; Rio Negro; Orinoco)

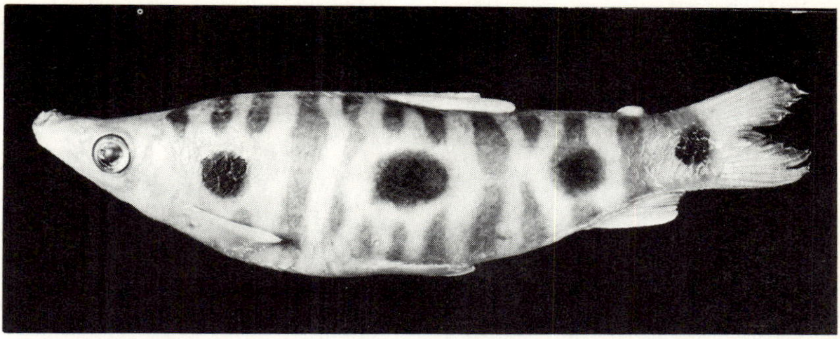

Anostomus spiloclistron. Photo from the Zoological Museum of Amsterdam, courtesy of Dr. H. Nijssen.

aa. 38-42 lateral line scales.

c. 4 roundish or slightly horizontally elongate spots on flank; longitudinal lines between rows of scales; about 12 bars across back, some extending onto flank between the spots . . . *A. (A.) plicatus* and *A. (A.) spiloclistron* (Guyana and Surinam)

cc. Body with large longitudinal bands on flank.

d. Color pattern formed of 2 narrow light longitudinal bands on a dark background; back striated; dorsal fin usually at midbody except in very large specimens; 12 or 13 predorsal scales (pectoral fins i, 14-15). . . . *A. (A.) ternetzi* (Guianas; Orinoco; Rio Araguaia and Rio Xingu)

dd. Color pattern composed of 3 dark longitudinal bands on light background; back uniformly brown, not striated; dorsal fin usually in front of midbody; 11 predorsal scales (pectoral fins i, 12-14).

e. Depth 3.33-3.75 and thickness of body 8-13 in standard length; transverse scales 5 or 5½/4 or 4½; snout relatively long, the eye slightly behind middle of head (snout-to-dorsal 1.05-1.20 in dorsal-to-caudal) . . . *A. (A.) brevior* (Mana and Camopi Rivers in French Guiana)

ee. Depth more than 3.8 and thickness of body 7.0-7.5 in standard length; transverse scales 4 or 4½/3½ or 4; eye in middle of head (snout-to-dorsal 1.10-1.25 in dorsal-to-caudal) . . . *A. (A.) anostomus* (an elongate subspecies, called *A. anostomus longus*, comes from the upper Amazon; the typical comes from the Guianas (except French Guiana ?), Orinoco and (?) Colombia)

Anostomus (*Laemolyta*) *taeniatus.* Photo by Harald Schultz.

A young *Anostomus* (*Laemolyta*) *proximus.* Photo by Harald Schultz.

Lisa of the *varius*-complex, possibly *Anostomus* (*Laemolyta*) *nitens.* Photo by Dr. Herbert R. Axelrod.

Anostomus (*Laemolyta*) *petiti* (or close to it). Photo by Harald Schultz.

Unidentified lisas, probably close to *Anostomus* (*Laemolyta*) *taeniatus*. Photo by Dr. Herbert R. Axelrod.

GENERA *SYNAPTOLAEMUS, SARTOR* AND *GNATHODOLUS*
SPECIALIZED HEADSTANDERS

These genera, each with only one species, are modified offshoots of an *Anostomus*-like ancestor or similar form. They are confined to very special biotopes, such as crevices in rocks or fallen trees, and they inhabit restricted areas (having been collected, up to now, only in the "bifurcation" of the Upper Orinoco and the Rio Casiquiare and in the Upper Xingu).

Synaptolaemus cingulatus (from both localities) has a striking coloration composed of seven narrow yellow to orange rings encircling the body on an almost black background; as in *Abramites* and *Leporinus desmotes*, the dorsal, ventral and anal fins are included, in part, within the dark pattern. Its teeth are the least specialized of the three: the median teeth of both jaws are not enlarged and simply truncate, and the lateral ones are slightly bi- or trifid. It eats only insects, judging from the stomach contents.

Sartor respectus (from above the falls in the Upper Xingu) has a quite trivial color pattern (6 or 7 inconspicuous transverse dark bands and longitudinal lines between the scales-rows) but a striking dentition, the lower median pair of teeth being enlarged and pointed upward and backward like a pair of tusks; likewise, the eyes are directed obliquely upward and backward. It has a mixed diet, eating algae, insects and sponges.

Gnathodolus bidens. Holotype from the Canal de Casiquiare (after Myers, 1950).

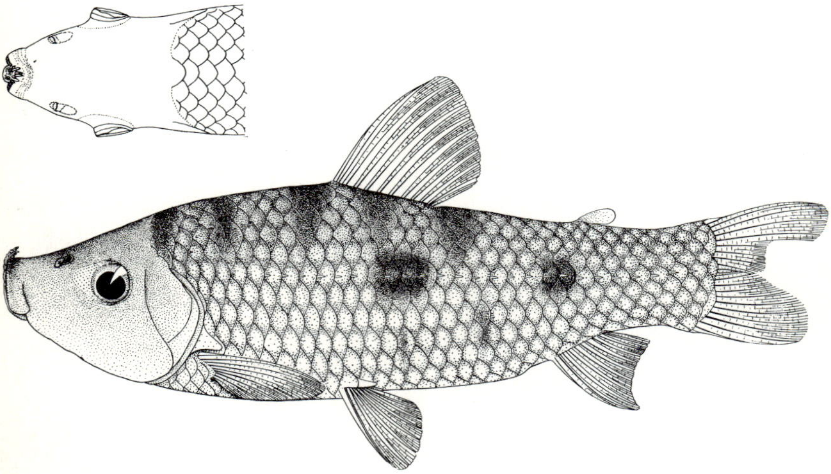

Gnathodolus bidens, from the bifurcation of the Orinoco and Casiquiare rivers, also has a quite common coloration (two diffuse round spots and a few bars on the back), but an even more striking dentition, the lower teeth having disappeared except for the median pair, which also protrude out of the papillae of the lips, but with a sickle-shaped form; the eyes are directed obliquely backward. Nothing is known of its diet.

Comparison of *Synaptolemus* with *Anostomus* and *Gnathodolus.* (a) Right premaxillary and right mandibular teeth; (b) position of teeth on jaws; (c) side view of head (after Myers, 1950).

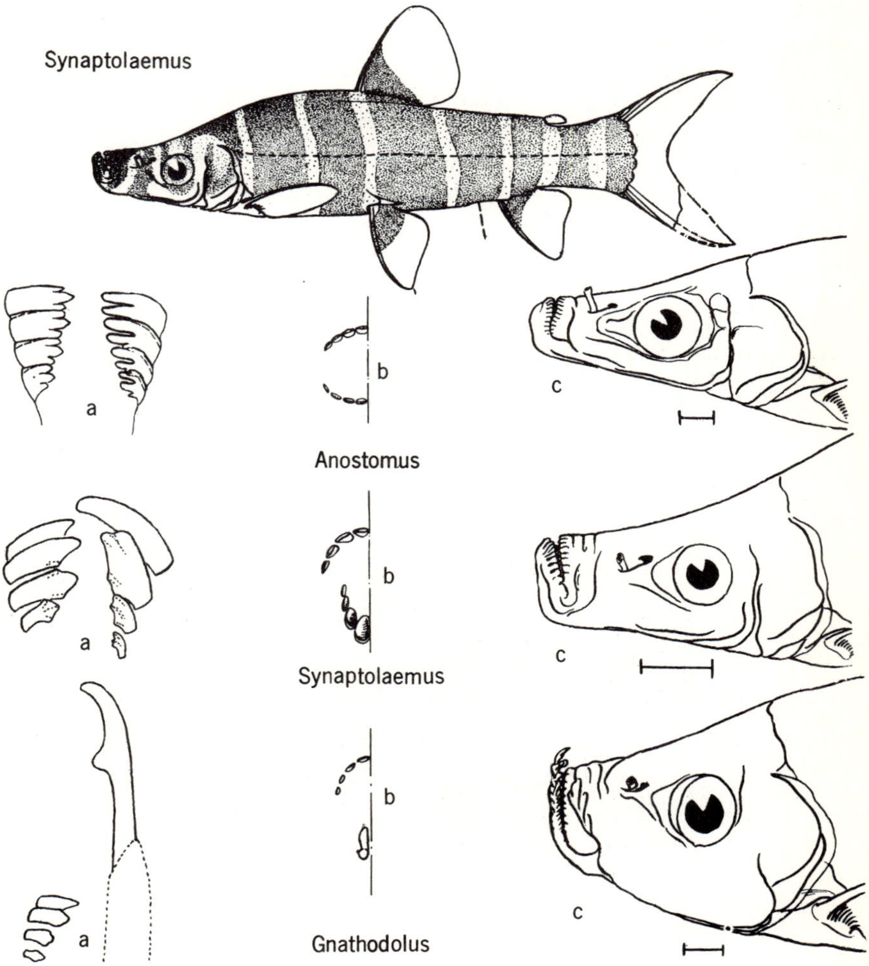

Anostomus (*Anostomus*) *ternetzi.* Photo by Harald Schultz.

Head of *Anostomus ternetzi.* Photo by Harald Schultz.

Anostomus (*Anostomus*) *anostomus.* Photo by R. Zukal.

Family Hemiodidae

Hemiodids and their relatives

These fishes, of rather small size with the body usually fusiform and the mouth subinferior, resemble somewhat the leporins and headstanders, but with a very different dentition: in most of the species the lower jaw is reduced and lacks teeth. On the upper jaw as well as sometimes on the feeble maxillary, the teeth are small, more or less cuspidate incisors which are loosely inserted into the bone or even into the gums. We will find this regressive tendency again in the chilodins and prochilodins which follow, whereas in the next (and last) step, the anodins and curimatins have completely lost their teeth, at least after the postlarval stage.

In the course of their evolution, the hemiodids have diverged rather considerably, according to their respective ecology (surroundings). At least three groups may be recognized: (1) species living in the rapids or open waters; (2) species living on (in) the sandy bottom; (3) species living on the bottom, usually on rocks, rarely on sand. They can be distinguished by the following characters:

a. Mouth subinferior, more or less protractile; no maxillary teeth; fontanels present; adipose lid present; paired fin not very developed
. . . Subfamily HEMIODINAE

b. Gill membrane free from isthmus; scales cycloid; mouth not much protractile; fontanel broad; gill-rakers normal; buccal cavity without papillae; live in open water.

c. Teeth unicuspid in both jaws; lower jaw with two rows of teeth in its anterior part . . . Tribe Micromischodini

cc. Teeth multicuspid; no teeth in lower jaw after the post-larval stage
. . . Tribe Hemiodini

bb. Gill membrane partly united to isthmus; scales cyclo-ctenoid; mouth strongly protractile; fontanel narrow; inner face of gill-rakers with numerous tubercles; buccal cavity with several valvulae or papillae; live on sand beaches (teeth very weak, tricuspid) . . . Tribe Bivibranchiini

aa. Mouth strictly inferior, with the snout scarcely movable; maxillary and mandibular teeth present but never numerous; fontanels closed; adipose lid absent; paired fins developed; bottom-fishes (teeth usually broad incisors, truncate or multicuspid; gill membrane not united to the isthmus; scales cycloid)
. . . Subfamily PARODONTINAE

SUBFAMILY HEMIODINAE

Tribe Micromischodini

The single species *Micromischodus sugillatus* is one of the most interesting discoveries of recent years. It was collected near Manaus by Dr. Tyson Roberts, who described it in detail in 1971.

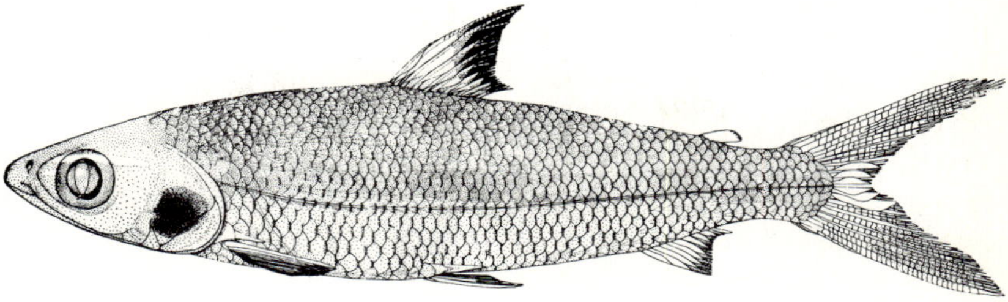

Micromischodus sugillatus (after Roberts, 1971).

It is the only hemiodid with numerous minute teeth on both jaws, and one of the few non-characids with two rows of teeth on the anterior part of the lower jaw (about 55 to 60 in each row), yet in the shape of the body it does not differ at all from a more "conventional" hemiodin such as *Argonectes*.

Tribe Hemiodini

Four genera comprise this group characterized by multicuspid teeth present only on the premaxilla (except at an early post-larval stage, where some mandibular teeth are to be seen, as in *Curimata*), the snout scarcely mobile except in *Argonectes*, and a few other technical characters to be seen in the following key to the genera:

a. Teeth with more than 3 cusps; adipose lid not entirely covering pupil; snout scarcely protractile; gill membrane not united with isthmus.
  b. Dorsal fin small; mouth usually subinferior, with the teeth almost horizontal, backward directed and almost parallel to premaxillary bone.
    c. Scales of back only a little smaller than those of abdominal region ... *Hemiodopsis*
    cc. Scales of back much smaller than those of abdomen; ratio of apparent height of scales below and above lateral line 1.8-2.2 in adult ... *Hemiodus*
  bb. Dorsal fin filamentous, at least 1.5 times longer than head; mouth almost terminal, with teeth almost vertical and inserted obliquely relative to premaxillary axis ... *Pterohemiodus*
aa. Teeth with 3 cusps, loosely inserted into the lips; adipose lid completely covering pupil; snout partly movable; gill membrane united forward to isthmus (an almost perfect link with the next tribe, the Bivibranchiini) ... *Argonectes*

Anostomus (Anostomus) trimaculatus. Photo by Harald Schultz.

Anostomus (Anostomus) plicatus. Photo by Harald Schultz.

Synaptolaemus cingulatus from the Upper Xingu. Photo by Harald Schultz.

Sartor respectus from the Upper Xingu. Photo by Harald Schultz.

GENUS *HEMIODOPSIS*

With at least 15 species, *Hemiodopsis* (formerly known as *Hemiodus*, that actually replaces *Anisitsia*) is the largest hemiodid genus. The following key was published in *Tropical Fish Hobbyist* (November, 1964); it is here somewhat modified:

a. Less than 80 lateral line scales.
  b. No definite transverse bands on sides; usually a longitudinal band and/or some spots.
    c. Less than 55 lateral line scales; transverse scales 7 or 7½/4 or 5; a longitudinal band (*gracilis*-group).
      d. Body slender; 42 lateral line scales . . . *H. gracilis* (Amazon, Guyana and Rio Sao Francisco)
      dd. Body less elongate; 44 or more lateral line scales.
        e. 44-46 lateral line scales.
          f. Longitudinal band beginning below dorsal fin level . . . *H. goeldii*, a species often confused with *H. semitaeniatus* (Guianas and Rio Xingu)
          ff. Longitudinal band continued anteriorly by a lighter band to opercle . . . *H. thayeria* (Rio Negro)
        ee. 50-53 lateral line scales (longitudinal band beginning below dorsal fin level) . . . *H. fowleri* (Amazon)
    cc. 55-72 lateral line scales (*semitaeniatus*-group).
      g. A longitudinal band; less than 64 lateral line scales.
        h. 9/56-60/5 scales; longitudinal band beginning below dorsal fin level . . . *H. semitaeniatus* (Rio Guapore)
        hh. 11/63/6 scales; longitudinal band continuous from eye to end of lower caudal lobe . . . *H. ternetzi* (Rio Tocantins)
      gg. No longitudinal band.
        i. No marks on flank; 10-11/69-72/6 or 7 . . . *H. immaculatus* (Rio Negro and middle Amazon basin)
        ii. A round spot on the middle of the flank; 17-18 (?)/65-70/11 scales . . . *H. rodolphoi* (Rio Parahyba)
  bb. Three transverse bands on side and one on caudal peduncle (*quadrimaculatus*-group).
      j. 42-45 lateral line scales.
        k. 9 branched anal rays; about 27 gill-rakers on lower arch . . . *H. quadrimaculatus quadrimaculatus* (Camopi River in French Guiana)
        kk. 8 branched anal rays; 16-21 gill-rakers on lower arch . . . *H. quadrimaculatus vorderwinkleri* (Upper Amazon; Guyana, Surinam)

jj. 50 or more lateral line scales.

l. 9½-10/50-52/5 or 6 scales ... *H. huraulti* (Surinam and French Guiana)

ll. 12/64-67/7½ or 8 ... *H. sterni* (Upper Rio Juruena)

aa. 80-130 lateral line scales (usually a round spot on the flank) (*microlepis*-group).

m. 20/80-83/12 scales ... *H. parnaguae* (Rio Parnagua; a similar form collected in Surinam, still undescribed, has 81-99 lateral line scales)

mm. More than 100 lateral line scales.

n. 24-25/110-112/14-15 scales ... *H. microlepis* (Amazon, Rio Negro, Rio Guapore; a similar form collected in Bolivia, still undescribed, has 27-28/21-23 transverse scales)

nn. 30-32/120-125/18 or 19 scales ... *H. argenteus* (Rio Orinoco)

Paratype of *Hemiodopsis huraulti.* Photo by the author.

Unidentified *Hemiodopsis* of the *gracilis*-group. Photo by Harald Schultz.

Hemiodus unimaculatus or *Curimata ocellata,* mimetic species from the vicinity of Humaita, Brazil. Photo by Dr. Herbert R. Axelrod.

Hemiodopsis gracilis. Photo by Hilmar Hansen.

Hemiodopsis of the *microlepis*-group from Humaita, Brazil. Photo by Dr. Herbert R. Axelrod.

GENUS *HEMIODUS* (synonym *ANISITSIA)*

In this small group the scales of the back of adult or semi-adult fishes are notably smaller than those above the ventral fin; apart from this character there is nothing which distinguishes *Hemiodus* from *Hemiodopsis*, and old descriptions (where the size of the scales is not mentioned) are unreliable in this respect.

Hemiodus unimaculatus, formerly called *Anisitsia notata*, is a fusiform fish strongly resembling a *Hemiodopsis* of the *microlepis*-group for example, with a round black spot astride the lateral line behind the dorsal fin level. It is common in the Guianas, the Rio Orinoco and probably in the state of Para (under the name *H. microcephalus*), and even in the Rio Guapore. To the south it is apparently replaced by a species with more lateral line scales (about 90 instead of 60-72), *H. orthonops*, from the Rio Paraguay.

GENUS *PTEROHEMIODUS*

The two known species, *P. atranalis* and *P. luelingi*, both from the Upper Amazon, have the body shape and the color pattern of a generalized hemiodin such as *Hemiodus unimaculatus*, but the dorsal fin very high and the teeth obliquely set.

GENUS *ARGONECTES*

This genus, together with *Atomaster*, forms a transition between the "true" hemiodins such as *Hemiodus*, and the "true" Bivibranchiini. Following recent studies, it seems that the two known species may be united into a single one, *Argonectes longiceps* (older name than *A. scapularis*), with a certain variability in the color pattern according to the localities (Rio Negro, upper Rio Orinoco, Lower Amazon and the Maroni River between Surinam and French Guiana). Specimens from the upper Orinoco (the form called *A. scapularis*) have a black, collar-like marking bordering the gill-opening, dark pigment on the dorsal fin, dark bands on the caudal lobes, and an elongate band behind the dorsal fin level. In specimens from the Maroni River, the mark on the flank is less elongate and resembles that of *Hemiodus unimaculatus*.

Tribe Bivibranchiini

The members of this group are hemiodins having adapted themselves to a diet composed at least partly of small animals living in the sand. They have thus developed a strongly protractile mouth to suck in the sand and bucco-pharyngeal papillae and valvulae to filter it. Two genera are known,

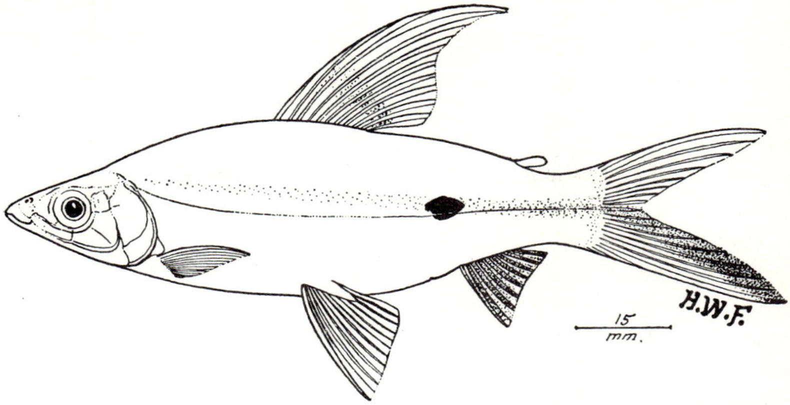

Holotype of *Pterohemiodus atranalis* (after Fowler, 1950) and *Bivibranchia protractila* (after Eigenmann, 1912).

Atomaster, which is still rather close to the hemiodins in the strict sense, and *Bivibranchia*, which is a very specialized form. The two genera differ in the following way:

a. Gill membrane free from isthmus; scales very numerous, their borders pectinate; valvulae on roof of mouth not much developed . . . *Atomaster*
aa. Gill membrane united to isthmus (on a small surface); scales not more than 62, crenulate only in adult; valvulae on roof of mouth well developed . . . *Bivibranchia*

The single species of *Atomaster*, *A. velox*, was collected in 1924 in the Rio Tocantins, state of Goyaz, Brazil. It has not been found again. It is straw colored with a longitudinal silvery band.

Bivibranchia protractila exists not only in the Rio Tocantins, but also along the sand beaches of the Branco and Negro rivers, near Santarem, and in Guyana. It has a plain color. Two other species have recently been found in collections from Surinam.

Hemiodopsis goeldii. Photo by Harald Schultz.

Hemiodopsis of the *gracilis*-group. Photo by Dr. Herbert R. Axelrod at Rio Negro.

Hemiodopsis sterni. Photo by Harald Schultz.

Hemiodopsis quadrimaculatus vorderwinkleri. Photo by Dr. Herbert R. Axelrod.

SUBFAMILY PARODONTINAE

The members of this group (that deserves full familial status according to certain ichthyologists) are all specialized for life on the river bottoms, mostly on rocky soil: the mouth is strictly inferior, the breast flat and broad, and the paired fins, chiefly the pectorals, are well developed and used to "walk" on the bottom. Other more technical characters have been given in an earlier key. Two genera are here recognized, as follows:

a. 8 or 10 multicuspid incisior teeth on a straight line along the premaxillary bones; usually 2 similar maxillary teeth; circumorbital bones not covering face ... *Parodon*

aa. 6 truncate incisior teeth, not in a straight line on the premaxillary bones; no maxillary teeth; circumorbital bones well developed, covering practically the whole face ... *Saccodon*

GENUS *PARODON*

Within this rather important unit, a few subunits may be found. They will here be called subgenera and may be distinguished as follows:

a. Pectoral fin with only one unbranched ray; ventral fin usually with i, 7 rays.

b. 1 to 3 small truncate teeth on sides of lower jaw; upper teeth usually with a straight edge ... Subgenus *Parodon* typical

bb. Lower jaw not toothed in adults; upper teeth usually with a rounded edge ... Subgenus *Apareiodon*

aa. Pectoral fin with two unbranched rays; ventral fin with i, 8 rays ... Subgenus *Parodontops*

Subgenus *Parodon*

The species of this subunit are perhaps the least specialized of the Parodontinae. Two species are somewhat independent from the rather generalized type. The first, for which Fowler proposed the generic name *Nematoparodon*, is *Parodon apolinari*, from the upper course of the Rio Orinoco in Colombia and Venezuela. It is remarkable in having a filamentous dorsal fin and tubercles on the snout. These peculiarities are generally considered as secondary sexual characters, and are much less common in the characoids than in the cyprinoids for example. *Parodon apolinari* is convergent to a *Labeo* in this respect. The other species is *Parodon guyanensis* from French Guiana and Surinam, which has rounded incisor teeth, as in *Apareiodon*, numbering 10 instead of 8 as in other species of the subgenus; it also has an elongate body. Apart from these characters which are probably of subgeneric rank, *P. guyanensis* is recognizable at first sight by a black bar on the dorsal fin and a zigzag mark along the lateral line.

The other 9 species, very close to each other, may be distinguished by means of the following key (adults only):

a. More than 20 cusps on middle upper teeth in adult; an irregular longitudinal band ... *P. buckleyi* (Ecuador)

aa. Less than 20 cusps on middle upper teeth in adult.

b. More than 12 cusps on middle upper teeth.

c. Depth usually 3.5-3.8 in standard length.

d. A series of vertical dark bars on sides; anal fin ii, 6; scales 5/39 or 40/4 or 5, 13½ or 14½ in a predorsal series ... *P. caliensis* (Colombia; an Andean species living at 1000-1100 m altitude)

dd. Usually one or two longitudinal bands; anal fin ii, 7; scales 4 or 4½/35-39/3 or 4, 11½ or 12½ in a predorsal series.

e. Scales 4/35-38/4; about 17-18 cusps on middle teeth; caudal lobes equal; head about 4.35 in standard length; one longitudinal band ... *P. suborbitale*, type of the genus (Venezuela; perhaps Colombia)

ee. Scales 4/39/3; about 15 cusps on middle teeth; lower caudal lobe more elongate than upper one; head about 5.15 in standard length; two longitudinal bands ... *P. bifasciatus*, a very rare species, apparently slightly different from *suborbitale* (Rio Branco basin)

cc. Depth usually more than 3.8 in standard length.

f. Depth 3.9-4.25 in standard length; pectoral fin usually i, 12-13.

g. A longitudinal band up to end of middle caudal rays; anal fin ii, 6 ... *P. carrikeri* (upper Rio Paraguay)

gg. A series of about 15 lozenge-shaped spots along lateral line; anal fin ii, 7 ... *P. gestri* (Mato Grosso)

ff. Depth 4.2-5.6 in standard length; pectoral fin i, 14 or 15 ... *P. tortuosus* (probably 2 subspecies, *P. t. caudalis* and *P. t. tortuosus*, sympatric, respectively, with *P. carrikeri* and *P. gestri* and having, respectively, the same pattern (sibling forms?))*

bb. 9-12 cusps on middle upper teeth.

h. A very conspicuous, well delimited black longitudinal band (depth 3.8-4.6 in standard length) ... *P. pongoense*, originally described as an *Apareiodon*, synonym *A. caquetae* (Peruvian Amazon, Ecuador and Colombia)

hh. A diffuse dark longitudinal band (depth 4.5-4.66 in standard length) ... *P. nasus* (probably 2 subspecies, *P. n. nasus* from Mato Grosso and *P. n. hilarii* from Minas Gerais, differing in the number of lower teeth and other details)

* (*P. pirassunungae* from S.E. Brazil is a synonym of *P. tortuosus*)

Bivibranchia sp. Photo by J. Elias.

A *Hemiodopsis* of the *microlepis*-group. Photo by R. Zukal.

Parodon pongoense. Photo by Dr. Herbert R. Axelrod.

Parodon sp. close to *P. affinis.* Photo by Klaus Paysan.

Subgenus *Apareiodon*

The species of this group, 8 or so in number and all from eastern and southeastern Brazil and the La Plata basin, curiously parallel the *Parodon* species in the evolution of their meristic characters and coloration. It is not impossible that, in the future, they will be distributed differently within the genus because the presence or absence of a few mandibular teeth as well as the more or less rounded edge of the upper teeth seem to be rather poor characters, at least at the generic or subgeneric level.

The following key is partly compiled from the literature and it is by no means intended as a revision of the subgenus:

a. Usually two maxillary teeth on each side (occasionally one in *P. piracicabae* and *P. davisi*).
  b. Rather deep bodied species, depth usually less than 4 in standard length . . . *P. davisi* (N.E. Brazil);

  bb. Elongate species, depth more than 4 in standard length.
    c. 35-36 lateral line scales and 10-11 predorsal scales; usually ii, 9 dorsal rays (a conspicuous black longitudinal band plus a dorsal band) . . . *P. itapicuruensis* (Rio Itapicuru, coastal river of Eastern Brazil near Rio S. Francisco basin)
    cc. 39-44 lateral line scales and 12-14 predorsal scales; usually ii, 10 dorsal rays.
      d. Dorsal fin with 2 black bars; 39 lateral line scales . . . *P. machrisi* (near Rio de Janeiro)
      dd. Dorsal fin plain; 40-44 lateral line scales.
        e. A conspicuous longitudinal band plus another one on back . . . *P. piracicabae* (basin of the Tiete-Parana rivers, S.E. Brazil)
        ee. A longitudinal line plus 6 transverse bands on back (usually less conspicuous) . . . *P. affinis*, synonym *P. paraguayensis* (La Plata, Parana, Paraguay basins)

aa. Only one maxillary tooth on each side (depth 4.33-5.0 in standard length).
  f. Less than 40 lateral line scales (34-39), 14 predorsal scales; a longitudinal band with 7-9 lozenge-shaped spots superimposed on it . . . *P. ibitiensis* (S.E. Brazil)
  ff. Usually 40 or more lateral line scales and only 10-12 predorsal scales.
    g. Head 4.5-5.0 in standard length; 4 or $4\frac{1}{2}/3$ transverse scales; a longitudinal line along body . . . *P. hasemani* (Rio S. Francisco)
    gg. Head 4.4 in standard length; transverse scales said to be 4/2; a series of bars or lozenge-shaped spots along body . . . *P. mogiguacuensis* (Rio Mogiguacu)

Saccodon wagneri, showing ventral view of head and thorax and teeth (after Eigenmann, 1922), and *Saccodon dariensis* (after Meek & Hildebrand, 1916).

Subgenus *Parodontops*

The single species *Parodon ecuadoriensis* (synonym *Parodon terminalis*) from the Pacific Coast of Ecuador represents one of the extremes of the specialization toward a bottom life and strongly parallels certain *Characidium.* It has the breast broad and flat and the pectoral fins very low, horizontal, and spread like those of an *Eleotris* (see footnote under *Saccodon*). Apart from the usual irregular longitudinal dark band, this species has an orange caudal fin barred with black.

GENUS *SACCODON*

With their armor-plated head, species of the genus *Saccodon* are strange fishes indeed, looking like remnants of an older age now surviving in some remote torrents of the Andes and Panama. According to Schultz and Miles, the species can be identified in the following way:

a. Pectoral rays ii, 15-16; 39-41 lateral line scales.
  b. Dorsal rays ii, 10; anal rays ii, 7; color pattern of elongate dark blotches arranged in three streaks along sides; caudal and anal fins spotted . . . *S. caucae* (Upper Rio Cauca)
  bb. Dorsal rays ii, 9; anal rays ii, 8; color plain; fins not spotted . . . *S. wagneri* (western slope of Ecuador)
aa. Pectoral rays ii, 12; 35-37 lateral line scales (color pattern much like that of *S. caucae*) . . . *S. dariensis* (Panama)

Parodon affinis (or close to it). Photo by Harald Schultz.

Parodon suborbitale. Photo by Dr. Fernandez-Yepez.

Two aspects of *Caenotropus labyrinthicus*. Upper photo by Harald Schultz, lower photo of a specimen from Puerto Gaitan, Colombia by Dr. Herbert R. Axelrod.

In an interesting paper just published, T. Roberts (1974) suggests that certain northwestern Parodontinae, so far separated into distinct species or even genera on the basis of the structure of their teeth (as it was done also here), are actually variant individuals ("morphs") of a *Saccodon*. These would be *S. dariensis* in Panama and Colombia (synonym *S. caucae*, according to Roberts), and *S. wagneri* in Ecuador. In other words the genus *Saccodon* would present a "dental polymorphism", having the teeth variable in form as well as in number.

Following this interpretation (of course to be tested against other hypotheses such as mimicry, sibling species and so on, by means of crossing experiments and/or study of the chromosomes, for example), *P. (Parodontops) ecuadoriensis* (see preceding section) would fall into the synonymy of *Saccodon wagneri* and *Parodontops* (type *ecuadoriensis*) thus would become a synonym of *Saccodon*.

Family Curimatidae

Curimatas

With the curimatids we reach the most regressed of the characoids, in a certain sense. Most of the curimatids have no teeth when adult, and they content themselves with food found in the mud, i.e. mostly algae and detritus like many carps; others have a mouth in the form of a sucking disc, with which they rasp their food from diverse substrata. All have the long intestine of typical herbivorous animals.

Four groups are here recognized within the curimatids, some of which are considered as distinct families by certain ichthyologists.* They can be diagnosed as follows:

a. Anal fin with 3 or 4 anterior unbranched rays; fourth branchial arch thickened; swimming position head down (7-10 branched dorsal rays; teeth present, few in number, bifid, set into the lips) . . . Subfamily CHILODINAE

aa. Anal fin with 2 anterior unbranched rays; fourth branchial arch normal; swimming position horizontal (8 or more rarely 9 branched dorsal rays).

b. Mouth protractile, forming a sucking disc, with two comb-like rows of teeth set into the lips; a predorsal spine (gill membrane united to the isthmus; body bream-shaped; mostly detritivorous) . . . Subfamily PROCHILODINAE

bb. Mouth scarcely protractile, not forming a sucking disc, lacking teeth after the post-larval stage; no predorsal spine.

c. Bottom fishes, with the body shape of a bream, very similar to that of a prochilodin, the breast and belly often flattened or, on the contrary, keeled; gill-rakers usually rudimentary, gill membranes united to the isthmus; adipose lid usually absent; maxilla reduced; mostly mud eaters when adult. . . . Subfamily CURIMATINAE

* T. Roberts (1973) presents some osteological evidences favoring relationships between Prochilodinae (which he treats as a family and calls Prochilodontidae, contrarily to the Rules) and Anostomidae, rather than Curimatidae. However the jaws, the teeth and the general organization of the two first named groups are markedly different.

cc. Pelagic fishes with the body fusiform, very similar to that of a hemiodin, the breast and belly rounded; gill-rakers long and numerous, gill membranes free from the isthmus; adipose lid covering the eye; maxilla relatively developed; plankton eaters ... Subfamily ANODINAE

SUBFAMILY CHILODINAE

Headstanders

The chilodins share with certain anostomids the faculty of swimming head down: this must be in connection with some special ecological needs and not due to close relationships. Phylogenetic connections are more likely to be looked for in the direction of the hemiodids (according to certain ichthyologists) or the curimatids.

The author published some years ago (*Tropical Fish Hobbyist*, May, 1964, pages 5-10 and 63-67) a simplified review of the headstanders. Although its exaggerated simplicity may be confusing, it is repeated below, no one having since added essential data.

The two genera here recognized may be distinguished as follows:

a. Mouth distinctly inferior, with thin lower lip, the lower jaw included; pharyngeal teeth rather large and relatively not numerous; scales crenulate in the adults; predorsal region rounded except just before dorsal fin; post-ventral region rounded; anal rays iii, 7i or 8, the fin with a concave border; rather large size (up to 185 mm) ... *Caenotropus*

aa. Mouth terminal or slightly upturned, the lips thick, the lower jaw slightly prominent when the mouth is closed; pharyngeal teeth rather small and numerous; scales cycloid; predorsal and postventral regions keeled; anal rays iii, 10 or 10i, the fin with a straight border; usually not more than 120 mm in length ... *Chilodus*

GENUS *CAENOTROPUS*

HEADSTANDERS

Two species are known: *Caenotropus labyrinthicus* and *C. maculosus*. *Caenotropus labyrinthicus*, from the Amazon basin, the Rio Orinoco and also the Rio Parnaiba, has a black longitudinal band along its body, extending from the tip of the snout to the end of the middle caudal rays, with a bulge shortly before the dorsal level; the first dorsal rays are dark. In the Rupununi River there is a close form characterized by a broken lateral band and a conspicuous bar on the dorsal fin.

Caenotropus maculosus, from the Guianas, has about the same color pattern, with the tip of the dorsal fin black. It differs in having fewer longitudinal scales, postdorsal scales, gill-rakers and branched caudal rays, a longer head, the adipose fin above the level of the last anal rays (instead of above the first ones as in *C. labyrinthicus*), and some teeth on the lower lip (instead of none or one or two very rudimentary ones as in *C. labyrinthicus*). The following table gives the principal characters of both species:

Caenotropus maculosus. Photo by Dr. Herbert R. Axelrod.

Chilodus punctatus, juvenile. Photo by Dr. Herbert R. Axelrod.

A pair (?) of *Chilodus punctatus.* Photo by H.J. Richter.

	labyrinthicus	*maculosus*
Standard length/length of head	3.6-3.8	3.4-3.5
Lateral line scales (counting all)	29-31	26-27
Postdorsal scales (up to adipose fin)	9-10	6½ or 7
Gill-rakers	19-23	14-16
Branched caudal rays	17	15-16

GENUS *CHILODUS*
SPOTTED HEADSTANDER

The single known species (at least the single one regularly discussed), *Chilodus punctatus*, from the Guianas, upper Amazon and upper Orinoco Rivers, is identifiable at once by the form of its cyprinoid mouth, somewhat resembling that of a Pekinese dog, its higher body, its spotted dorsal fin, and a different color pattern of the body; the longitudinal line, conspicuous from snout to opercle, is only "suggested" posteriorly by a series of brown spots at the center of each scale. During spawning (which is a very rare event in captivity), the spotted pattern disappears and is replaced by a broad humeral spot.

The spotted headstander is a rather variable species as regards the body shape and color pattern. This is why *Chilodus zunevei*, from the upper Maroni in French Guiana (whose types are lost), has to be considered, until it is collected again, as particularly deep bodied specimens of the spotted headstander (body depth about 2.55 in the standard length, versus at least 2.75 in the "normal" headstander). On the other hand, there are indications that an undescribed headstander does live in the upper Amazon basin, more precisely in the Rio Japura. This species has the upturned mouth, the long anal fin and spotted dorsal fin of *Chilodus punctatus*, but the elongate body and the longitudinal band of *Caenotropus* (either *labyrinthicus* or *maculosus*).

SUBFAMILY PROCHILODINAE
Bocachicos

This is a rather important group of no less than 25 species, all reaching a fairly large size (30 to 45 cm). The biology of some, which are of economic importance in Brazil or Colombia, has been rather extensively studied by local ichthyologists, and the group has been recently reviewed from a systematic point of view by Dr. Mago, with an accent on the

Venezuelan species. According to this review, three genera can be defined as follows:

a. Predorsal spine forked in front; edge of labial teeth rounded; snout moderate in length, jaws equal or upper one not very prominent, the mouth not inferior.
   b. Scales cycloid or crenulate, the postdorsal ones in a regular series from dorsal to adipose fin, finely striated; caudal fin with conspicuous bars ... *Semaprochilodus*
   bb. Scales ctenoid, at least in adults, the postdorsal ones usually not in a regular series and not striated, similar to those of flanks; caudal fin plain or spotted, the spots sometimes arranged in wavy lines but never forming conspicuous bars ... *Prochilodus*

aa. Predorsal spine blunt in front and not forked; labial teeth pointed; snout long, upper jaw prominent with the mouth inferior (scales cycloid; caudal fin plain) ... *Ichthyoelephas*

GENUS *SEMAPROCHILODUS*
BOCACHICOS

The genus *Semaprochilodus* (possibly only a subgenus of *Prochilodus*) is better characterized by the very peculiar color pattern of the caudal and anal fins of all species than by the structure of the scales. They can be identified as follows:

a. Elongate species, body depth 2.8-3.3 in standard length (scales crenulate, 10 or more in a transverse series from dorsal fin to lateral line; 3 or 4 oblique bars on each caudal lobe).
   b. 66-76 lateral line scales and 12-13/12-14 or 15 (up to ventral fin) transverse scales; body depth about 2.8-2.9 in standard length ... *S. taeniurus* (middle Amazon basin)
   bb. 48-50 lateral line scales and 10/10 transverse scales; body depth 3.0-3.33 in standard length ... *S. binotatus* (Rio Branco; a dubious species)

aa. Deep species, body depth 2.0-2.7 in standard length.
   c. $7\frac{1}{2}$-9 scales in a transverse series above lateral line; scales with border feebly crenulate.
      d. 3-4 oblique bars on each caudal lobe; 45-49 lateral line scales ... *S. kneri* (Rio Orinoco)
      dd. 4 (juvenile)-7 oblique bars on each caudal lobe; 38-41 lateral line scales ... *S. insignis* (Rio Branco; Guianas)
   cc. 9-13 scales in a transverse series above lateral line.
      e. 7-9 bars on each caudal lobe in adult, rather indistinct in old specimens; edge of scales scarcely crenulate (body depth 2.0-2.6; 47-57 lateral line scales, 10-11/10-$11\frac{1}{2}$ transverse scales) ... *S. laticeps* (Rio Orinoco; sympatric with *S. kneri*)
      ee. 3-4 bars on each caudal lobe in adults; edge of scales distinctly crenulate.
         f. 47-52 lateral line scales, 9-10/9-10 transverse scales (body depth

Unidentified *Chilodus* from Rio Japura in Brazil. Photo by Harald Schultz.

Semaprochilodus taeniurus. Photo by Dr. Herbert R. Axelrod.

Semaprochilodus theraponura from Rio Urubu. Photo by Harald Schultz.

2.35-2.7 in standard length, possibly 2.8-2.85 in very large individuals, or in certain populations) . . . *S. theraponura*, synonym *S. amazonensis* (middle and upper Amazon)

ff. 54-61 lateral line scales, 12-13/12½-13 transverse scales (body depth 2.0-2.4 in standard length) . . . *S. squamilentus* (eastern Brazil)

GENUS *PROCHILODUS*
BOCACHICOS

The bocachicos ("small mouths") of the genus *Prochilodus* are among the least variable of all characoids. They lack typical characters permitting their sure identification, and the present state of their systematics is poor. The color pattern of their caudal and dorsal fins, as well as their scale formulas, seem the only actually reliable characters, if associated with the body depth (provided that it is measured on a good series of adult specimens). The following is a very crude estimate of what may be the real situation.

Two artificial groups may be found within the genus, depending on the coloration of the caudal fin, which may be spotted (the spots somewhat arranged in wavy lines, but never forming conspicuous bars as in *Semaprochilodus*), or plain, often quite dark, without any pattern.

The first group (with the caudal spotted) may be called the *nigricans*-group (from the oldest species name). Mostly according to the scarce data of the literature, the species can be identified as follows:

Key to the *Prochilodus nigricans*-group (caudal fin spotted):

a. Usually less than 50 lateral line scales.

b. Body depth 2.45-2.8 in standard length.

c. 43-46 lateral line scales and 7-9/5½-7 transverse scales . . . *P. rubrotaeniatus* (Guianas, Venezuela, Rio Negro and Rio Branco, upper Amazon)

cc. 48 lateral line scales and 11/7 transverse scales . . . *P. beni* (Bolivia)

bb. Body depth 2.8-3.5 in standard length.

d. Snout moderate in length, upper jaw not projecting (body depth 2.8-3.2 in standard length).

e. 37-43 lateral line scales (8/7 transverse scales) . . . *P. corimbata*, possible synonyms *P. oligolepis* and *P. cearaensis* (eastern and southeastern Brazil)

ee. 44-49 lateral line scales (8-10/7-8 transverse scales) . . . *P. nigricans* and *P. ortonianus* (two sibling species (?) from the Amazon; *P. cephalotes* may be the young of one of these) dd. Snout elongate, upper jaw projecting, resembling that of genus *Ichthyoelephas*.

f. Body depth 2.9-3.0 in standard length; 40-43 lateral line scales and 6½-7½/5½ transverse scales . . . *P. maripicru* (Guianas)

ff. Body depth 3.5 in standard length; 47 lateral line scales and 8/7-8 transverse scales . . . *P. caudifasciatus* (Rio Perene in Peru, a poorly known species)

aa. 49-65 lateral line scales (body depth 2.3-3.1 in standard length and 10-12/8-10 transverse scales) . . . *P. mariae* (Rio Orinoco; a species very close to *P. brama*)

The second group, with the caudal fin plain (*Prochilodus* typical or *P. argenteus*-group), is even less well known than the preceding. Certain species, such as *Prochilodus vimboides* with *P. steindachneri* and *P. hartii*, *Prochilodus reticulatus* with *P. magdalenae*, *Prochilodus lineatus* with *P. affinis*, etc., form groups, usually sympatrics, which can be sibling species or simply synonyms. All species except *P. reticulatus* (with *P. magdalenae*) are from eastern and southeastern Brazil, some extending south to Argentina.

The following key, like the preceding one, is mostly compiled from the literature.

Key to the *Prochilodus argenteus*-group (caudal fin plain):

a. Dorsal fin spotted.

b. Body depth 2.25-2.4 in standard length.

c. 37 lateral line scales . . . *P. steindachneri* (Rio Parahyba; may be the young of *P. vimboides*)

cc. 41-43 lateral line scales . . . *P. brevis* (Bahia)

bb. Body depth 2.6-3.5 in standard length.

d. 6/5-6 transverse scales (35-42 lateral line scales) . . . *P. vimboides* and *P. hartii* (two sibling species (?) from Parahyba)

dd. 8-9/6-8 transverse scales.

e. Snout moderate in length, upper jaw not projecting (usually more than 47 lateral line scales) . . . *P. lineatus-affinis* complex (Rio de la Plata, Rio Paraguay and Rio Sao Francisco)

ee. Snout elongate, upper jaw projecting, resembling that of genus *Ichthyoelephas*.

f. Usually less than 47 lateral line scales; species from northwestern South America . . . *P. reticulatus* (Venezuela)-*magdalenae* (Colombia, synonym *P. beani*)-complex

Young fishes of the genus *Semaprochilodus*, probably *S. taeniurus*. Photos by Dr. Herbert R. Axelrod.

Unidentified *Prochilodus* of the *nigricans*-group. Photo by Harald Schultz.

Semaprochilodus sp., close to *S. theraponura.* Photo by Dr. Herbert R. Axelrod.

ff. 47-48 lateral line scales; species from southeastern South America ... *P. scrofa* (Rio de Janeiro, Rio Paraguay and (?) Argentina)

aa. Dorsal fin plain in adults.

g. Less than 60 lateral line scales.

h. 41-45 lateral line scales; body depth 2.5-2.7 in standard length in adults ... *P. argenteus* (type of the genus, a synonym of *P. marggravi* which is a suppressed name; synonyms *P. lacustris* and *P. costatus*; Rio Sao Francisco basin, Rio Parnahyba)

hh. About 48-50 lateral line scales; body depth about 3 in standard length ... *P. dobulinus* (Amazon) and *P. platensis* (a dubious species from Argentina, which may be synonymous with *P. lineatus*)

gg. About 60 lateral line scales (body depth 2.45 in standard length) ... *P. brama* (Rio Tocantins)

Profile of the heads of two bocachicos (*Prochilodus*) and of the patalo (*Ichthyoelephas longirostris*) showing the lengthening of the snout of the latter. *Prochilodus maripicru* seems to be intermediary in this respect. Original drawings by the author.

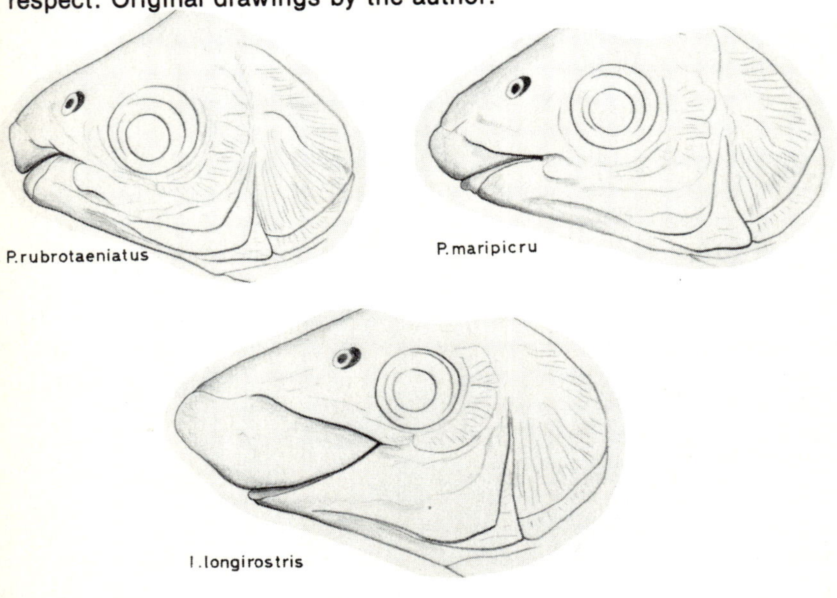

Prochilodins (called bocachicos in Peru and jaraquis, among other names, in Brazil) migrating upwards and jumping the falls. Photo by Harald Schultz.

A catch of prochilodins by the Indians. Photo by Harald Schultz.

A young *Prochilodus*. Photo by Dr. Herbert R. Axelrod.

Prochilodus nigricans (or *P. ortonianus*) from Solimoes. Photo by Dr. Herbert R. Axelrod.

Curimata cyprinoides from Rio Purus. Photo by Dr. Herbert R. Axelrod.

Prochilodus sp. about 5 inches in length. Photo by Aaron Norman.

GENUS *ICHTHYOELEPHAS*

PATALOS

The two species of the genus, *I. longirostris* from Colombia and *I. humeralis* from Ecuador, are closely related to each other; they differ in the size of the eye as well as in a humeral spot (present in *I. humeralis*). Both are relatively large species (more than 40 cm when adult), and easily recognizable thanks to their very elongate snout.

SUBFAMILY CURIMATINAE

Curimatas

Unlike the bocachicos, which have been only partly reviewed, the curimatas have been the object of a rather extensive revision in 1948. However, this work has been much discussed, due to the extreme splitting of the group (the number of genera having almost tripled from 10, several already superfluous, to 27, with no less than 17 new genera!).

The present survey will take into account only a few of these genera. Some of the others are perhaps worthy of consideration, but this would lead us too far astray, owing to the considerable number of species involved (perhaps 100-110). Most of the species, like most bocachicos, lack "good" systematic characters. They are very much alike except for the

The most common *Curimata* from the Guianas and Amazon belongs to the *C. spilura*-complex. Photo by Dr. Herbert R. Axelrod, Rio Aguaro.

Curimatopsis sp., possibly *C. macrolepis.* Photo by Timmerman.

following features: number of scales (there are nevertheless important overlaps), form of the preventral region (and associated changes in the predorsal and postventral regions, rarely that of the postdorsal region) and, in some cases, color pattern.

All curimatas look like bocachicos without the predorsal spine and without a sucking disk. The adults are toothless; the very young ones, which probably have a very different diet, have conical teeth.

For simplicity, the curimatas will be divided into 2 groups: large species with many scales (usually 48-110 lateral scales), and more evolutionarily divergent species, probably derived from them. The latter are smaller, with fewer scales, and sometimes evolve like the tetras, with the appearance of forms with scaled caudal fins, pigmy forms with an incomplete lateral line, etc.

GENUS *CURIMATA*—typical and specialized forms
LARGE CURIMATAS

This group includes mostly forms living in large rivers, in which the structure of their ventral region has become adapted to their way of life (a flat breast often meaning a life near the bottom, a keeled breast an adaptation to pelagic life, and so on). From the least specialized to the most specialized forms (apparently), we have one large group of species with not

Curimata ciliata. Photo by Dr. Herbert R. Axelrod.

Curimata rhomboides. Photo by Dr. Herbert R. Axelrod.

Curimata isognatha, juvenile. Photo by Dr. Herbert R. Axelrod.

Curimata (*Gasterostomus*) *latior.* Photo by Dr. Herbert R. Axelrod.

very many scales (48-70 lateral line scales), a short anal fin as in most curimatids, and the predorsal line usually not modified; and a small group of 4 species, placed in 3 different genera (or subgenera, depending on the authors), with many scales (85-110 lateral line scales), 15-17 anal rays (except in *C. planirostris*), and the predorsal line naked.

The first sub-group includes species closely allied to the type of the genus *Curimata*, *C. cyprinoides*, and several related forms, as follows:

a. Postventral keel, if any, not spinous.
  b. Rather elongate species, body depth usually more than 2.6 in standard length.
    c. 50-54 strongly ctenoid lateral line scales (probably no postventral keel) ... *C. rutiloides* (Mato Grosso and Rio Negro) and *C. aspera* (Rio Huallaga)
    cc. 56-70 feebly crenulate lateral line scales ... *C. vittata* (Mato Grosso and Rio Negro) and *C. macrops* (Sao Paulo)
  bb. Rather deep species, body depth usually less than 2.6 in standard length.
    d. Predorsal region trenchant; lateral line relatively low, with only 6 or 7 transverse scales to ventral fins ... *C. simulata* (Amazon), *C. cisandina* and *C. alleni* (sympatric species from Peruvian Amazon)
    dd. Predorsal region rounded; lateral line along axis of body, with 8-11 transverse scales to ventral fins.
      e. Preventral region more or less rounded; scales strongly ctenoid or ciliate (50-61 lateral line scales) ... *C. ciliata* (synonyms *C. essequibensis* (?), Guyana, and *C. isognatha*, Amazon, Sao Paulo ?)
      ee. Preventral region flat, the angles well marked; scales never strongly ctenoid ... *C. cyprinoides*-group, comprising the nominal form *C. cyprinoides* (from lower Amazon), *C. cyprinoides schomburgki* (Guianas), *C. copei* (Surinam), *C. kneri* (Rio Negro) and *C. falcata* (Rio Xingu)
aa. Postventral keel spinous ... *C. rhomboides*-group (Amazon basin); the 2 or 3 spp. (?) of this group are often classified as *Psectrogaster*, a subgenus of *Curimata* or a distinct genus

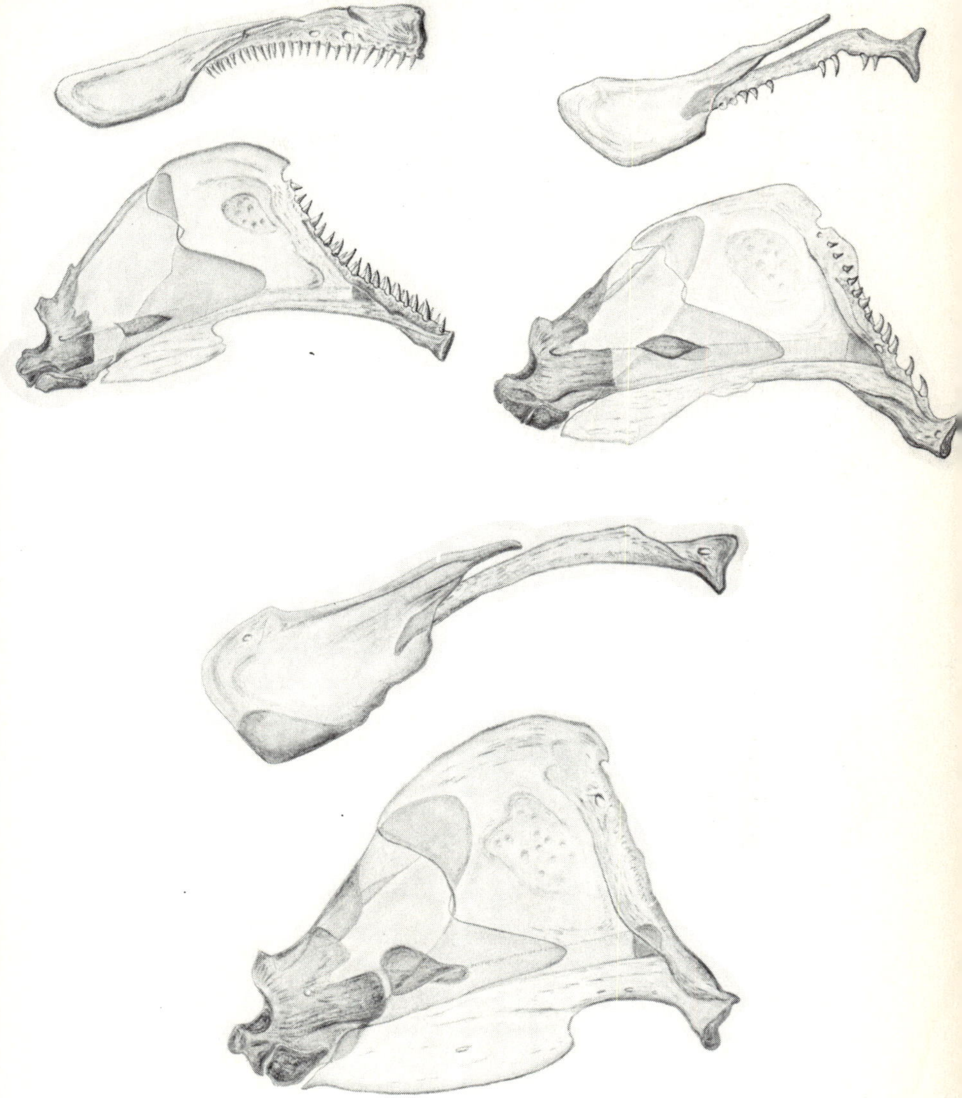

Regression of the teeth in a young, unidentified species of *Curimata* from Guyana. The first figure was drawn from a post-larva 15 mm in standard length; the second from a post-larva 17 mm in standard length (some teeth are already lost); and the third from a young individual somewhat more than 20 mm in standard length, with the toothless jaws characteristic of adult curimatins. Original drawings by the author, all of the same scale.

Curimata of the *elegans*-group. Photo by Dr. Herbert R. Axelrod.

Curimata vittata from the vicinity of Humaita, Brazil. Photo by Dr. Herbert R. Axelrod.

Curimata sp., possibly *C. metae.* Photo by Dr. Herbert R. Axelrod.

Curimata spilura. Photo by Dr. Herbert R. Axelrod.

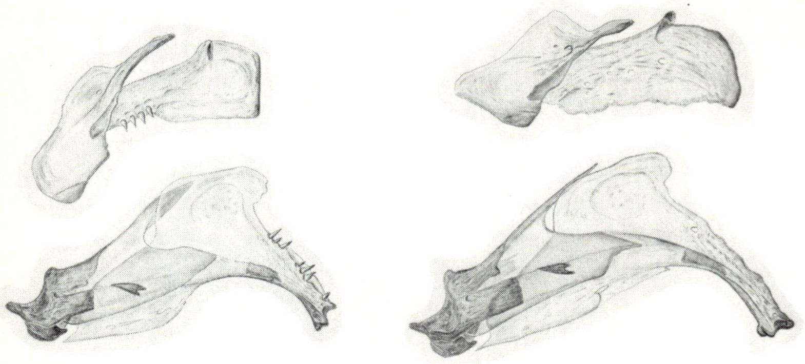

Regression of the teeth in another species of *Curimata*, possibly *C. helleri*. The first figure drawn from a post-larva 20 mm in standard length; the second from a slightly larger individual (23 mm SL). Original drawings by the author.

The second subgroup includes only a few specialized species, all with a postventral keel; they may be identified as follows:

a. Preventral region rounded or flattened as in preceding subgroup.
  b. Postventral keel not spinous (genus or subgenus *Semitapicis*).
    c. 82-90 lateral line scales and 11-12 anal rays (total) . . . *C. (S.) planirostris* (Amazon, Rio Negro)
    cc. 95-110 lateral line scales and 15-17 anal rays (total) . . . *C. (S.) laticeps* (Amazon, Venezuela)
  bb. Postventral keel spinous (genus or subgenus *Potamorhina*) . . . *C. (P.) pristigaster* (Amazon, Rio Negro)
aa. Preventral region forming a continuous keel with postventral region (genus or subgenus *Gasterotomus*) . . . *C. (G.) latior* (Amazon and La Plata basins)

SMALL CURIMATAS—evolved forms

This is the largest group, with some 80-90 species. It may be split into curimatas with a scaled caudal (corresponding to the tetra genus *Moenkhausia*, for example), curimatas with an incomplete lateral line (corresponding to the tetra genus *Hyphessobrycon*, for example), and "ordinary" unmodified curimatas (corresponding to the tetra genus *Astyanax*, for example). The appropriate generic name for the last group is debated. *Curimata* (in the broad sense) will be used here.

According to certain authors, most of the small curimatas fall into the following divisions:

a. Preventral region not very flattened, usually rounded (not as in the type species *C. cyprinoides*, for example).

b. Postventral region rounded, predorsal region more or less angular.

c. Mouth terminal or subinferior: about 25 or more species, for which the generic names *Cyphocharax* (oldest), *Allenina, Curimatorbis* and *Pseudocurimata* have been applied. The most common species are *C. argentea, C. elegans* (which have a black spot at dorsal fin base), *C. gilberti, C. spilura* (which have a more or less conspicuous caudal spot), *C. microcephala*, etc.

cc. Mouth inferior: about 10 species for which the generic name *Cruxentina* has been suggested; the best known species is perhaps *C. nasa*.

bb. Postventral region keeled, predorsal region rounded: 2 or 3 species with terminal mouth, which have been called by the generic name *Bondia*.

aa. Preventral region flattened as in the type species *C. cyprinoides*.

d. Mouth terminal or subinferior: a few species (3 or 4) sometimes called *Rivasella*.

dd. Mouth inferior: two species, *C. hypostoma* and *C. trachysteta*, for which the generic name *Steindachnerina* has been proposed.

The 20 or more species which are not included above would probably fall within section (c) when the form of their ventral region (usually not described) is studied. Also, a specimen without an adipose fin, perhaps an abnormal individual but possibly a good species, has been called by the generic name *Curimatoides*.

Type of *Curimata multilineata*. Original drawing by Mr. Pablo Bravo in 1930, published for the first time by permission of Dr. G.S. Myers.

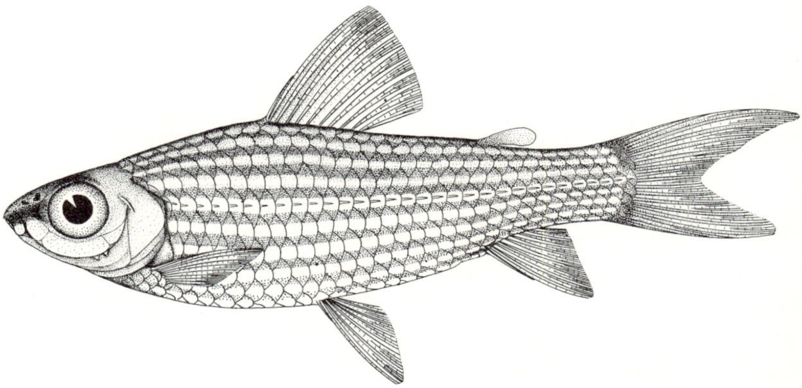

Curimata lineopunctata, a rare species from Rio Calima, Colombia. Photo by Dr. Herbert R. Axelrod.

Unidentified, parasitized species of *Curimata* from Rio Aguaro, possibly *Curimatella alburna.* Photo by Dr. H.R. Axelrod.

Curimatopsis macrolepis, male. Photo by Dr. Herbert R. Axelrod.

Unidentified species of *Curimata,* close to *C. microcephala.* Photo by Harald Schultz.

GENUS *CURIMATELLA*

CURIMATINS WITH SCALED CAUDAL

Apart from the numerous scales on the caudal lobes, this genus cannot be told apart from *Curimata*. The species can be identified by means of the following key:

a. More than 6 transverse scales above lateral line; 43-46 lateral line scales.
  b. 9/7 transverse scales; body depth 2.5-2.66 in standard length . . . *C. lepidura* (Rio S. Francisco)
  bb. 7½/6½ transverse scales; body depth 3 in standard length . . . *C. xinguensis* (Rio Xingu)

aa. 5 or 6 transverse scales above lateral line; 28-42 lateral line scales.
  c. 39 to 42 lateral line scales; body depth usually more than 3 in standard length (no peduncular spot; a plumbeous longitudinal band) . . . *C. meyeri* (Amazon and Rio Negro)
  cc. 30 to 38 lateral line scales; body depth usually less than 3 in standard length.
    d. A peduncular spot.
      e. About 35 lateral line scales; preventral region flattened; body elevated, with an acute angle just in front of the dorsal fin . . . *C. dorsalis*, possible synonym is the subspecies *C. alburna caudimaculata* (Amazon)
      ee. 30 to 32 lateral line scales; preventral region rounded or not much flattened; predorsal region not angular . . . *C. rehni* (Mato Grosso)
    dd. No peduncular spot (35 to 38 lateral line scales; preventral region flattened) . . . *C. alburna*, possible synonyms *C. immaculata* and (?) *C. lineata* (very common in the Amazon and Guianas)

GENUS *CURIMATOPSIS*

CURIMATINS WITH INCOMPLETE LATERAL LINE

As pointed out above, certain curimatas have followed parallel evolutionary lines with the tetras, and some of them have a very short lateral line. The group is a difficult one. A few small species (which have been put into the dubious genus *Hemicurimata*) may well be the young of large curimatas, the lateral line usually not being complete at the postlarval stage. The *Hemicurimata* species (*esperanzae, maculata, saladensis*, and possibly others) are just like curimatas, with the lateral line perforating the first 6-10 scales only.

The *Curimatopsis* species are more peculiar. They differ from *Curimata* not only in their short lateral line, but also in their relatively large maxillary bone, which is free from the antorbital (whereas it is partially covered by

Curimatella alburna (above) and *C. meyeri* (below). Photos by Dr. J. Gery.

the antorbital in the other curimatins), as well as by a curious sexual dimorphism: the males, when they reach maturity, have a much deeper caudal peduncle than the females. Two species have been described, the well-known *Curimatopsis macrolepis* from the Amazon and the Guianas (with a dubious subspecies, *C. macrocephalus*, in the lower Amazon), and the more recently discovered *C. evelynae* (named for Mrs. H. R. Axelrod), from the upper Rio Meta in Colombia. The males of the latter species, which are nicely colored with pastel tints, have no caudal spot (or a very faint one), and are more elongate than the males of *C. macrolepis*.

SUBFAMILY ANODINAE

This curious group includes only 3 species of pelagic fishes, strongly resembling a hemiodid but lacking teeth, which are presently in an uncertain systematic position (the resemblance with the hemiodids may be due to a common origin* or, on the other hand, it may be due to convergence, the same pelagic way of life ending in the same adaptive structures).

* This has just been suggested by T. Roberts (December, 1974), too late to be taken into account in the present book.

Curimatopsis macrolepis female. Photo by Dr. Herbert R. Axelrod.

Curimatopsis sp. from the Rio Negro, possibly *C. evelynae.* Photo by Dr. Herbert R. Axelrod.

Two male *Curimatopsis evelynae.* Photo by Dr. Herbert R. Axelrod.

Anodus elongatus from Tefe, Brazil. Photo by Dr. Herbert R. Axelrod.

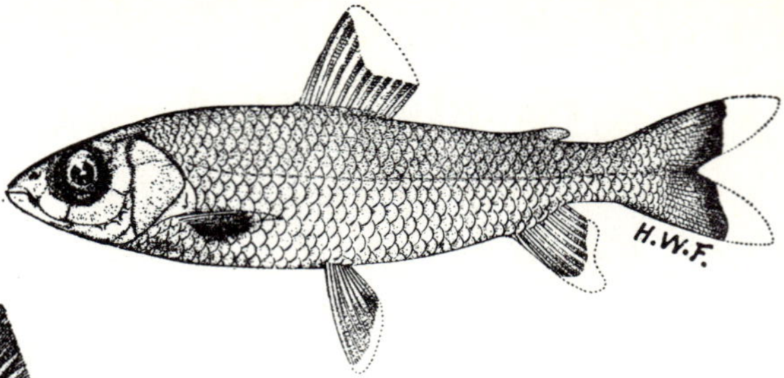

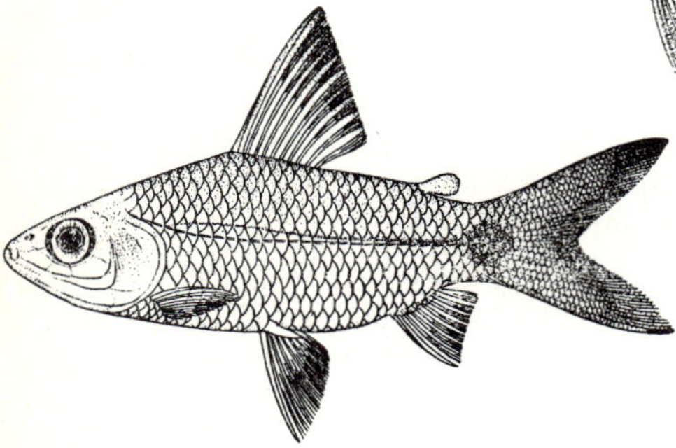

Curimatella meyeri and *C. rehni,* type (after Fowler, 1950); and *Eigenmannina melanopogon* (photo by Dr. J. Gery).

Anodus elongatus (synonyms *A. steatops* and *A. jordani*) from the large rivers in the Amazon basin (it is not found in the Guianas), and *Eigenmannina melanopogon*, this also restricted to the Amazon, are elongate, fusiform fishes (the body depth more than 4 in the standard length), with many small scales (95 and more), the gill membrane free from the isthmus, a more or less developed adipose eyelid, and very numerous, long and slender gill-rakers (55 to 100 on the lower arch), attesting to a planktonic diet. The species differ in the form of the head, *A. elongatus* having a normal head with the mouth terminal, whereas *E. melanopogon* has a very large head as compared with the rest of the body (which tapers caudally), the top of the head is concave, and the mandible is produced, with the mouth somewhat upturned. Both species are silvery, and *Anodus elongatus* has a dark spot on the flanks (irregular in size and form), which adds to its resemblance with a hemiodid, where this mark is common. A third species, *A. orinocensis*, was described by Steindachner in 1888 from the Orinoco. It has not been rediscovered since.

Family Gasteropelecidae

Hatchetfishes

The hatchetfishes, or flying characins, are surface fishes identified at once by their rounded, trenchant belly and their very large pectoral fins. The keel is formed by the joined coracoid bones (part of the "shoulder" of the fish), which are much flattened and enlarged and used to cleave the water surface. Very powerful muscles capable of setting the wing-like pectoral fins in motion at a high speed (similar to that of the well known humming-birds) are connected with these bones.

The technique of the flight is not entirely known. It is known that the fish rises progressively above the surface, in the manner of the ships called hydrofoils. But one is not certain whether it really leaves the water. However, owing to the fluttering of their pectoral fins, the hatchetfishes can be considered as true flying fishes, in contrast with pantodons and marine "flying fishes," which only soar, and with several jumpers such as the characoid *Triportheus* (which is nevertheless capable of covering several meters above water). As for the purpose of the specialization, no one knows. It has been suggested that the fishes can eat small aerial insects or may be able to escape predators. The author favors the latter hypothesis, altogether without much argument.

As concerns the other characters, the hatchetfishes have most in common with the Characidae, chiefly with certain tetras. In addition, they have the usual adaptations to life at the surface (straight outline and superior mouth).

Dr. Stanley Weitzman, one of the best comparative anatomists of our time, has studied, among other characoid groups, the hatchetfishes.* He recognizes two groups, which may be considered as subfamilies within the family: the genus *Thoracocharax*, which is the most primitive, on the one hand (subfamily Thoracocharacinae) and the genera *Gasteropelecus* and *Carnegiella*, on the other (subfamily Gasteropelecinae). *Thoracocharax* is distinguished at once by its impressive keel.

Technically the genera may be identified as follows:

a. An adipose fin.

b. 19-22 longitudinal line scales; coracoid bones very enlarged (3 outer pre-maxillary teeth; anal fin with iii, 31-42 rays)
. . . *Thoracocharax*

bb. 28-37 longitudinal line scales; coracoid bones less developed (0-1 outer pre-maxillary teeth; anal fin with iii, 19-33 rays)
. . . *Gasteropelecus*

aa. No adipose fin (otherwise with many characters of *Gasteropelecus*)
. . . *Carnegiella*

* See for example: *Tropical Fish Hobbyist*, January, 1970, pages 38-64.

Thoracocharax stellatus. Photo by Harald Schultz.

Thoracocharax securis? Photo by Klaus Paysan.

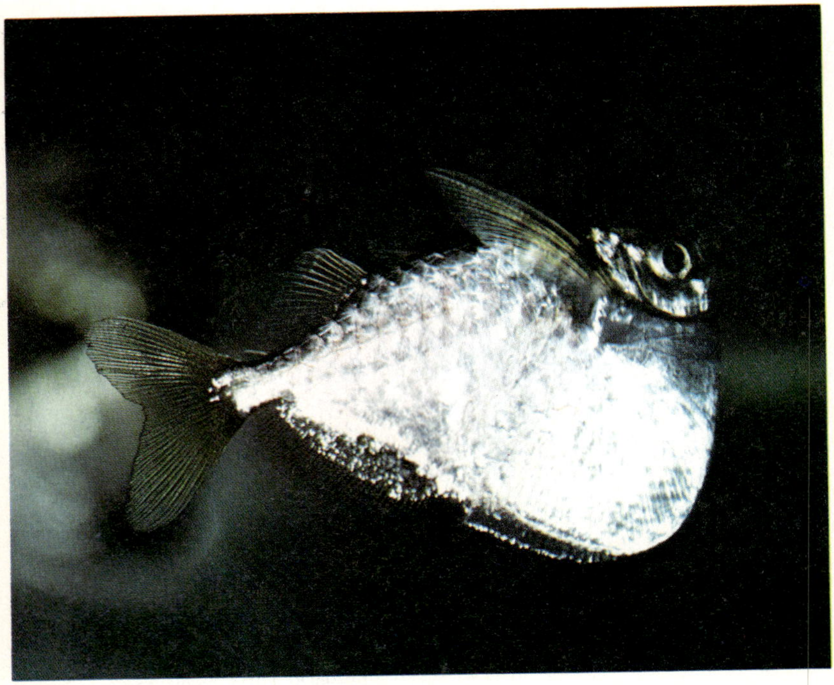

Carnegiella myersi. Photo by Klaus Paysan.

Gasteropelecus sternicla. Photo by Dr. Herbert R. Axelrod.

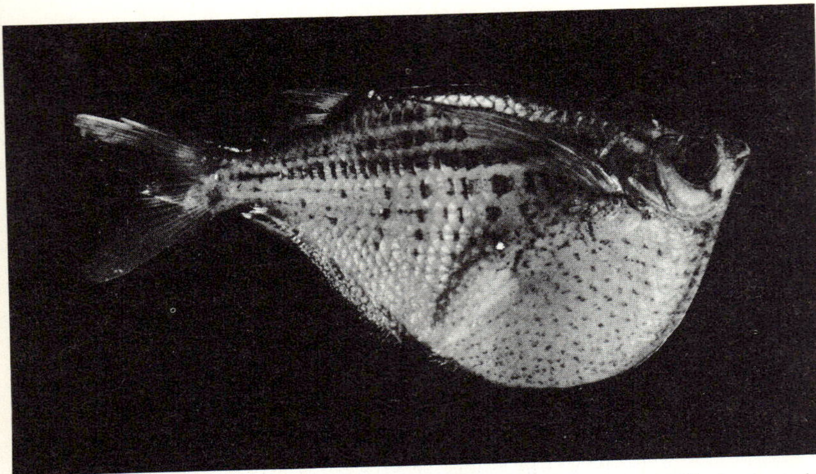

Gasteropelecus maculatus, preserved specimen about 2½ inches in length. Photo by Dr. Stanley H. Weitzman.

GENUS *THORACOCHARAX*
SILVER HATCHETFISHES

As stated above, the silver hatchetfishes belonging to the genus *Thoracocharax* can be distinguished from *Gasteropelecus* by their more prominent belly which forms almost a right angle with the dorsal outline. Two species are known: *Thoracocharax stellatus*, an ubiquitous fish found from Rio Paraguay to Colombia and Venezuela, and *Thoracocharax securis* (also called *pectorosus*, a junior synonym), which is found mostly in the middle-upper parts of the Amazon basin. *T. stellatus* has a black spot on the base on the dorsal fin (as well as 2 or 3 rows of scales on the anal fin base which must be counted with a lens), whereas *T. securis* has a plain dorsal fin except at its tip (as well as 5 or 6 scale rows on the anal fin base).

GENUS *GASTEROPELECUS*
SILVER HATCHETFISHES

This genus, formerly called *Pterodiscus*, includes 3 species which are very similar in appearance. They can be distinguished from each other mostly by anatomical details, as follows:

a. Maxillary not toothed; small size (one outer premaxillary tooth; sixth circumorbital bone present) . . . *G. levis* (Lower Amazon)

aa. Maxillary toothed; larger size.

b. Depth usually 1.8 or less in standard length; body spotted; outer premaxillary tooth and sixth circumorbital bone present as in *G. levis* . . . *G. maculatus* (Venezuela, Colombia, and Panama)

bb. Depth usually 1.85-1.95 in standard length; body plain; no outer premaxillary tooth; sixth circumorbital bone lacking ... *G. sternicla* (Guianas and Trinidad; Mato Grosso in Brazil; Peruvian Amazon)

The latter species is by far the commonest species, at least in the aquarium trade. Almost nothing is known of its biology.

GENUS *CARNEGIELLA*
MARBLED AND PIGMY HATCHETFISHES

The species of this genus, characterized by the lack of an adipose fin, are the smallest and prettiest of the hatchetfishes. They may be identified rather easily by their color pattern, as well as by some technical details, as follows:

a. Body marbled with broad bands extending from keel to dorsum ... *C. strigata* (Guianas and Amazon Basin)

aa. Body with small spots or lines.

b. Keel with narrow parallel lines, sometimes with a black border; pectoral fins with a black band; body rather deep; circumorbital bones lacking; anal fin with less than 31 rays; maxilla with less than 3 teeth ... *C. marthae* (Venezuela; Upper Amazon basin; Rio Madeira)

bb. Keel with a black border, plain or with some dots; pectoral fins plain; body relatively elongate (as concerns a hatchetfish); 2nd to 5th circumorbital bones present; 33-36 anal rays; maxilla with 4-5 teeth ... *C. myersi* (Peruvian and Bolivian Amazon)

In contrast to the other species of hatchetfishes, the species of *Carnegiella* are what zoologists call polytypic, which means that 2 or more distinct populations can be recognized within the species concepts of *Carnegiella marthae*, *C. myersi* and *C. strigata*, respectively. As concerns *Carnegiella marthae*, sometimes called the black-winged hatchetfish, there

Carnegiella myersi, live individual about 1¼ inches in length. Photo by Dr. Stanley H. Weitzman.

Carnegiella strigata strigata, the form called *vesca* by aquarists. Photo by Dr. Herbert R. Axelrod.

Carnegiella strigata fasciata. Photo by Harald Schultz.

Carnegiella marthae, form with thoracic lines. Photo by Dr. Herbert R. Axelrod.

Carnegiella marthae, form with thoracic spots. Photo by Dr. Herbert R. Axelrod.

is certainly a difference in the number of anal rays between specimens from the Rio Orinoco and Rio Negro, and specimens from the Peruvian Amazon and Rio Madeira: usually 22-25 for the former and 26-30 for the latter (total counts; based on 17 specimens from middle-upper Rio Negro, and 18 specimens from Igarape Preto on the Amazon at the Brazilian-Peruvian border, and Guayaramirim on the Rio Madeira). The typical form from the Orinoco-Negro Rivers has to be called *Carnegiella marthae marthae*, and the form from the Peruvian Amazon, which has already been described, may be called *Carnegiella marthae schereri*.

The pigmy hatchetfish, *Carnegiella myersi*, seems to be polytypic too. In Bolivia there is a population which is clearly deeper bodied than the typical form from the Peruvian Amazon.

Finally, the present author has recently shown that the well known polymorphism of the marbled hatchetfish *Carnegiella strigata* (which has different patterns of the thoracic bands depending on the localities) may be explained by the existence of two forms (probably subspecies) which perhaps intergrade, at least in some places in the Peruvian Amazon. Thanks to infrared photography of the types in the British Museum, it has been put into evidence that these types have the color pattern called *vesca* by most ichthyologists and aquarists. Therefore the name *C. vesca* is a synonym of the older name *C. strigata*, and the fishes having the *vesca*-pattern are rightly known as *Carnegiella strigata*. The fishes with the "old" *strigata* pattern would be given the next available name, *C. fasciata* or more precisely *Carnegiella strigata fasciata*, since it is probably a subspecies and not a "full" species.

In the nominate subspecies, *C. strigata strigata*, the foremost band, which runs obliquely from the middle of the thoracic convexity toward a point behind the pectoral root, is divided into two branches almost from its origin. This pattern is seen in fishes coming from places situated north of the Amazon, in the Guianas, upper Amazon, and also in the Rio Purus. In the subspecies *C. strigata fasciata*, the thoracic band is not divided until it reaches the level of the pectoral fin. This form occurs in the Amazon basin (lower, middle and upper Amazon) and somewhat south of it, as well as in the Rio Caqueta in Colombia, north of the Amazon.

Both forms have been collected in the middle Amazon, in the upper Amazon and in the Rio Purus. It is not known whether they live together or isolated from each other; some samples* are intermediate in certain characters and may prove to be intergrades. Finally, within each subspecies there is a variability in the color pattern as well as in the body proportions and in the anal ray counts.

* In the Rio Meta, a tributary of the Rio Orinoco in Colombia, as well as in the Rio Negro and (?) Upper Amazon.

Family Serrasalmidae

Pacus, Silver Dollars and Piranhas

Considered by some as a subfamily within the family Characidae (strictly a matter of opinion and of no great importance), the family Serrasalmidae (=serrated salmons) is well characterized by a number of unique, or almost unique, features. The body is compressed, usually deep and in the form of a disc, somewhat like the famous discus fishes of the family Cichlidae, and with a trenchant belly armed with spines. These spines, usually called serrae or scutes, vary in number from 6 (genus *Acnodon*) to 70 or so (certain species of the genus *Colossoma*). There is usually a single spine in front of the dorsal fin (except in *Mylossoma* and *Colossoma*), and the dorsal fin itself is more elongate than in most characoids. The scales are small and numerous, the maxillary bone reduced and never provided with teeth (except in *Colossoma bidens*?), and the other teeth are very variable depending upon the eating habits of the fish. The specializations of the alimentary system can be used to divide the Serrasalmidae into three quite natural groups: (1) the pacus, etc. (Myleinae) which are vegetarians, (2) the piranhas and pirambebas (Serrasalminae) which are carnivorous, and (3) a species forming by itself the third subfamily (Catoprioninae), which, probably having evolved from a predatory ancestor, has degenerated into a semi-parasitic life, eating the scales of other fishes. The following key will serve to distinguish between the three subfamilies:

a. Upper jaw with a double row of teeth, the inner row molariform; lower jaw usually with a pair of conical teeth behind the main series
. . . subfamily MYLEINAE

aa. Both jaws with a single series of usually sharp teeth.
b. Teeth close-set, interlocking, forming a continuous saw
. . . subfamily SERRASALMINAE

bb. Teeth spaced, tuberculate, everted . . . subfamily CATOPRIONINAE

The Serrasalmidae are inhabitants of tropical and subtropical South America with the exception of the coastal rivers of southwestern Brazil and west of the Andes. This distribution indicates that we are dealing with a relatively recent group which probably appeared, possibly from a tetra-like ancestor, after the Andean formation (i.e. in Tertiary times),

SUBFAMILY MYLEINAE

Pacus et cetera

Compared with their ferocious cousins the piranhas, the pacus are similar in shape and have the same general organization, but with a very

Head study of *Colossoma* sp. Photo by Dr. Herbert R. Axelrod.

Colossoma sp. from Rio Aguaro, Venezuela. Photo by Dr. Herbert R. Axelrod.

Colossoma oculus, young specimen from Iquitos. Photo by Dr. Herbert R. Axelrod.

A somewhat older *Colossoma oculus*. Photo by Dr. Herbert R. Axelrod.

different dentition. They have two rows of teeth on the upper jaw, the front row being composed of 3 (on each side) more or less incisor-like teeth and the inner row composed of 4 broad teeth, more or less in the form of molars. The molar teeth are capable of crushing large leaves, fruits or seeds falling into the river. Certain species, such as *Colossoma* spp. or *Myleus pacu*, attain a large size and are an important source of food for the Indians. Seven genera are usually recognized within the group. Two of them, *Colossoma* and *Mylossoma*, are exceptions within the family in not having acquired the predorsal spine which is characteristic for the Serrasalmidae* (together with the Prochilodinae and the Stethaprionini, but in these last cases the spine is different and was formed from other bones by so-called "convergent evolution"). For this and other reasons, *Colossoma* and *Mylossoma* are considered by students of the group as the most primitive genera, linking (so to speak) the Serrasalmidae with some other characoids such as the Tetragonopterinae. The principal characters which separate the genera are presented in the following key:

KEY TO THE MYLEINE GENERA

a. Supposedly primitive forms without predorsal spine.
- b. 6 or more teeth on each side of lower jaw; anal fin with less than 30 rays, not scaled except on its base; very large size ... *Colossoma*
- bb. 4 teeth on each side of lower jaw; anal fin with more than 35 rays, scaled; large size, disciform ... *Mylossoma*

aa. More specialized forms with predorsal spine.
- c. Preventral region usually keeled and serrated; normally a pair of small conical teeth behind the main series on the lower jaw.
  - d. Adipose fin shorter than distance from end of dorsal fin to beginning of adipose fin; anal fin strongly falcate in females and bilobed in males.
    - e. Serrae well developed and numerous in front of as well as behind ventral fins ... *Myleus*
    - ee. Abdominal serrae not well developed in front of ventral fins ... *Utiaritichthys*
  - dd. Adipose fin longer than dorsal-adipose distance; sexual dimorphism not striking, the male never (or rarely ?) with a bilobed anal ... *Metynnis*
- cc. Preventral region flat, not serrated.
  - f. Teeth like those of *Myleus* and *Metynnis*, without a pair of small conical teeth behind main series on lower jaw, which has 4 or 5 teeth on each side ... *Acnodon*
  - ff. Teeth of front row of both jaws flattened, trilobed; a pair of small conical teeth behind main series on lower jaw, which has 8 or 9 teeth on each side ... *Mylesinus*

GENUS *COLOSSOMA*
PACUS

This genus, well characterized by the features mentioned previously in the key, could be split into two groups (subgenera) on the basis of

* Phylogenetic methods applied to the group show they are likely to have lost the serrasalmid spine.

Colossoma (Piaractus) brachypomum. Photo by Gerhard Budich.

differences in the internal anatomy: the subgenus *Piaractus* with at least 3 species and the nominate subgenus *Colossoma* with probably a single species, *Colossoma oculus* (if *nigripinnis* is synonymous with *oculus*, which is not yet quite proved). These species are difficult to identify. Most of them are very large when adult. They have an economic importance in the Amazon basin, where they are called pacus (in the Guianas, where the genus *Colossoma* does not exist, this vernacular name belongs to the largest *Myleus*, chiefly to *Myleus pacu*).

Colossoma (Piaractus) brachypomum is the type of the subgenus *Piaractus* (synonyms *Colossoma mitrei* and *Colossoma canterai*??). In this species the front part of the gas-bladder is smaller than the posterior one and there is a single muscular band. Dorsal fin ii or iii, 12-13; anal fin iii, 21-23; scales 110-120; serrae 65-70; gill-rakers 17 or 18 on lower part of the first arch; uniformly dark or marbled when adult (600 mm), the body spotted and fins dark edged in the young. This species has a very rare feature: its adipose fin is rayed when the fish is adult.

Colossoma (Piaractus) macropomum is like the preceding species, but the head is longer (3 times in the standard length instead of 4 or thereabouts) and broader, the bones of the cheek smooth (instead of rugose) and the largest suborbital almost complete (instead of not covering the cheek).

A young *Colossoma bidens*? Photo by Harald Schultz.

Colossoma sp., probably also *C. bidens.* Photo by Dr. Herbert R. Axelrod.

A young *Mylossoma* sp. Photo by Dr. Herbert R. Axelrod.

A young *Colossoma,* probably *C. oculus.* Photo by Aaron Norman.

Colossoma (Piaractus) bidens: like *C. brachypomum*, but with fewer scales (85-95), fewer serrae (50-60), and the adipose fin not rayed. Characteristic of the species is a large blotch on the opercle.

Colossoma (Colossoma) oculus: in this species, often called *C. nigripinnis*, the front part of the gas-bladder is larger than the hind one and there are two muscular bands. Dorsal fin iii or iv, 13-16; anal fin iii or iv, 20-24; scales 60-80; serrae 45-55; gill-rakers numerous, 35-45 on lower part of the first arch; adipose fin rayed in adults. Grows to 400 mm in length. Young spotted, adults plain with the fins dark-brown.

GENUS *MYLOSSOMA*
POMPANOS, SILVER DOLLARS, PACUS

The 5 or so species of the genus *Mylossoma* have a deep, compressed body, almost as disciform as the *Metynnis* species, chiefly when young. During growth, the depth of the body becomes comparatively less than its length, as was mentioned by L. P. Schultz concerning a Venezuelan species. This growth phenomenon is called negative allometry. *Mylossoma* is easily distinguished from *Colossoma* by its shape, as mentioned, as well as by its long anal fin, almost entirely covered with minute scales.

There are two common species in the Amazon basin (as well as in the Orinoco), *Mylossoma aureum* and *Mylossoma duriventre*, which can be identified as follows:

a. 10 to 16 serrae behind ventral fins, last one not close to first anal ray; 28-34 branched anal rays; base of adipose fin 3.75-4.25 in dorsal fin base
. . . *M. aureum* (synonyms *M. herniarius* and *M. unimaculatus* ?)

aa. 18 to 22 serrae behind ventral fins, last one very close to first anal ray; about 37 branched anal rays; base of adipose fin 2.6-3.66 in dorsal fin base
. . . *M. duriventre* (synonyms *M. albiscopus*, *M. ocellatum* and *M. argenteum*)

Adults of both species are plain silvery and attain a rather large size (up to 250 mm or perhaps more). They live in the deeper and broader part of the rivers, for example in the Rio Negro facing Manaus and down to the barra (confluent with the Rio Solimoes), where they are fished for with seines and then sold in the Manaus market. Young fish are much more colorful and are good aquarium subjects (although rarely seen); when very young, they have an ocellus on the flanks which probably acts as a recognition signal within the school of fishes. The other species are less well known. Norman, in 1929, recognized three other forms:

(1) *Mylossoma paraguayense*, from the Rio Paraguay, which resembles *M. aureum* but with fewer post-ventral serrae (as in *M. duriventre*);

(2) *Mylossoma acanthogaster*, from Lake Maracaibo in Venezuela, is best identified by its short anal fin base length, which goes a little more than 3 times into the length (instead of about 2 in *aureum* and *duriventre*); in life, its sides are silvery, its eye and anal fin red;

(3) *Mylossoma ventriosa* is a very poorly known species, known from a single specimen with a different shape of the hinder part of the body than that of the other four species.

GENUS *MYLEUS*
PACUS

Like *Colossoma*, this genus can be subdivided. Four subgenera could be recognized owing to the quite different structure of the teeth, as follows:

a. Front teeth in upper jaw not very compressed, somewhat conical, set apart from each other as well as from second series; 21-27 branched dorsal rays; distance from end of dorsal fin to adipose fin 9 to 10 in standard length.

b. Front teeth in upper jaw follow curve of the bones and are moderately set apart from middle teeth of second series.

c. A pair of small conical teeth always present behind front series on lower jaw; 32-43 branched anal rays (23-27 branched dorsal rays; 31-46 serrae) . . . subgenus *Myloplus*

cc. No conical teeth behind front series on lower jaw; 29-31 branched anal rays (21-26 branched dorsal rays; 35-47 serrae) . . . subgenus *Paramyloplus*

bb. Front teeth in upper jaw pointing forward, forming a "V" and not following curve of the bones, the foremost ones set very far apart from middle teeth of second series (which are somewhat behind their followers in this series); adipose fin relatively long, dorsal fin relatively short, with 21-24 branched rays; 28-32 branched anal rays; 28-47 serrae, the number in correlation with size of specimen . . . subgenus *Prosomyleus*

aa. Front teeth in upper jaw compressed, incisor-like, appressed against each other as well as against teeth of second series; conical teeth behind main lower series often very small or even lacking; dorsal fin short, with 18-21 branched rays; distance from dorsal to adipose 7.8 to 9.4 in standard length . . . nominate subgenus *Myleus*

Subgenus *Myloplus*

The systematics of this group is in a poor state, the species, with few exceptions, having never been revised. Certain species, such as "*Tomete*" *maculatus* Amaral Campos for example, have been described on the basis of characters like bilobed anal fin, filamentous dorsal rays and red spots on body which are secondary sexual characters common to all *Myleus* at reproduction time. It is almost unidentifiable.

The best defined species is probably *Myleus (Myloplus) torquatus* from the Branco, Solimoes, Madeira and Xingu rivers, because of its distinctive coloration with a conspicuous black border on the anal and chiefly on the caudal fin. It has the following meristics: dorsal fin ii, 21-23; anal fin iii,

A young *Mylossoma* from Tefe. Photo by Dr. Herbert R. Axelrod.

An adult *Mylossoma* from Tefe. Photo by Dr. Herbert R. Axelrod.

Mylossoma aureum? from Boca de Tapaua, Rio Purus. Photo by Dr. Herbert R. Axelrod.

Mylossoma duriventre. Photo by Harald Schultz.

Myleus torquatus **(female). Photo by Harald Schultz.**

27-30; serrae 30-39; at least immature individuals of about 100-125 mm in length are quite deep (depth 1.3-1.45 in standard length) with a narrow interorbital. However, even this species poses a problem: the specimens captured after the types do not show the narrow humeral band described by Kner.

The other species, three, five or perhaps more in number, could be grouped into a complex of species (the "*rubripinnis*-complex") which is rather well defined as a whole as compared with *torquatus* (no black border on unpaired fins; dorsal rays usually more than 23, anal rays usually more than 30), but difficult to split into workable taxonomic units. There are at least three forms in the Guianas: *Myleus (Myloplus) rubripinnis rubripinnis* (synonym *M. asterias?*) in Guyana (probably) and in Surinam, and *Myleus (Myloplus) rubripinnis luna* together with *Myleus (Myloplus) kneri* in the Maroni River at least; one form in the Rio Paraguay, *Myleus (Myloplus) tiete* (synonym *M. levis*); in the Amazon basin it is not certain whether the commonest species should be called *rubripinnis* (or a subspecies of it) as in the Guianas or *gurupyensis* (of which *M. maculatus* is perhaps a synonym). Apart from the two last named forms, the species of the subgenus *Myloplus* may be tentatively distinguished by means of the following key:

a. Broad black border on caudal and anal fins (see also *Utiaritichthys*); dorsal fin ii, 21-23; anal fin iii, 27-30; serrae 30-39; lateral line scales 75

. . . *M. torquatus*

aa. No black border on caudal fin; dorsal fin usually with 23 or more branched rays; anal usually with 30 or more rays.

b. Anal fin iii, 30-33 (depth 1.33-1.6; interorbital 1.66-1.75; dorsal fin iii, 24-25; serrae 32-47; anterior part of anal fin with a black margin)

. . . *M. tiete* (synonym *levis*) Paraguay

bb. Anal fin iii, 32-43

c. Depth 1.6 (in the adult male); serrae 31 (interorbital 1.85; dorsal fin iii, 26 anal fin iii, 36-37) . . . *M. kneri* (Maroni River)

cc. Depth 1.33-1.55, serrae 33-46.

d. Anal fin iii, 35-43; serrae 37-46 (dorsal fin ii, 23-26; depth 1.4-1.55; interorbital 1.75-1.9) . . . *M. rubripinnis rubripinnis* (synonyms *asterias, ellipticus*)

dd. Anal fin iii, 32-38; serrae 33-39 (dorsal ii, 23-27; depth 1.33-1.55; interorbital 1.85-2.13) . . . *M. rubripinnis luna*

It can be seen from the preceding key that color pattern does not play a large role in species recognition (at least little is known about the coloration except in *M. torquatus*). Thus the identification of the figures is often questionable. However, it is rather certain that young *rubripinnis* have the front part of the anal fin red.

Myleus rubripinnis (juvenile). Photo by Franz Werner.

Young *Myleus pacu* or *M. rhomboidalis.* Photo by Dr. Herbert R. Axelrod.

Young *Myleus* from Rio Tocantins, possibly *M. tiete.* Photo by Harald Schultz.

Myleus rubripinnis luna, adult male in breeding colors, from Igarape Preto, Alto Solimoes. Photo by Harald Schultz.

Subgenus *Paramyloplus*

Only one species, from the Guianas, is known, with two subspecies; *Myleus (Paramyloplus) ternetzi ternetzi* and *M. (P.) t. goslinei*. They diverge in certain meristic characters such as gill-rakers, scales of the lateral line and number of dorsal rays. Externally there is nothing which distinguishes these forms from species of the subgenus *Myloplus*. The nominate subspecies may have, in life, a dark blotch on the anal which would extend onto the flanks: this mark is indicated by its first describer.

Subgenus *Prosomyleus*

This is a distinct unit, not only by its tooth structure, as mentioned in the key, but also by its broad adipose fin contrasting with a rather short dorsal fin, the former contained 3 to 5 times in the latter. Two species belong to this sub-group: the type species, *Myleus rhomboidalis*, and *Myleus schomburgki*. The latter has a very conspicuous vertical black band on its body which is very characteristic and serves to easily distinguish it from the former.

As far as meristic characters are concerned, the two species do not differ much from each other, except that the anal fin is somewhat longer in *M. schomburgki*. There is some evidence that the latter species is more complex than believed (this is often the case with animals whose color pattern makes them recognizable at first sight). What we call *M. schomburgki* is possibly composed of 2 (or more?) species. In the Araguaia and Xingu rivers, the band on the flanks is shorter than in the typical form and the anal fin rays are more numerous (about 34 to 36 branched rays versus 31 to 33).

Subgenus *Myleus*

The pacus of this sub-group attain a length of 500 to 600 mm, which is larger than the other species of *Myleus* (which are distinguished in French Guiana by the vernacular "pacoucines," meaning little pacus). They are identified at once by the aspect of the upper teeth, which are exposed, rather sharp cutting incisors appressed against the inner molariform teeth.

At least two species are known, *Myleus pacu*, the type species from the Guianas and the Amazon Basin, and *Myleus micans* from the Rio Sao Francisco Basin. *Myleus pacu* has an impressive number of synonyms *(setiger, doidyxodon, divaricatus, trilobatus, unilobatus)* owing to the great variability in the form of its dorsal and anal fins (which are secondary sexual characters). *Myleus micans* is less common, being apparently restricted to the Sao Francisco Basin; *M. altipinnis* with a similar restricted distribution is probably the same species. *Myleus micans* has more rays and abdominal spines than *M. pacu*, as follows:

	M. pacu	*M. micans*
Dorsal rays	ii, 18-21	ii, 22-24
Anal rays	iii, 30-33	iii, 34-35
Serrae	28 (juveniles) to 44	50 to 54

M. micans is also said to have "a general appearance of the head distinctively sheep-like."

Both species are plain when adult, except at the breeding season when they have large red spots on the body. The young pacus are nicely brown-colored, flecked when very young, which is a protective coloration also seen in the fry of the cichlids. When motionless, these young fishes are invisible against the background. Another peculiarity is the enormous migrations undertaken by the pacus (and most Serrasalminae) at the breeding season. These migrations (upstream, to the small "creeks" of the upper course of the rivers) are comparable in importance with those of certain marine species, even if the fresh-water milieu is much more restricted.

GENUS *UTIARITICHTHYS*

The single species *Utiaritichthys sennaebragai* from the Tapajos, Tocantins and Araguaia rivers, looks just like a *Myleus* that might have lost part of its abdominal serration; the anterior serrae are weak, few in number (about 9) and the preventral region is flattened, whereas the post-ventral region, as well as the other features such as the teeth, is exactly like that of the real pacus. In my opinion, *Utiaritichthys* could enter into *Myleus* as a subgenus, for the benefit of simplicity, but the present book is not an appropriate place to make such a change. In coloration *U. sennaebragai* has much in common with *Myleus torquatus*, both having a conspicuous black border to the anal and caudal fins.

Utiaritichthys sennaebragai (after Gosline, 1951).

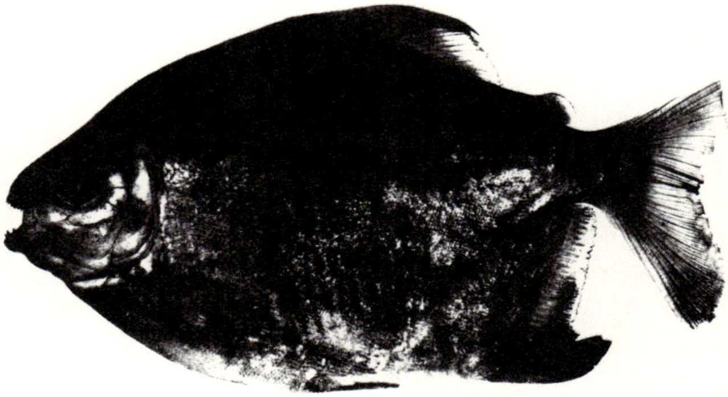

Myleus rubripinnis luna from Alto Juruena, Brazil.

Myleus ternetzi from Alto Solimoes. Photo by Dr. Herbert R. Axelrod.

Myleus schomburgki, adult female from Alto Solimoes. Photo by Harald Schultz.

Adult female *Myleus schomburgki* from Rio Araguaia. Photo by Harald Schultz.

GENUS *METYNNIS*
SILVER DOLLARS

Differing from all other serrasalmins by its long adipose fin, this genus is a good example of the inconsistency of the classification of certain groups of animals: it would include more than 20 species, according to one author (E. Ahl), and only six according to other zoologists. The truth probably lies between, but a precise number is presently impossible to give, the species being very difficult to identify. One species, *Metynnis luna* (synonym *M. guaporensis*?), which is the type of the genus, diverges from the other species by its very numerous gill-rakers (60-62) and the low number of serrae (about 23).

The other species (at least 5, but probably more) may be ranged within the subgenus *Myleocollops* (type species *M. goeldii*, which is perhaps a synonym of *M. lippincottianus*). In this subgenus, as contrasted with the preceding (nominate subgenus *Metynnis*), the species have never as many as 60 gill-rakers or as few as 23 serrae.

Two species, *M. lippincottianus* and *M. hypsauchen*, are common. The other ones, such as *M. maculatus*, *M. argenteus* (or *altidorsalis*) and *M. mola* (maybe others) are much less known. All other described species (some of them, such as *M. roosevelti* and *M. schreitmuelleri*, are often cited in aquarium literature) would be synonymous with one of the above cited forms, as follows (in alphabetical order):

altidorsalis	=	*argenteus* ?
anisurus	=	*argenteus*
calichromus	=	*hypsauchen*
dungerni	=	*argenteus*
ehrhardti	=	*hypsauchen*
eigenmanni	=	*argenteus*
fasciatus	=	*hypsauchen*
goeldii	=	*lippincottianus*
heinrothi	=	*argenteus*
orbicularis	=	*lippincottianus*
orinocensis	=	*hypsauchen*
otuquensis	=	*mola*
roosevelti	=	*lippincottianus*
schreitmuelleri	=	*hypsauchen*
seitzi	=	*lippincottianus*
snethlageae	=	*argenteus*

Although most modern authors agree that the *Metynnis* species have to be lumped as far as possible, the present author does not assume the responsibility of the above synonymy. For example *M. hypsauchen*, the *M. argenteus*-complex and the *M. lippincottianus-maculatus*-complex appear

Metynnis sp. Photo by Harald Schultz.

to be composed of several different forms, which could be subspecies or sibling species.

The following key will help to identify the best defined species of the subgenus *Myleocollops:*

a. Serrae projecting beyond abdomen, the forked serration behind ventral origin extending vertically and parallel with the other ones
. . . *M. mola* (Rio Paraguay basin)

aa. Serrae not projecting beyond abdomen, the forked serration behind ventral origin "flaring out into a double ended anvil" (Gosline).

b. Supraoccipital process very long, usually less than 2.4 in distance from base of occiput to origin of dorsal fin; adipose fin very long and low, the dorsal-adipose distance usually more than 1.5 in its base (about 35 gill-rakers)
. . . *M. hypsauchen* (Guianas, Amazon and Paraguay basins)

bb. Supraoccipital process not very long, usually more than 2.6 in distance from base of occiput to origin of dorsal fin; adipose fin not very long and low, the dorsal-adipose distance usually less than 1.5 in its base.

c. Usually 36-41 serrae; predorsal spine serrated; a vertical humeral bar (21-25 gill-rakers; 47-56 predorsal scales) . . . *M. maculatus* (Rio Guaporé basin)

cc. Usually 29-37 serrae; predorsal spine usually not serrated; humeral spot, if present, not vertically elongate.

d. Gill-rakers 21-25; body roundish (depth about 66 to 82 per cent of the standard length) . . . *M. lippincottianus* (Amazon basin)

dd. Gill-rakers 17-23; body lozenge-shaped (depth about 76 to 88 per cent of the standard length) . . . *M. argenteus* (Guianas and Amazon basin)

Unidentified young *Myleus.* Photo by Dr. Herbert R. Axelrod.

Metynnis maculatus or *M. lippincottianus.* Photo by H.J. Franke.

Metynnis maculatus or *lippincottianus*, often called *M. roosevelti* by aquarists. Photo by H.J. Franke.

Metynnis sp. Photo by Dr. A. Fernando-Yepez.

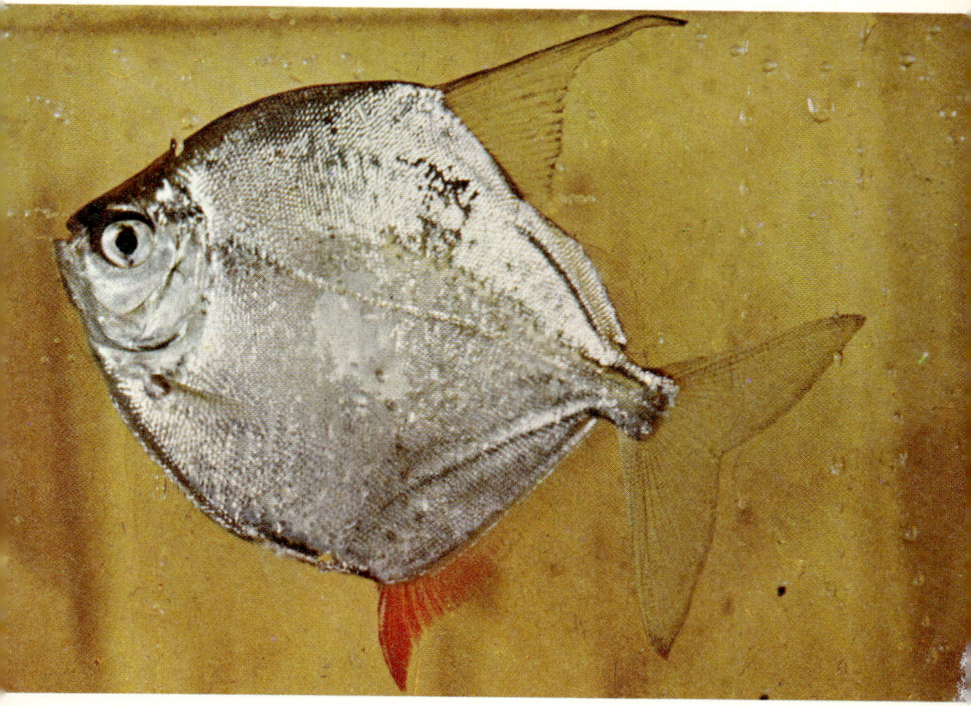

The silver dollars are the mylein serrasalmids most often seen in aquariums. Two species are relatively common: the spotted silver dollar called *roosevelti* by the aquarists *(M. lippincottianus* or eventually *maculatus)* and the silver dollar called *schreitmuelleri* by the aquarists *(M. hypsauchen)*, which is unspotted, strictly speaking, but may have a humeral spot and a few bars on the dorsum. Both forms grow to a relatively large size (much smaller than the other pacus) and readily eat aquarium plants. Both have reproduced in captivity.

GENUS *ACNODON*
SHEEP-PACUS

Only two species are known in this interesting genus which, unlike most other pacus (except *Utiaritichthys* and the next genus, *Mylesinus*), has the breast flat and not serrated; other peculiarities are the very long predorsal spine as well as the aquiline "nose" of the type-species, *Acnodon oligacanthus*. The two species, which have about the same proportions, may be identified as follows:

Acnodon normani and *Mylesinus schomburgki* (after Gosline, 1951).

a. Color pattern half-barred when adult; dorsal fin high; anal lobe elongate; eye small (about 3.5 in head); snout rather pointed, profile straight; outer upper teeth ranged as in *Myleus (Prosomyleus)* . . . *A. normani* (Alto Xingu, Alto Tocantins)

aa. Color pattern plain (silvery) when adult; dorsal fin low, anal lobe relatively short; eye large (about 2.65 in head in a specimen 180 mm in standard length); profile of snout strongly hooked; outer upper teeth ranged as in *Myleus (Myleus)* . . . *A. oligacanthus* (Guianas)

Obviously the two species have adapted themselves to a different diet (different seeds or fruits?), and *A. oligacanthus* may lead a crepuscular life (at least judging from the very large eye).

GENUS *MYLESINUS*

This genus, with the single species *Mylesinus schomburgki* (not to be confused with *Myleus schomburgki*, a quite different form) may be considered as extinct in the Guianas. Not a single specimen has been collected for more than a century. Gosline, in 1951, was fortunate enough to study a specimen coming from the Rio Tocantins; the main characters of this rare species are mentioned in the key. In general appearance, as well as in the counts of the fin rays, *Mylesinus schomburgki* looks much like *Myleus pacu* and may be a modification of this form.

SUBFAMILY SERRASALMINAE

Piranhas and Pirambebas

The Serrasalminae, with the single genus *Serrasalmus*, are unique among the characoid fishes in having sharp interlocking cutting teeth forming a continuous saw on both jaws, well adapted to taking a good piece of flesh out of fishes or even mammals (see more concerning their biology and behavior in *The Piranha Book*, edited by G. S. Myers, TFH Publications, 1972). The teeth are so well interlocked that an entire set on one side has to be replaced when a tooth is worn (unlike most of the characoids where the replacement occurs tooth by tooth).

According to the structure of the teeth, five subgenera may be recognized within the genus. They are about at the same classification level as the subgenera of *Myleus* and are defined as follows:

a. Teeth symmetrical, usually with 5 cusps; base of anal fin covered with no more than 2 scale rows; no teeth on palate . . . subgenus *Pygopristis* (a single species)

aa. Teeth not symmetrical, with no more than 3 cusps; base of anal fin covered with more than 2 scale rows.

b. Head and jaws narrow, profile behind eyes more or less concave, interorbital space more than 2 times in length of head; usually some teeth on palate (pirambebas).

Another *Metynnis* of the *maculatus-lippincottianus* complex, possibly a new form from the Alto Xingu. Female. Photo by Harald Schultz.

Metynnis luna. Photo by Dr. Herbert R. Axelrod.

Young *Metynnis.* Serrae protruding beyond abdomen are similar to those of *M. mola,* but possession of such protruding serrae may be characteristic of other juveniles as well. Photo by Dr. Herbert R. Axelrod.

Metynnis sp., probably *M. hypsauchen,* is generally called *schreitmuelleri* by aquarists. Photo by Kremser.

c. Teeth on palate obtuse, few in number (3 or 4), eventually lacking; suborbital narrow, leaving a wide naked zone on cheek even in the adult; body disciform, depth usually 1.5 to 1.66 in standard length in adults ... subgenus *Pristobrycon* (at least 4 species)

cc. Teeth on palate well-formed, acute, numbering 8 or 10 (except in the very old where they may be completely worn out); suborbital covering at least 3/4 of the cheek, usually complete in the adult; body not very high, depth usually more than 1.66 in standard length in adults ... nominate subgenus *Serrasalmus* (10 or so species, maybe more)

bb. Head and jaws broad and heavy, snout flat and mandible prominent; profile behind eyes more or less convex, interorbital space less than 2 times in length of head; palate never toothed (true piranhas).

d. Adipose fin not rayed ... subgenus *Taddyella* (2 or more species ?)

dd. Adipose fin rayed in adult ... subgenus *Pygocentrus* (one species restricted to the Rio Sao Francisco)

GENUS *SERRASALMUS*

Subgenus *Pygopristis*

Pirambeba

The species *Serrasalmus denticulatus* is the type of this subgenus, which is the least evolved toward predation of the serrasalmins. The young ones resemble tetras, with the body less deep than the next subgenus, and the adipose fin, as well as the caudal fin, is large. The species has a palate not toothed, the teeth at least 5-cuspidate and the suborbitals very weak, as are the muscles of the jaws. There is no spine after the anus and only one or two rows of scales on the anal fin which, unlike that of most piranhas, is bilobed in the male. It is a nicely colored fish which has the dorsum blue, the ventral part of the body as well as the base of the caudal fin yellow or orange, and the rest of the body bright red (eye, cheeks and fins, in their distal part for the dorsal and caudal fins). This strictly harmless fish is mostly Guianan, perhaps also Amazonian.

Fernandez-Yepez has described another species from Venezuela, *Pygopristis antoni*, which looks much more like a *Pristobrycon* (next subgenus).

Subgenus *Pristobrycon*

Pirambebas

This group comprises the less harmful *Serrasalmus* with the body flat and disciform, the mouth small and the armature of the cheek weak. They probably do not live in large schools, but instead more or less stay around the schools of more aggressive species and apparently profit from their hunting.

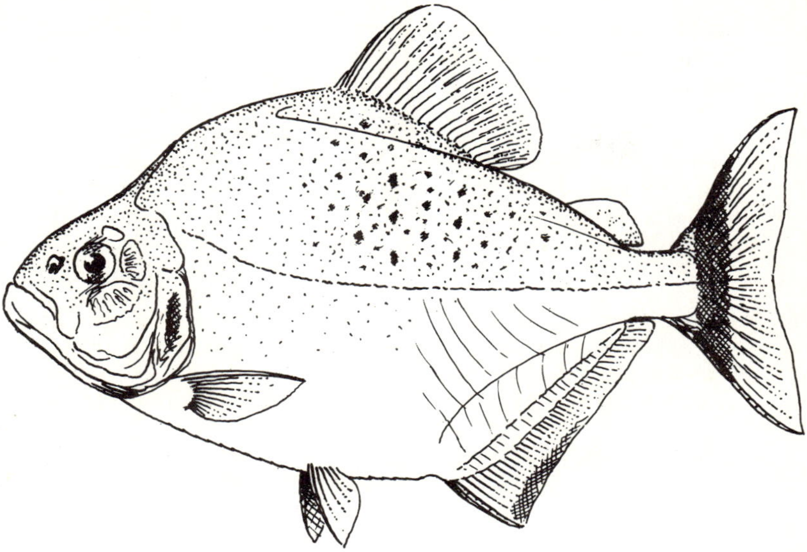

Pygopristis antoni. Illustration by Dr. A. Fernandez-Yepez from Myers, 1972.

The species are extremely close to each other and may be classified by the following key, which is compiled from a recent revision of the subgenus:

a. No humeral spot; dorsal fin at midbody; usually 84-86 scales in lateral line; depth 1.55 to 1.85 in standard length; snout 4.2 to 5.0 in length of head; naked zone on cheek contained 2.2 to 3.0 times in width of suborbital.

b. Head about 3.2 in standard length; no teeth on palate; body sometimes striped; no serrae after anus; dorsal-adipose distance 1.4-1.55 in base of dorsal fin (base of dorsal fin 2.45-2.9 in depth of body; anal fin rays 30-32, its base 2.8-3.0 in standard length; serrae 28-29) . . . *S. striolatus*, Lower Amazon and Guianas (a species very close to the preceding subgenus)

bb. Head 3.3-3.5 in standard length; probably a few blunt teeth on palate; body plain; probably a pair of serrae after anus; dorsal-adipose distance 1.0-1.1 in base of dorsal fin (base of dorsal fin 2.9-3.4 in depth of body; anal fin rays 31-34 (?), its base 2.5-2.7 in standard length; serrae 29-33) . . . *S. aureus*, Amazon and Guyana (synonym *gymnogenys*)

aa. A humeral spot, body often spotted; dorsal fin behind midbody; 70-80 or 81 scales in lateral line; depth 1.45-1.65 in standard length; snout short, 5 to 6.5 in length of head; naked zone on cheek contained 1.2 to 2.5 times in width of suborbital.

Serrasalmus (Pygopristis) denticulatus. Photo by Harald Schultz.

A form from the *Metynnis argenteus* complex from Rio Aguaro, Venezuela. Photo by Dr. Herbert R. Axelrod.

Amazonian pirambeba of the subgenus *Pristobrycon*, probably *S. serrulatus*. Photo by Dr. Herbert R. Axelrod.

Serrasalmus (*Pristobrycon*) sp., possibly young of *S. striolatus*, from the Marmelos, a tributary of the Madeira River. Photo by Dr. Herbert R. Axelrod.

c. Caudal fin with a black border; dorsal-adipose distance 1.2 to 1.25 in base of dorsal fin; naked zone on cheek about 1.2 in suborbital; anal rays 29-30 (serrae 34-35) . . . *S. calmoni*, Lower Amazon, Guyana and Venezuela (synonyms *bilineatus* and *coccogenis*)

cc. Caudal fin without a black border; dorsal-adipose distance 1.0-1.15 in base of dorsal fin; naked zone on cheek 2.0-2.5 in suborbital; anal rays 31-33 (serrae 27-34) . . . *S. serrulatus*, Amazon and Guianas (probable synonym *scapularis*)

Ramirez has described a species resembling a *Pristobrycon*, *S. altuvei*, which I have never seen (not even the original description). All that can be said is that it strongly resembles *S. calmoni*, which, under the name *coccogenis*, has already been described from Venezuela. As concerns *Pygopristis antoni*, it would not differ much from *S. gymnogenys*, which is synonymous with *S. aureus*.

Nominate subgenus *Serrasalmus*
Pirambebas

Here we have the "true" *Serrasalmus*, with strong teeth and jaws, but not true piranhas; there is not a single record of serious injuries made by members of this group (except occasionally a finger or toe cut away). Technically the nominate *Serrasalmus* species which are more elongate than

Serrasalmus elongatus. Photo by Dr. Herbert R. Axelrod.

the *Pristobrycon* species have well-formed, acute teeth on the palate, as well as, usually, a rather strong armature of the cheek. Exceptions are: *S. hollandi*, *S. elongatus* and *S. sanchezi*, which may have as few as 5 pterygoid teeth (average 7 in individuals of *S. rhombeus* 130-240 mm long); the *humeralis*-group, which has a naked zone bordering the large suborbital; and very old specimens of some species (chiefly *S. rhombeus*, but probably others) which, at the size of 300 mm or so, begin to lose their pterygoid teeth.

The description of the subgenus may well start with the *humeralis*-group which is still near the preceding subgenus owing to its relatively feeble armature of the cheek. *S. humeralis*, for example, is quite difficult to tell from *S. serrulatus*, in the same way as *S. striolatus* is still very close to *S. (Pygopristis) denticulatus*: this similarity is one of the reasons which prevent a splitting of the genus *Serrasalmus*. The *humeralis*-group (large suborbital leaving a naked zone on cheek which is included about 3 times in the width of the bone) is probably composed of only 3 species, which can be distinguished as follows:

a. Body depth 1.75 or 1.80 in standard length ... *S. humeralis* in the broad sense (the type of *humeralis* differs from all other specimens called *humeralis* by authors in having 38 or 39 serrae instead of 26-35; it may be an exception, in which case the following species may all be synonymous or at least geographical forms: *iridopsis*, *eigenmanni* and *nalseni* (this last form was not seen; its unusual number of dorsal rays, 20, would place it close to *S. caribe*, an uncertain species from the Orinoco))

aa. Body depth more than 2 times in standard length (serrae 32-37).

b. Body depth about 2.15; 14 branched dorsal rays ... *S. hollandi* (south of Amazon basin)

bb. Body depth about 2.6; 13 branched dorsal rays ... *S. elongatus*, Amazon basin (*S. pingke* is perhaps synonymous with *elongatus*, although it would have a larger suborbital)

These 3 species form an interesting secondary evolutionary line towards an elongate body. The end of the line, *S. elongatus*, is supposed to have

A pirambeba resembling *Serrasalmus serrulatus.* Photo by Harald Schultz.

A pirambeba resembling *Serrasalmus calmoni.* Photo by Harald Schultz.

Serrasalmus elongatus from the area of Humaita, Brazil. Photo by Dr. Herbert R. Axelrod.

A giant pirambeba from the Alto Xingu, possibly an old specimen from the *Serrasalmus serrulatus* group. Photo by Harald Schultz.

adapted itself to a fin-eating habit, similar to the African *Phago* and allies.

The second or *rhombeus*-group is based on the oldest species, the type of the genus *Serrasalmus* which is the type genus of the family. In this species the suborbital covers the cheek when the fish is still young, at the size of 100 mm or slightly more. Some of the species grow to a large size (450 mm), at which time they are almost black, with the front part of the body quite heavy: they look like a true piranha and they have often been confounded with species of the subgenus *Taddyella*, inasmuch as they have the palate toothless at this age, as pointed out above. However, they are always solitary in contrast with the piranhas. The identification of the species is difficult and their classification is controversial. Seven or so forms may be recognized; certain of them, such as *S. marginatus* and perhaps *S. medinai* (not seen by the author), are very close to *S. rhombeus* and probably only subspecies of this common species.

All have 13-16 branched dorsal rays and 28-36 branched anal rays. They can be distinguished mostly by their coloration and a few proportions:

a. A black border on caudal fin.

b. Black border to end of rays; usually no triangular humeral spot; probably no red on belly when alive.

c. Interorbital width about 2.4-3.0 in length of head . . . *S. rhombeus* (Guianas and Amazon basin) (*S. rhombeus* elongate form, has body depth 2.2 instead of 1.75-1.95, with available names *S. gibbus* and (?) *S. humeralis gracilior; S. medinae* from Venezuela, with depth about 2.1, may belong here)

cc. Interorbital width usually a little more than 3 in length of head . . . *S. rhombeus marginatus* (Rio de la Plata basin)

bb. Black border not reaching tips of the rays; usually triangular humeral spot; throat and belly red when alive.

d. Serrae 40-42; teeth on palate quite rudimentary . . . *S. sanchezi* (Upper Amazon)

dd. Serrae less than 40; teeth on palate well formed . . . *S. spilopleura* (Amazon basin)

aa. Border of caudal fin hyaline or grey . . . *S. brandti* (Rio Sao Francisco) (*S. fernandezi* from Venezuela may belong here)

The species most commonly met with in aquaria (at least in Europe, since piranhas and pirambebas are forbidden in the USA), are probably

A typical activity in the spawning act of a pair of piranhas (after Ledecky, 1966).

A young *Serrasalmus rhombeus*. Photo by Harald Schultz.

Serrasalmus hollandi from Alto Guapore. Photo by Harald Schultz.

An old *Serrasalmus rhombeus* (often called *S. niger*). Photo by Harald Schultz.

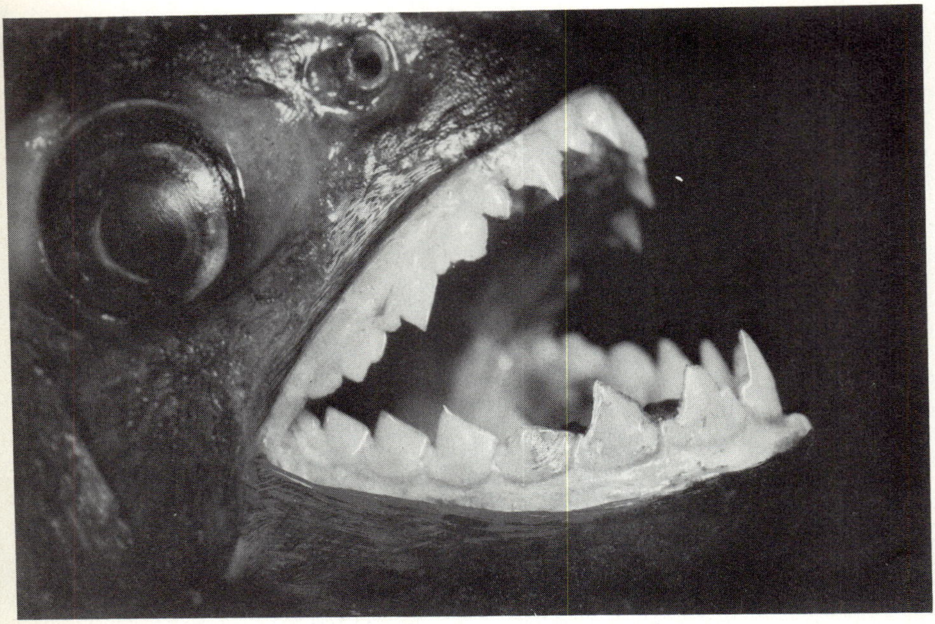

The cutting teeth of one of the dangerous piranhas. Photo by Harald Schultz.

S. rhombeus or "white piranha" and *S. spilopleura* which may be one of the red piranhas, and which I would call "diamond piranha" owing to its frequently bronze spotted scales (as in *Moenkhausia pittieri*). The spawning of both has been achieved, and it is now a known fact that the piranhas, at least the most specialized ones, guard their eggs and young. This is a highly evolved habit, quite unexpected among such a primitive group as the characoids.

Subgenera *Pygocentrus* and *Taddyella*

True Piranhas

The species of dangerous piranhas are few in number and do not differ much from each other. For example, the only feature which distinguishes *Pygocentrus* from *Taddyella* is the presence of some rudimentary rays in the adipose fin of the former. The species *Serrasalmus (Pygocentrus) piraya* occurs only in the Rio Sao Francisco basin, where it has been much feared.

The members of the subgenus *Taddyella,* actually the only dangerous piranhas (when trapped into a portion of river and hungry), are perhaps only two in number, the piranha from the Rio Paraguay, *S. ternetzi,* and the red piranha, *S. nattereri,* common in the entire Amazon basin as well as in the Orinoco (if *S. notatus* is a synonym of it, which is quite probable). They differ only in the form of the body, *S. ternetzi* having the anterior profile more arched and the dorsal-adipose distance shorter than that of *S. nattereri.* All other published names *(altus, notatus,* etc.) are probably referable to *S. nattereri. S. niger,* never rediscovered since it was collected at the beginning of the last century, is probably (in the opinion of the author at least) an abnormal specimen of a very old *S. rhombeus. Pygocentrus manueli* is not known by the author, even by its description. *

* This is perhaps the place to remind readers that predators, in all groups of animals, play a fundamental role in population control, keeping the other species (as well as their own) in balance. It is not only stupid but dangerous for the very fragile ecological equilibrium to suppress predators. Man has suppressed his own predators (in the large sense, including all selective pressures), thus provoking the demographic explosion we are just beginning to be conscious of.

An old *Serrasalmus rhombeus* from the Japura River. Photo by Harald Schultz.

Serrasalmus spilopleura male from the Cincinnati Aquarium. Photo by Dr. Herbert R. Axelrod.

Serrasalmus spilopleura female from the Cincinnati Aquarium. Photo by Dr. Herbert R. Axelrod.

Serrasalmus nattereri. Photo by Dr. Herbert R. Axelrod.

Wimple piranha, *Catoprion mento.* Photo by Harald Schultz.

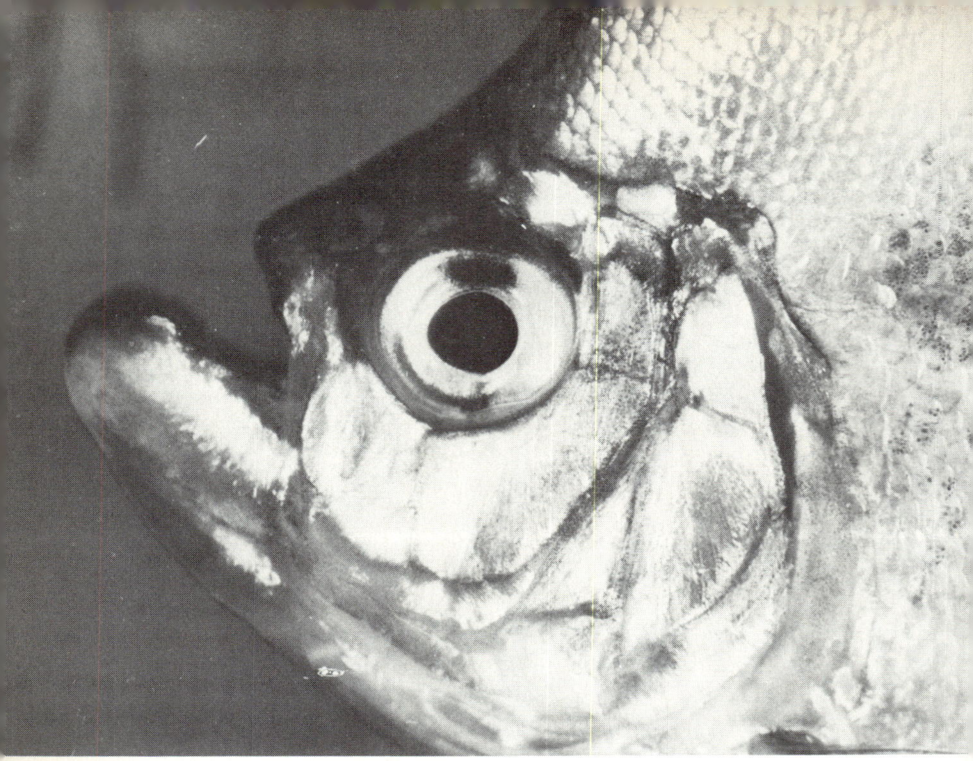

Catoprion mento. Photo by Harald Schultz.

SUBFAMILY CATOPRIONINAE

Scale-eater or wimple Piranha

The single genus *Catoprion* with the single species *Catoprion mento* looks very much like a piranha, but with the lower jaw even more prominent and armed with curious everted teeth. These teeth serve to remove rows of scales from other fishes. According to Ladiges, who observed the phenomenon, this takes place vertically in a reversed manner, from bottom to top. However, the fish can live on other food, preferably alive.

Another peculiarity is the very long, almost filamentous dorsal fin. Otherwise it has the relatively long dorsal and anal fins, as well as the predorsal spine and the serrae of its cousins, the piranhas. The wimple piranha occurs in certain parts of the Amazon basin (where it is not very abundant), in correlation with some ecology of its own, which is not necessarily in relation to its scale-eating habit.

Part III
American Tetras and Characins: Family Characidae

Family Characidae
American Tetras and Characins

This is the most important characoid group: it has more species (about 700) than the other 13 families combined. These species are mostly characterized by a long anal fin (with a few exceptions in the tetras), the maxillary bone not very small and usually toothed (contrasting with that of the African tetras) and the dorsal fin with 10-13 rays. The body shape, as well as the disposition of the teeth, is very variable. For classification purposes, although there are many overlaps between the groups within this large family, it has to be split into several subfamilies as follows:

a. Predators with teeth usually conical or even caniniform; maxillary bone long, often entirely toothed.

  b. Body elongate and compressed, clupeoid; a ventral keel present.

    c. Scales relatively large, of the Tetragonopterine-type; certain teeth with cusps; no palatine teeth . . . AGONIATINAE (1)

    cc. Scales small, of the Acestrorhynchine-Serrasalmine-type; all teeth conical or caniniform; a patch of palatine teeth . . . RHAPHIODONTINAE (2)

  bb. Body shape variable, but never elongate and compressed at the same time; a ventral keel only in 2 instances (*Gilbertolus* and *Gnathocharax*) . . . CHARACINAE (3)

aa. Omnivorous fishes with teeth rarely purely conical, never caniniform; maxillary bone rarely very long and entirely toothed.

    d. Mandibular teeth in two rows (the inner row composed of a pair of small conical teeth behind the main series, as in African tetras and certain serrasalmids); premaxillary teeth usually in more than two rows . . . BRYCONINAE (4)

Hydrolycus scomberoides. Photo by Dr. Herbert R. Axelrod.

Head of *Hydrolycus scomberoides* from Rio Aguaro, Venezuela.

Rhaphiodon vulpinus. Photo by Dr. Herbert R. Axelrod.

Head of *Rhaphiodon vulpinus.* Photo by Dr. Herbert R. Axelrod.

dd. Mandibular teeth uniserial; premaxillary teeth very rarely in 3 rows.

e. Body more or less elongate and compressed, clupeoid or siluroid; dorsal fin behind anal origin.

f. Scales large; body shape very similar to that of an *Agoniates*, by convergency or real affinity . . . CLUPEACHARACINAE (5)

ff. Scales small as in the Rhaphiodontinae; body shape less clupeoid . . . PARAGONIATINAE (6) (some forms not easily separable from certain Aphyocharacinae)

ee. Body never elongate and compressed at the same time; dorsal fin never behind anal origin except in *Piabucus*.

g. Postorbital bone very large (teeth conical or tricuspid, only one premaxillary series as in the Paragoniatinae) . . . APHYOCHARACINAE (7)

gg. Postorbital bone never very developed; teeth variable.

h. A caudal gland in the mature males . . . GLANDULOCAUDINAE (8) (certainly an artificial group)

hh. No caudal gland.

i. A predorsal spine (body deep and compressed, usually with a ventral keel) . . . STETHAPRIONINAE (9)

ii. No predorsal spine.

j. Usually 2 series of premaxillary teeth, rarely 3 (in *Creagrutus*) or one (in certain iguanodectins).

k. Form of teeth and length of maxillary bone not varying with age . . . TETRAGONOPTERINAE (10)

kk. Premaxillary teeth uniserial in young and biserial in adults; maxillary bone growing considerably with age . . . RHOADSIINAE (11)

jj. Premaxillary teeth usually in one series, more or less regularly set . . . a complex of forms comprising the CHEROIDONTINAE in the strict sense, as well as *Megalamphodus, Aphyodite et al., Grundulus et al., Henochilus et al.,* and *Probolodus* (12)

SUBFAMILY (1): AGONIATINAE

The two or three species forming the subfamily, *Agoniates halecinus, A. anchovia* and ? *A. ladigesi*, are very odd, ribbon-like fishes inhabiting the broad rivers of Guyana and those of the Amazon region. They are rare and perhaps even extinct in Guyana, where not a single specimen has been found for more than 100 years.

The largest known example (the type of *A. ladigesi*, a species very close to *A. anchovia*, if not the same) is only 170 mm in total length. All specimens

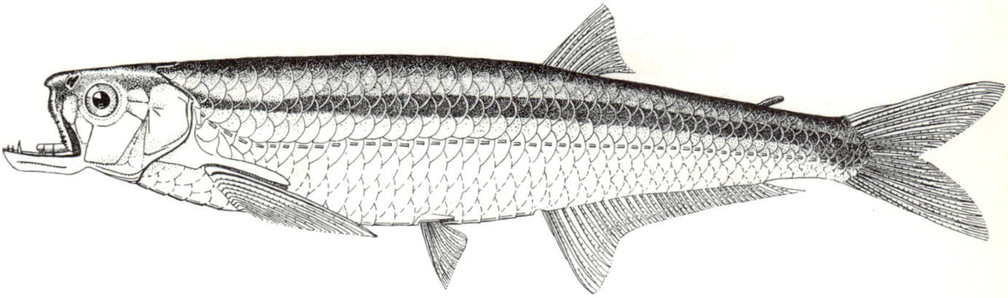

Agoniates anchovia, type (after Eigenmann, 1916).

are silvery and resemble certain deep-sea fishes. Their teeth are mostly conical or even caniniform, very acute, except that the frontal ones on both jaws show some remnants of cusps. With this type of dentition, they are likely to eat small fishes or shrimps.

SUBFAMILY (2): RHAPHIODONTINAE*

Like the Agoniatinae, the Rhaphiodontinae are a small group of odd-looking fishes. But they have quite different scales and different cranial structures, and they may not be as close as the key suggests.

The anal fin is usually quite long, the mouth broad, oblique and superior, with one row of numerous, acute, caniniform and conical teeth in each jaw; the gill-rakers are spinous as in *Acestrorhynchus*. These fishes, which reach a large size (some specimens are said to attain 70 cm in total length), have a strong ventral keel associated with developed pectoral fins and reduced ventral fins, as in certain other characoid fishes (Gasteropelecidae, *Triportheus*, certain Characinae and Glandulocaudinae) but, as far as known, they are unable to "fly."

Four species are known, two in the genus *Hydrolycus* and two in the genus *Rhaphiodon*, both genera differing in the position of the dorsal fin relative to the anal fin (in front of the anal fin in *Hydrolycus*, at the same level or behind it in *Rhaphiodon*).

Hydrolycus pectoralis inhabits the Rio Negro, the upper Amazon region and Bolivia. It has 90 lateral line scales and 43-46 branched anal rays. The caudal fin is hyaline, but the tips of the pectoral fins are black. *Hydrolycus scomberoides*, from the Guianas and Amazon, Orinoco and Paraguay basins, has 100-125 lateral line scales and 31-40 branched anal rays. The tip of the caudal fin is black and the pectoral fins are hyaline.

*After the setting of this book, G. J. Howes 1976 (Bull. Brit. Mus. (N.H.), Zool., 29(4): 201-248) has published a paper on this and the next subfamilies (treated as tribes), which offers interesting views on their functional anatomy.

Post-larval *Rhaphiodon vulpinus*. Photo by Dr. Herbert R. Axelrod.

Unidentified *Cynopotamus*, possibly a new species. Photo by Harald Schultz.

Charax tectifer. Photo by Harald Schultz.

Charax gibbosus. Photo by Dr. Herbert R. Axelrod.

Rhaphiodon vulpinus, from the Amazon and Paraguay-Parana basin, is elongate, with 125-130 lateral line scales and 40-45 branched anal rays. It possesses a median caudal lobe, like certain *Brycon* and a few other characoids. *Rhaphiodon gibbus* (sometimes called *Cynodon gibbus*), in contrast with the preceding, has the middle third of the body much shortened (as if the fish had lost its middle vertebra). This gives to it a bizarre profile.

The four species are apparently carnivorous (probably ichthyophagous). The size of certain individuals and their sharp teeth make their bite more than a simple incident, as reported by certain naturalists.

SUBFAMILY (3): CHARACINAE

This important group, which is close in certain respects to the preceding subfamily, is composed of fewer than 20 genera, including *Charax* which (being the oldest name) is the origin of the names Characinae, Characidae and Characoidei. It is usually split into 2 tribes, the Characini, which are an heterogeneous assemblage of several genera, certain of which are very specialized, and the Acestrorhynchini with only 3 genera which are monophyletic. Between the two groups, or forming a linking ring with the Tetragonopterinae, is a curious fish, *Bramocharax*, from certain Central American lakes, in a separate tribe. It resembles an *Astyanax* but with a large, characin type mouth with caniniform teeth (as well as pluricuspid ones).

Tribe Characini

True Characins

Like the Rhaphiodontinae, most of the Characini have a large mouth, obliquely directed, a long, entirely toothed maxilla, the top of the head almost horizontal and the dorsal profile humped. To this type correspond most of the forms of the genera *Cynopotamus*, *Charax* et al. and *Roeboides*, among others. A few forms have developed a terminal mouth (or still have it), some having a rather generalized tetra-like body shape.

Quite artificially, they may be split into species with elongate conical teeth, often caniniform, and species with mammilla-like teeth or tusks, usually outward (forward) directed, forming an outer series (the inner row being "normal") or even being the only "dentition" of the species, which is adapted to the scale-eating habit. A transition may be found in the poorly known genus *Eucynopotamus*.

The genera can be recognized as follows:

a. Predators armed with conical or canine-like teeth, never with mammilla-like tusks (mouth oblique).
  b. Scales ctenoid; rather large size . . . (1) *Cynopotamus*
bb. Scales cycloid; usually small size.

c. Body shape characteristic, the top of the head horizontal and the dorsal profile usually humped.

d. Preopercular angle not pointed; pectoral fins not very developed.

e. Lateral line complete; premaxillary teeth in 2 series . . . (2) *Charax*

ee. Lateral line incomplete; premaxillary teeth in 1 series . . . (3) *Asiphonichthys*

dd. Preopercular angle acute, often spinous; pectoral fins usually very developed.

f. No ventral keel.

g. Premaxillary teeth in two series; mouth oblique, the mandible moderately oblique and flattened (a single series of mandibular teeth) . . . (4) *Acanthocharax*

gg. A single series of teeth on the upper jaw; mouth superior, the mandible almost vertical, strongly flattened (2-3 small conical teeth behind the main mandibular series, at least in one species) . . . (5) *Roestes* (synonym *Xiphocharax*)

ff. A ventral keel, the pectoral fins very developed.

h. Lateral line complete; 58-68 longitudinal scales; 40-50 branched anal rays . . . (6) *Gilbertolus*

hh. Lateral line incomplete; 33-35 longitudinal scales; 26-32 branched anal rays . . . (7) *Gnathocharax*

cc. Very small species; body form generalized, like that of a tetra.

j. Preopercular angle without a real spine; opercle not spinous; clavicle notched; predorsal line scaled; lateral line complete; first pectoral ray soft.

k. A single series of teeth on upper jaw . . . (8) *Heterocharax*

kk. Premaxillary teeth in two series . . . (9) *Lonchogenys*

jj. Preopercular angle with a strong spine; opercle with 2 strong spines; clavicle notched; predorsal region naked; lateral line incomplete; first pectoral ray modified into a strong spine (as in most siluroids) . . . (10) *Hoplocharax*

aa. Scale-eaters (at least partly) with some or all of their teeth transformed into mammilla-like, more or less outward pointing, tubercles.

l. General aspect of a *Charax*, the mouth oblique; only a part of the dentition mammilla-like .

m. A few frontal teeth on both jaws mammilla-like; no teeth pointing forward or outward except the last maxillary teeth . . . (11) *Eucynopotamus*

mm. Mammilla-like processes pointing forward and outward on both jaws; "normal" teeth in two series on the upper jaw . . . (12) *Roeboides*

ll. General aspect of a generalized Tetragonopterine (except the mouth structures), the mouth not oblique; all the teeth transformed into mammilla-like tubercles, at least in upper jaw.

Gilbertolus alatus? These specimens have not been studied and may represent a population of *Gnathocharax.* Photo by Dr. Herbert R. Axelrod.

Gnathocharax steindachneri. Photo by Dr. Herbert R. Axelrod.

Gnathocharax steindachneri. Photo by Dr. Herbert R. Axelrod.

Gnathocharax steindachneri from a different locality (Rio Negro) than the above. Photo by Dr. Herbert R. Axelrod.

n. Mouth terminal; mandible with an inner series of acute, conical "real" teeth ... (13) *Exodon*

nn. Mouth strongly inferior, profile of head resembling that of a shark; 8 or so tubercles on mandible, the foremost ones pointing outward, without inner series ... (14) *Roeboexodon*

THE GENUS *CYNOPOTAMUS**

DIENTUDOS

Cynopotamus (meaning "river dog") is the largest characin genus after *Roeboides*, all other genera having few species or being monotypic. The most typical *Cynopotamus* species look exactly like a big *Charax* with rough scales; certain species are more elongate and have recently been classified within a distinct subgenus, as follows:

a. Body depth 3.33-3.9 in standard length, without a hump; clavicle usually feebly notched; large suborbital bone narrow; 2 mandibular rows of teeth, inner row composed of 6-9 conical teeth ... Subgenus *Acestrocephalus* (synonym *Galeocharax*)

aa. Body depth usually less than 3.3 in standard length, dorsal profile much elevated, forming a sort of hump.

b. Clavicle only feebly notched; large suborbital almost complete; inner mandibular row composed of 3 or 4 conical teeth ... Subgenus *Hybocharax*

bb. Clavicle with a distinct notch; large suborbital narrow; no conical teeth behind the mandibular canines ... Subgenus typical *Cynopotamus* (synonym *Cyrtocharax*)

Six species of the subgenus *Acestrocephalus* are known: the type species *Cynopotamus anomalus* from the Rio Magdalena, the rather common *C. humeralis* and the closely related *C. kneri* from Rio de la Plata and Rio Paraguay, *C. gulo* from the upper Amazon, and 2 lesser known Amazonian forms, *C. sardina* (not very far from *C. anomalus*) and *C. goeldii* (very close to *C. gulo*). The largest specimens of *C. humeralis* reach 150-200 mm and are sold in the fish markets of the Rio de la Plata.

The subgenus *Hybocharax* has only two species: *C. bipunctatus*, the type species, from Venezuela, and *C. magdalenae*, sympatric with *C. anomalus* in the Rio Magdalena.

The subgenus *Cynopotamus* has 6 species, 3 in the Rio de la Plata and Rio Paraguay (sympatric with *C. humeralis*) and 3 in the Amazon and north of the Amazon. The former group, which has 10 branched dorsal rays, includes a pair of species, *C. argenteus*, the type species, and *C. calliurus*, which are possibly siblings, as well as *C. kincaidi*, more recently discovered and probably rare.

The second group, which has 9 branched dorsal rays, is composed of *C. amazonus* (synonym *C. limaesquamis*) from the Amazon and *C.*

* In a still more recent study, Dr. Naercio Menezes has raised the subgenera to generic status. At the same time he synonymized *Hybocharax* but resurrected *Galeocharax*.

essequibensis from Guyana. A third related species, *C. atratoensis*, has two slightly different forms (subspecies ?) in Colombia and in Venezuela.

All species are probably carnivorous. All are much alike, and certain technical characters such as the number of scales and fin rays, not easy to count with the naked eye, are necessary to separate them.

GENERA *CHARAX* AND *ASIPHONICHTHYS*
TRUE CHARACINS OR DIENTUDOS

Like the *Cynopotamus* species, the *Charax* species, which are very similar in shape and color pattern but usually have humeral and caudal spots, can be split into 3 subgenera, as follows:

a. No teeth on palate (ectopterygoid bone) . . . nominate subgenus *Charax*
aa. 10 or so teeth on each ectopterygoid bone.
b. Dorsal region normally scaled; maxilla long, reaching beyond middle of eye . . . Subgenus *Moralesia*
bb. Dorsal region naked; maxilla short, reaching to middle of eye . . . *Charax rupununi*, representing a new subgeneric type

There are only two species in the subgenus *Charax*. *Charax gibbosus*, one of the oldest characoid species, dating from Linnaeus, is relatively abundant in the Guianas as well as in the Amazon basin. *Charax pauciradiatus*, a poorly known Amazonian species, differs from the preceding in the naked predorsal line and perhaps in the lower number of anal rays.

Asiphonichthys stenopterus (top) and *Acanthocharax microlepis* (bottom). Photos by the author.

Hoplocharax goethei (after J. Gery, 1966).

A recently discovered glass characin, *Asiphonichthys condei.* Photo by Harald Schultz.

Two aspects of *Heterocharax macrolepis* from Mai Juana near Manacapuru. Photo by Dr. Herbert R. Axelrod.

Likewise, 2 species of the subgenus *Moralesia* are known: *Charax tectifer* (synonym *copei*), from the Upper Amazon basin and Ecuador, and *C. metae*, from Upper Rio Meta, tributary of the Orinoco. The two forms differ only in the size and position of the humeral spot.

The *Asiphonichthys* species, which are *Charax* with a single more or less regular premaxillary tooth series and an incomplete lateral line, number three: *A. stenopterus*, the largest species (the size of a *Charax*, i.e. 100 mm or so) from La Plata basin, *A. hemigrammus*, known from a single post-larval specimen from Guyana, and a third just described species from the Rio Negro which is completely translucent in life (like certain *Roeboides* and *Phenacogaster*), *A. condei*.

GENERA *ACANTHOCHARAX* AND *ROESTES*

These genera seem rather close to each other and look as if they were two stages of an adaptation toward life near the surface, with their mouth becoming superior, their breast deep and their pectoral fins very large. This adaptation culminates in the next genera, the "flying" characins *Gilbertolus* and *Gnathocharax*, which have a true ventral keel and very developed pectoral fins.

The species *Acanthocharax microlepis*, from the Essequibo and Potaro Rivers in Guyana, is a small fish (about 80 mm in standard length), still resembling a *Charax* but with a spinous preopercular angle, a very large eye and a paddle-like maxillary, together with a naked predorsal region (as in *Charax pauciradiatus* and *C. rupununi*).

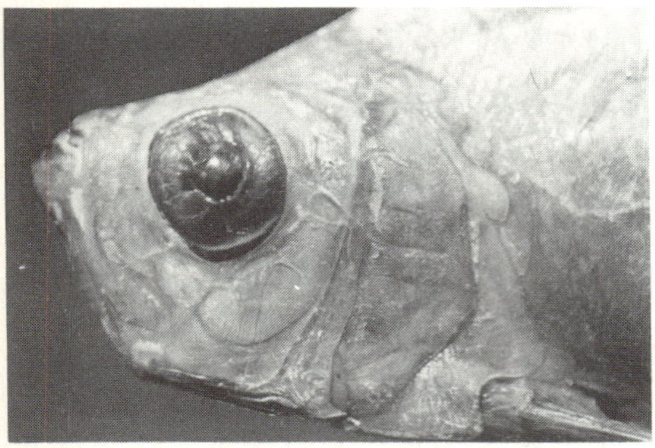

Type of *Roestes ogilviei* and a close-up of its head. Photos by the author.

Until very recently, the genus *Roestes* was poorly known. Its type species, *Roestes molossus*, had been collected only once, in the Rio Guapore, more than a century ago, and never seen since. Dr. Naercio Menezes was fortunate enough to discover some specimens from the Jurua and Livramento Rivers which agree fairly well with the original description, and presented an osteological study of them. Menezes found that *R. molossus* has much in common with a species from the Rio Branco, known up to now as *Xiphocharax ogilviei*, and therefore put the genus *Xiphocharax* into the synonymy of *Roestes* (this action is adopted here, but, of course, has to be checked by examination of the types of *Roestus molossus*, when possible).

The type of *Roestes ogilviei*, which the present author has seen in the Academy of Natural Sciences in Philadelphia, represents a species apparently restricted to the Rio Branco basin. It is a curious fish 173 mm in standard length with an almost vertical mandible forming an angle with the lower part of the head, the whole being perfectly flat as seen from the front.

Roestes molossus (not seen by the author) would differ from *R. ogilviei* in having 2 or 3 small conical teeth behind the main mandibular series (the type of *R. ogilviei* has the lower teeth uniserial) and fewer scales in the lateral line (77-80 instead of 82-84), but not in the dorsal profile, as stated by Menezes; the dorsal profile is slightly convex, without gibbosity, in both species.

GENERA *GILBERTOLUS* AND *GNATHOCHARAX*

Gilbertolus alatus and *Gnathocharax steindachneri* are apparently the only species in their respective genera. The former has been placed, like *Xiphocharax*, into the synonymy of *Roestes* by Menezes (who followed Steindachner) owing to its overall resemblance to the two species just discussed. However, its distinctive ventral keel and other adaptations seem to have full generic value, at least according to current standards in characid systematics. It is restricted to extreme northern South America, with the nominal form in the Rio Magdelena and two subspecies (according to certain ichthyologists) in the Rio Atrato, also in Colombia, and in coastal rivers of the Maracaibo basin (Venezuela). It reaches about 120 mm in length.

Gnathocharax steindachneri, which is much smaller, inhabits certain biotopes in the whole Amazon-Negro basin, as well as in the Upper Orinoco and Guyana. It looks like the young of *Gilbertolus*, with its small size, lack of a dorsal hump, and incomplete lateral line. In life, the upper part of the eye is bright red and the fins are yellow, with rosy and black spots on the caudal fin. The color pattern is quite variable.

Roeboides sp. Photo by Harald Schultz.

Roeboides sp., also from the Amazon basin. Photo by Dr. Herbert R. Axelrod.

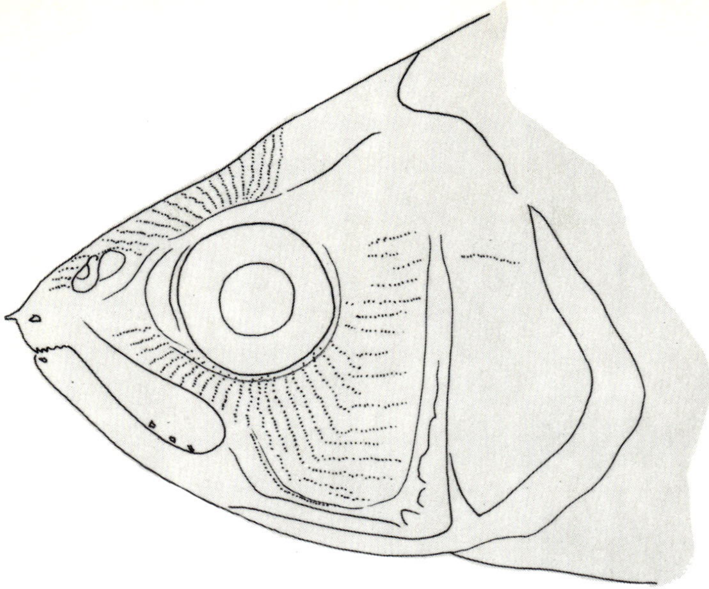

Pit-lines on the head of a species of *Roeboides* (*R. caucae*) (after Gery, 1966).

Exodon paradoxus. Photo by Dr. Herbert R. Axelrod.

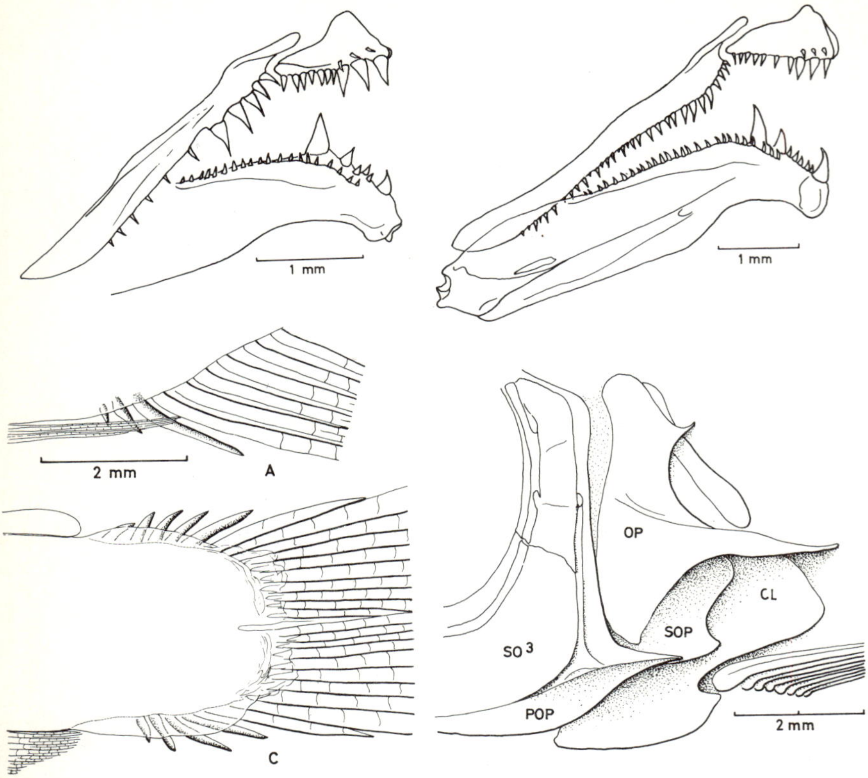

Jaws of *Hoplocharax goethei* (left) and *Heterocharax macrolepis*; accessory spinous caudal rays and opercular region of *Hoplocharax goethei* (after Gery, 1966).

GENERA *HETEROCHARAX, LONCHOGENYS* AND *HOPLOCHARAX*

These 3 genera, monotypic like the preceding ones, have been reviewed by me rather recently (*Ichthyologica*, June-September, 1966). The principal characters of these pigmy characins are given in the key.

Hoplocharax goethei, from the Rio Negro, is very interesting from an evolutionary point of view, being a museum in miniature of defensive specializations, with a collection of spines of various origin: one on the preopercle, two on the opercle, one on the pectoral fin, and several on the beginning of the anal and caudal fins.

Heterocharax macrolepis, from Guyana, Rio Negro and Middle Amazon basin, and *Lonchogenys ilisha*, somewhat larger, from Upper and Middle Rio Negro, are rather close to each other (the latter derived from the former ?) and have more generalized structures, including only a weak preopercular spine.

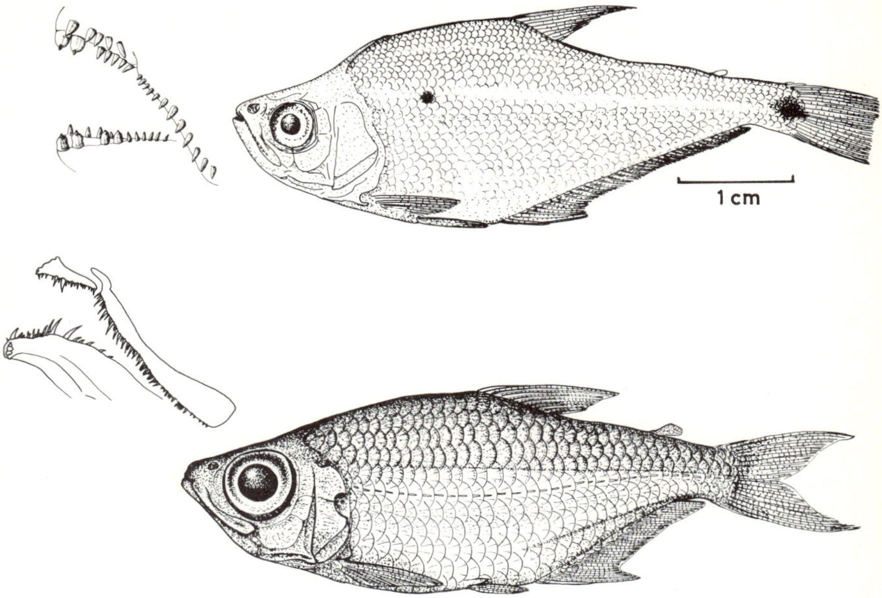

Cotype of *Eucynopotamus biserialis* with disposition of the teeth; type of *Lonchogenys ilisha* with disposition of its teeth. *L. ilisha* from original drawing by Mr. Pablo Bravo used by permission of Dr. G.S. Myers; other drawings by the author.

GENERA *EUCYNOPOTAMUS* AND *ROEBOIDES*

Eucynopotamus biserialis, from the Amazon, is a curious species resembling a low "humped" *Charax*, with the teeth more or less intermediate between a true characin and a *Roeboides*, with only a few mammilla-like tubercles, which do not point forward. It is possible that some of the tooth structures are juvenile (the largest known specimen is only 50 mm in standard length). On the whole the fish reminds one also of a *Phenacogaster*, which is actually classified within the true tetras. *Eucynopotamus* remains mysterious, and its present position as a precursor of *Roeboides* is to be considered as a working hypothesis. Dr. Roberts even suggested that it may be the young of a *Roeboides*.

In contrast with the preceding, *Roeboides* is a large genus containing 20 or so forms (subspecies included). Most of them are characterized by the body shape of a *Charax*, with a spine on the clavicle, usually a flat, pointed head, a gibbosity and a long anal fin, all in association with external mammilla-like teeth, cervical pit-lines (in some species at least), as well as a translucent body like that of *Asiphonichthys* and *Phenacogaster* in certain species when alive. Finally, the genus has scale-eating habits.

Head of *Oligosarcus* sp. Photo by Harald Schultz.

Oligosarcus sp., possibly new to science. Photo by Harald Schultz.

Acestrorhynchus of the *microlepis*-group. Photo by Dr. Herbert R. Axelrod.

Acestrorhynchus of the *falcatus*-group, possibly a new species. Photo by Harald Schultz.

Most of the species are in need of revision. As such a revision is not the purpose of the present book, the following key will only give outlines of the classification: *

a. Body shape rather generalized, quite elongate (body depth more than 3 in standard length), the hump scarcely pronounced . . . *R. bicornis* (upper Amazon) and *R. salvadoris* (San Salvador)

aa. Body shape *Charax*-like, body depth 2.15-3.0 in standard length, the hump easily visible.

b. 110 lateral line scales . . . *R. microlepis* (Amazon), the type species

bb. Less than 96 lateral line scales.

c. 85-95 lateral line scales . . . *R. myersi* (Amazon) and *R. bonariensis* (Rio de la Plata)

cc. 52-82 lateral line scales.

d. No humeral spot . . . *R. prognathus* (Mato Grosso and Rio Paraguay) and 3 northern species: *R. guatemalensis* (Central America), *R. magdalenae* and *R. romeroi* (Colombia)

dd. A humeral spot . . . a complex of species comprising *R. descalvadensis* (Mato Grosso), *R. xenodon* (or *R. francisci*, Rio Sao Francisco), *R. affinis* (upper Amazon) and *R. thurni* (Guianas), as well as 5 northern species, certain ones with 2 or more subspecies: *R. occidentalis* (Panama and Colombia), *R. caucae* and *R. meeki* (Colombia); *R. dayi* (Venezuela and Trinidad) and *R. bouchellei* (Nicaragua)

GENERA *EXODON* AND *ROEBOEXODON*

These two monotypic genera are among the most curious characoids.

Exodon paradoxus, from certain biotopes all over the Amazon basin (including the Rio Branco drainage in Guyana), can be locally very abundant when the surroundings closely fit its peculiar ecology (rivers with

*Since the writing of the key, Miller and Carr (Copeia No. 1, 1974, p. 121) have synonymized *R. salvadoris* and *R. bouchellei* with *R. guatemalensis*.

1

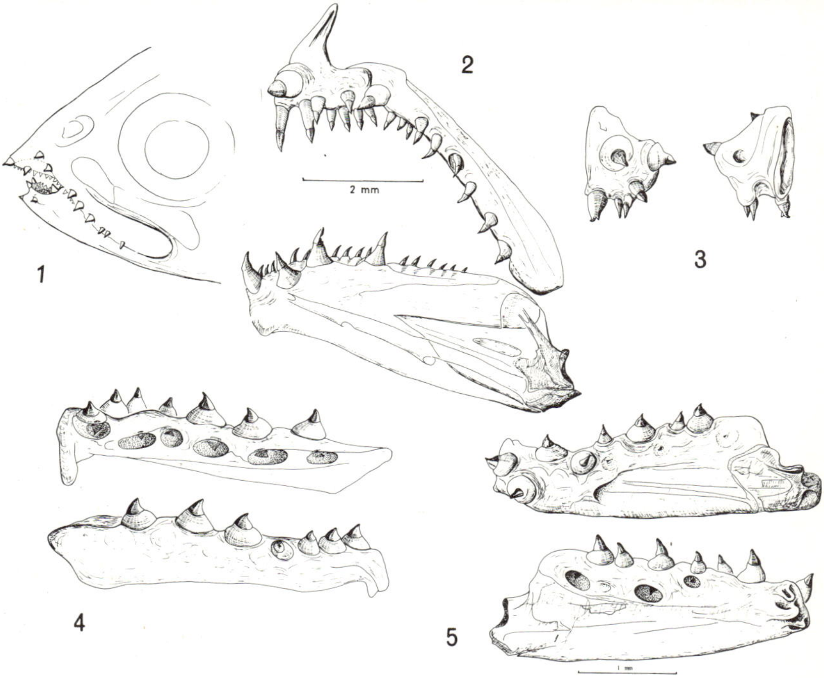

Roeboides caucae and dentition of *R.* cf *affinis* (1); jaws of *Exodon paradoxus* (2); neotype of *Roeboexodon guyanensis* with illustrations of the left premaxilla (3), left maxilla (4), and left mandible (5). Photos and illustrations by the author.

3

Acestrorhynchus sp., possibly *A. falcirostris.* Photo by Dr. Herbert R. Axelrod.

Holobrycon pesu. Photo by Harald Schultz.

Brycon of the *falcatus*-group, possibly *B. brevicauda.* Photo by Dr. Herbert R. Axelrod.

Another aspect of a *Brycon* of the *falcatus*-group (possibly also *B. brevicauda*). Photo by Harald Schultz.

sandy bottoms, mostly in the savannah regions, where it is frequently associated with the puffer *Colomesus asellus*). Its mouth is terminal and scarcely oblique; its premaxilla, which has an ascending process, has 3 outer tusks and 6 inner acute teeth; the maxilla bears a similar tubercle near its joint and 9 or 10 acute teeth, the latter ones pointing outward. The mandible has 4 external tubercles, the foremost one pointing forward, and an inner series of about 17 small conical teeth of the characin type.

Its stomach contents consist of piles of partly digested scales from other fishes. These little fish are quite brilliantly colored, with the body and the caudal fin yellow and the other fins red or reddish. The two very large spots, the humeral one in front of the dorsal fin level and the one covering the whole height of the caudal peduncle, are characteristic.

Roeboexodon guyanensis (synonym *R. geryi*), from the Guianas and the Serra do Roncador (upper drainage of the Araguaia-Xingu rivers, where it seems to be of a slightly different type), is even more striking than *Exodon paradoxus*. The mouth is completely inferior and the snout is prominent like that of a shark. Provided that the two fishes are from the same phylogenetic line, which is likely, one may say that *R. guyanensis* has lost the ascending process of its premaxillary bone, together with several of its teeth, having kept only two external tusks and four mammilla-like teeth. The tubercles on the maxilla and mandible no longer resemble any known characin tooth. In contrast with the extreme modification of the teeth, the color pattern of the species is more common than that of *Exodon* and consists of two "ordinary" humeral and caudal spots of a good, but not exaggerated, size.

Tribe Bramocharacini

It is not known whether *Bramocharax bransfordi* (and certain other forms), from Nicaraguan and Costa Rican lakes and rivers, belongs to the Acestrorhynchini or is to be classified in a separate tribe.* The latter choice is followed here. Indeed, this genus is a puzzle of tetragonopterine and characine features, having the body shape and the color pattern (the usual double spot, humeral and caudal) of an *Astyanax*, the head, including the long concave maxillary, of an Acestrorhynchine, and an association of multicuspid and caniniform teeth inside the mouth. There are one large and 3 small canines on the premaxilla, mixed with 4 pentacuspid teeth; one very large and one medium sized canines on the mandible associated with a large pentacuspid tooth immediately behind them; and 9 much smaller penta- or tricuspid teeth in a series which reminds one of most of the Tetragonopterinae; and finally, a concave maxilla armed with 14 or so pentacuspid teeth, again of the tetragonopterine type.

* Or even to be put into the Tetragonopterinae, as has been done recently by Dr. Donn Rosen.

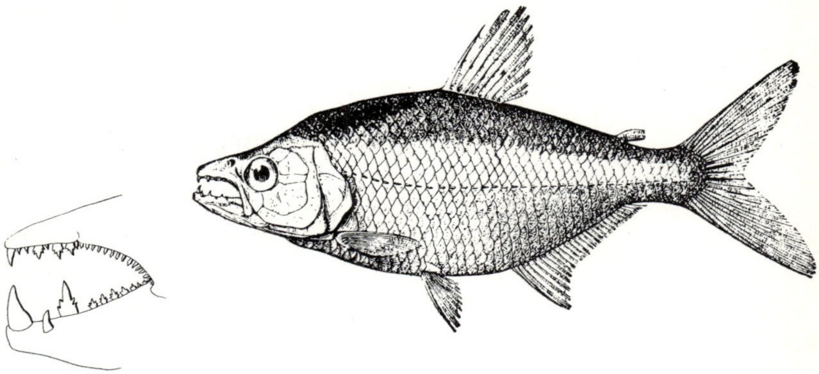

Bramocharax bransfordi (after Eigenmann and Ogle, 1907) and a semi-diagramatic illustration of its teeth (by the author).

Scissor macrocephalus, known from a single specimen of uncertain origin and classified by Eigenmann within his Tetragonopterinae, may belong here.

Tribe Acestrorhynchini

This group is one of the best-known characid groups, having been recently revised by Naercio Menezes. This Brazilian ichthyologist recognized 3 genera.

Paroligosarcus is a monotypic genus from S.E. Brazil which is like *Bramocharax* in having tri-to pentacuspid teeth, but unlike it in lacking canines.

Oligosarcus, which is like *Paroligosarcus* (but with canine teeth and no pentacuspid teeth), has a variable snout. One group of species (*Oligosarcus* in the strict sense) has a short snout, and the second group (called *Acestrorhamphus* by classical authors) has a long snout, grading into that of *Acestrorhynchus*. *Acestrorhynchus* differs from the most specialized *Oligosarcus* by its spinous gill-rakers (an adaption to predation), the absence of any cuspid teeth, the presence of strong canines on the maxilla, and by its Guianan-Amazonian distribution.

GENUS *PAROLIGOSARCUS*

Paroligosarcus pintoi inhabits the state of Sao Paulo in Brazil. It is a small fish (largest specimens about 85 mm in standard length) resembling an *Astyanax*, but with a different dentition and a long maxillary. It has a humeral spot, a lateral band and a caudal spot extended onto the middle caudal rays. It eats mostly insects fallen on the water surface.

Brycon melanopterus. Photo by Dr. Herbert R. Axelrod.

Brycon sp., probably close to *B. melanopterus.* Photo by Dr. Herbert R. Axelrod.

Unidentified *Brycon* from Rio Aguaro, Venezuela. Photo by Dr. Herbert R. Axelrod.

Brycon sp. from northwestern South America. Photo by Dr. Herbert R. Axelrod.

GENUS *OLIGOSARCUS*

Seven species, mostly from S.E. Brazil and the Paraguay-Uruguay basin, none in the Amazon basin (except at its periphery where there exists a still undescribed species), are known. As mentioned above, they may be ranked into two subgroups (at the subgeneric level ?) according to the relative length of their snout.

In the first group, which may be considered as primitive and near the stem of the tribe, the dorsal outline of the head is convex, the snout relatively short (about the diameter of the eye in projection) and the first upper canine is included in the mandible when the mouth is closed. *Oligosarcus (Oligosarcus) argenteus*, of unknown origin, *O. bolivianus* from Rio Lipeo in Bolivia, and *O. meadi*, from Rio das Velhas, belong to this group. They have about the same coloration and body shape as *Paroligosarcus*, and they differ (between themselves) in the number of scales.

In the second group *(Acestrorhamphus* auct.), the species are more elongate; they have the dorsal outline of the head straight (very slightly convex in *O. jenynsii*), the snout pointed and longer than an eye's diameter, and the first upper canine outside the mandible when the mouth is closed (owing to the lengthening of the upper jaw). Four species are known: *Oligosarcus (Acestrorhamphus) hepsetus* and *O. jenynsii*, two sibling species (sympatric) from southeastern Brazil and the La Plata basin, *O. robustus* (which is sympatric with *O. hepsetus*) in the Uruguay River, and the small *O. macrolepis* (80 mm), sympatric with *O. jenynsii* in the Jequitinhonha River. Again, they are all alike in habitus and differ only in the number of scales. They eat mostly fishes.

GENUS *ACESTRORHYNCHUS*

The pike-like fishes of this homogeneous and clear-cut genus grade from rather "dumpy" predators such as *A. lacustris* and *A. altus* (body depth usually less than 4 in the standard length) to very elongate "pikes" such as *A. falcirostris*, *A. minimus* and *A. nasutus* (body depth about 6 in the standard length), the bulk being composed of *A. falcatus* and its forms, *A. heterolepis* and *A. britzkii*, on the one hand and the siblings *A. microlepis-guianensis* on the other, the former with body depth about 4.0-4.75, the latter 4.75-5.5 in the standard length. The length of the upper jaw usually follows the lengthening of the body. All forms are ichthyophagous.

The species can be identified by means of the following key (mostly compiled from Menezes, 1969):

a. Anteriormost upper tooth caniniform, larger than the following one (body depth usually less than 6 times in standard length, except in *A. minimus*).
 b. Humeral spot large, conspicuous.
  c. Humeral spot round; body depth about 3.5-3.65 in standard length (89-115 lateral line scales).

d. 27-34 transverse scales above lateral line (usually more than 100 lateral line scales) . . . *A. altus* (Amazon and Paraguay basins)

dd. 25-26 transverse scales above lateral line (usually less than 100 lateral line scales) . . . *A. lacustris* (sympatric with the preceding)

cc. Humeral spot oval or triangular, vertically elongate; body depth usually more than 4 in standard length (18-24 transverse scales above lateral line, 80-96 lateral line scales) . . . *A. falcatus*-complex (Guianas, Amazon and Orinoco basins); probably more than 2 forms

bb. No humeral spot.

e. 130-145 lateral line scales; sensory canal branched on each scale (2 branches); a small spot on uppermost part of opercle . . . *A. heterolepis* (upper Amazon and Rio Orinoco)

ee. Less than 130 lateral line scales; sensory canal a single tube on each scale; no spot on upper part of opercle.

f. 92-122 lateral line scales.

g. A small spot just behind opercle; no black lateral band.

h. Body depth about 4.8 in standard length; 92-106 lateral line scales . . . *A. guianensis* (Guyana; Amazon and Orinoco basins)

Skull of an *Acestrorhynchus* of the *falcatus*-group. Photo by the author.

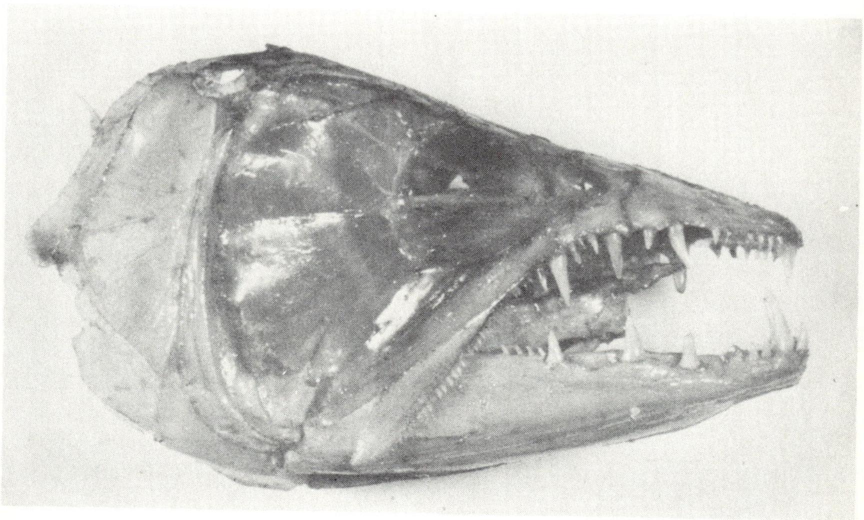

A young *Brycon* sp. from Colombia. Photo by Dr. Herbert R. Axelrod.

A young *Brycon*, possibly *B. melanopterus*. Photo by Dr. Herbert R. Axelrod.

Chalceus macrolepidotus, the pink-tailed chalceus. Photo by Harald Schultz.

Chalceus sp. Photo by Dr. J. Gery.

hh. Body depth about 5.5 in standard length; 108-122 lateral line scales ... *A. microlepis* (sympatric with *A. guianensis* except in Orinoco basin, and much more common than it; *A. cachorro*, known from a single specimen much deeper bodied than *A. microlepis*, may be a distinct species)

gg. No small spot just behind opercle; a black lateral band (body depth about 4.65 in standard length) ... *A. britzkii* (Rio Sao Francisco)

ff. 74-85 lateral line scales (a narrow lateral stripe; depth usually more than 6 in standard length; small size; greatly resembling *A. nasutus*) ... *A. minimus* (Amazon basin)

aa. Anteriormost upper tooth much smaller than the following one which is caniniform (body depth about 5.9-6.2 in standard length).

i. Very large size (up to 38 cm in standard length); very small scales (140-175 lateral line scales) ... *A. falcirostris* (Amazon and Orinoco basins; Guyana)

ii. Small size (up to 70 mm in standard length); relatively large scales (78-82 in standard length) (a double longitudinal dark stripe) ... *A. nasutus* (Guyana and Amazon basin)

SUBFAMILY (4): BRYCONINAE

The characins of the subfamily Bryconinae, with the exception of the genus *Triportheus*, resemble rather closely the trouts, a resemblance which is reinforced by the common possession of an adipose fin and a rather similar ecology, including their behavior when they are hooked by sportsmen. Needless to say, the resemblance of body shape and ecology is secondary, and the convergence of the two groups can be easily explained by evolutionary factors such as selective pressure (with the exception, in the present case, of the adipose fin, which is still not well explained).

The Bryconinae, which are technically characterized by the premaxillary teeth in 2, 3 or more rows, the mandibular teeth usually biserial and the absence of pterygoid teeth, have followed 3 different evolutionary lines, as follows:

a. Salmonoid body (trout-like); breast and abdomen not forming a keel; pectoral fins low and moderately developed.

b. Teeth conical or very feebly tricuspid, numerous, in 2 rows on both jaws; maxilla partly covered by the antorbital bone in *Salminus*; scales, of a very simple structure (like those of *Acestrorhynchus*); lateral line usually at mid-body ... Tribe Salminini

bb. Teeth pluricuspid, in 3 rows at least on upper jaw, in 2 rows on mandible with usually a pair of conical teeth behind main mandibular series; maxillary less covered by antorbital bone; scales relatively large, of the *Alestes*-type, usually with supernumerary scales like those of the Salminini; lateral line low ... Tribe Bryconini

aa. Body more clupeoid than salmonoid owing to the keel formed by enlarged breast and abdomen; pectoral fins high and developed (teeth and scales *Brycon*-like) ... Tribe Triportheini

Tribe Salminini

South American Salmons

The genera *Catabasis* and *Salminus* are mostly confined to the northwestern and southeastern part of tropical South America, being excluded from the Guianan basins and from most of the central parts of the Amazon basin. This is an argument in favor of their antiquity, which is enhanced by the many primitive features of their skull as shown by T. Roberts who studied their anatomy.

Catabasis acuminatus, the single species of the genus *Catabasis*, is known from a specimen 165 mm long captured before 1900 in the Rio Tiete (near Sao Paulo in Brazil). This does not mean the species is extinct. It differs

Salminus hilarii (after Castelnau, 1855).

Chalceus erythrurus from Purus. Photo by Dr. Herbert R. Axelrod.

Triportheus sp. Photo by Dr. Herbert R. Axelrod.

Triportheus sp., possibly *T. angulatus.* Photo by Dr. Herbert R. Axelrod.

Triportheus angulatus. Photo by Dr. Herbert R. Axelrod.

Salminus maxillosus (after Castelnau, 1855).

from *Salminus* in the less numerous, strictly conical teeth, as well as by a more pointed snout, giving it somewhat the profile of an *Oligosarcus* of the subgenus *Acestrorhamphus*. It may be near the stem from which the Characinae and *Brycon* diverged millions of years ago, as is probably the next genus.

Salminus, called dourados by the Brazilians, is apparently one of the most primitive characoid genera (together, according to S. Weitzman and T. Roberts, with *Hoplias* and *Hepsetus*, all in different lines however). Despite its importance as a food fish in S.E. tropical America and as a game fish in N.W. South America (together with certain *Brycon*), its classification is rather poor and the last review of the 4 known species dates from 1916.

They are mostly distinguished by their scales:

S. maxillosus, from La Plata basin (and Amazon ?) has 90-105 lateral line scales (14-18 transverse scales from dorsal fin to lateral line) as contrasted with the other three species, *S. brasiliensis*, *S. hilarii* and *S. affinis*, which have only 66-79 (10-12 transverse) scales.

S. brasiliensis (*S. brevidens* of most authors) is in turn different from the last two in having more scales (77-79 versus 66-73) and more anal rays (27-30 versus 24-26). It is found mostly in Rio S. Francisco and Rio Parana.

S. affinis, the northwestern form, and *S. hilarii*, found (if all the records are reliable, which is doubtful) in the same rivers as well as in the peripheral Amazon basin and Rio Apure, differ very slightly from each other in meristic characters, the former having slightly more scales than the latter.

Some of the species attain a fairly large size (60 cm for *S. affinis*) and, as stated above, they are locally of economic importance.

Tribe Bryconini

South American Trout

The numerous *Brycon* species (including *Holobrycon*) have followed an evolutionary line parallel to that of the *Roeboides* species. Like these forms, they speciated actively in northern South America as well as in Central America, with the same result from the systematic point of view: a pretty mess in their classification.

Holobrycon pesu, from the Guianas and Amazon basin, is the single species to lack a fontanel, except in the young. (The fontanel is a gap between the "parietal" and frontal bones on the top of the head, covered by the skin, exactly like in a baby). This species reaches about 150 mm in length. It is iridescent blue with a dusky humeral spot, a characteristic black adipose fin and black margin of the caudal fin, with the rest of the fin, as well as the other fins, being rusty.

Apart from *Holobrycon* (which is perhaps to be included within *Brycon*), some other generic names have been published, their author claiming that the type species was close to *Brycon* but different from it: *Chalcinopsis*, whose type species (*striatulus*) is said to have the preventral region keeled and the pectoral fins very developed, thus forming a link between the Bryconini (the type species of *Brycon*, *B. falcatus*, has the belly rounded and the pectoral fins moderately developed) and the Triportheini (a strong preventral keel); *Megalobrycon (cephalus)*, *Bryconodon (orthotaenia)*, *Othonophanes (labiatus)*, said to lack an inner pair of conical teeth behind the main mandibular series; and *Triurobrycon (lundii)* which has the middle caudal rays developed into a pointed accessory lobe. Some of these genera may be valid, at least judging from the description of the characters of their respective type species. But as most of the descriptions of the other species are incomplete, it is impossible to know which species is to be included within each supposed genus.

Chalcinopsis is the least questionable, if the type species (not seen) has the same keel as that of *alburnus*. Other species which may or may not be included are *C. bahiensis*, *C. behreae*, *C. chagrensis* and *G. hilarii*.

Myers and Weitzman (1960) cite some species which would belong to the "poorly defined subgenus *Megalobrycon*": *cephalus*, *atrocaudatus* (under the name *ecuadoriensis*, which is synonymous), *erythropterus* (which may be identical to *cephalus*), *ferox* (not cited by Myers and Weitzman, but likely to belong here), *melanopterus*, *microlepis*, *moorei*, *whitei* and *petrosus* (together with the complex, or synonyms, *argenteus-guatemalensis-obscurus*).

Bryconodon (with a single species, *orthotaenia*) and *Othonophanes* (with only *labiatus* and *bolivarensis*), seem to be defined solely by the absence of

Triportheus albus, juvenile. Photo by Dr. Herbert R. Axelrod.

Triportheus albus, adult. Photo by Dr. Herbert R. Axelrod.

Young *Triportheus,* possibly *T. rotundatus.* Photo by Dr. Herbert R. Axelrod.

Post-larval *Triportheus,* possibly *T. rotundatus.* Photo by Dr. Herbert R. Axelrod.

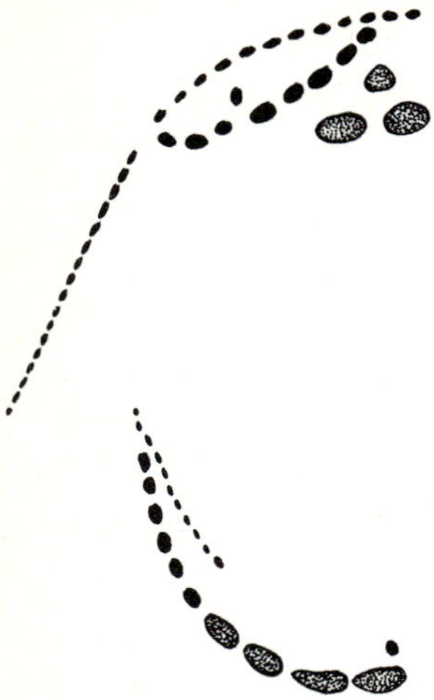

Dentition of a typical *Brycon* (*B. sibenthalae*) showing conical tooth behind main mandibulary series (plastiline cast, right side only). Drawing by the author.

a pair of conical teeth behind the main mandibular series (a character which is shared by certain *Megalobrycon* at least, including the type, *cephalus*). They may be synonymous with the last named genus (or subgenus), but the impression persists that some of the species have to be redistributed.

Finally the pointed median lobe of *Triurobrycon lundii* (together with *T. matrinchao* and *T. travassosi* ?) is scarcely a generic character.

Facing the above difficulties, the reader will not be surprised then if the following key goes down not to the 50 or so *Brycon* species (only one generic name is retained here), but refers only to "groups of species," based on number of scales and anal rays (and not on the characters already discussed, which would lead, if well known, to a more satisfying assemblage).

Key to artificial groups of *Brycon* spp:

a. A broad, V-shaped or crescent-shaped black band at base of caudal fin . . . *B. falcatus*-group (*Brycon* nominate) with *B. falcatus* in the Guianas and north of Amazon, *B. brevicauda* in the lower Amazon basin and southern tributaries, and 2 poorly-known species; *B. stuebeli* from upper Amazon and *B. matrinchao* from Rio Parnaiba

aa. No such caudal pattern.

b. Scales numerous, 65-85 in lateral line.

c. About 84 lateral line scales . . . *B. microlepis*, upper Paraguay basin

cc. 65-83 lateral line scales.

d. Northwestern South American and Central American species with a long anal fin (30-37 anal rays) . . . *B. striatulus*-group, with 3 Panamanian species, *B. chagrensis, B. striatulus* and *B. behrae*, and 3 Colombian species, *B. meeki, B. rubricauda* and *B. tovari*

dd. Amazonian and southern forms (except *B. whitei* which is from the upper Orinoco basin) with only 26-28 anal rays . . . *B. hilarii*-group, with *B. hilarii* in the Rio Sao Francisco, *B. cephalus* and *B. pellegrini* in the Amazon, and *B. whitei*

bb. Scales less numerous, 43-65 in lateral line.

e. Usually more than 48 lateral line scales (48 or 49 in *B. medemi*)

f. Usually more than 32 anal rays . . . *B. dentex* (Ecuador and Central America) and *B. guatemalensis* (Central America)

ff. Usually less than 32 anal rays.

g. 27-31 (rarely 32) anal rays . . . *B. orbignyanus*-group, the largest group, with 2 spp. from Ecuador, *B. atrocaudatus* and *B. acutus*, 3 from Colombia, *B. labiatus, B. moorei* and *B. medemi*, 2 from Central America, *B. petrosus* and *B. obscurus*, and the rest mostly from the south and S.E.: *B. devillei* from Bahia, *B. ferox* from S.E. Brazil, *B. lundii* from Rio das Velhas, *B. orbignyanus* and *B. orthotaenia* from La Plata basin

gg. 21-26 anal rays . . . *B. carpophagus*-group, with 2 spp. from Colombia, *B. henni* and *B. bolivarensis*, 2 from the Orinoco basin, *B. longiceps* and *B. bicolor*, 3 Guianan-Amazonian spp.: *B. carpophagus, B. melanopterus* (perhaps the most common Amazonian *Brycon*), and *B. capito*, as well as 2 southern species: *B. reinhardti* (Rio das Velhas) and *B. nattereri* (Parana)

Prionobrama filigera, adult pair. Photo by Harald Schultz.

Unidentified *Prionobrama,* possibly *P. paraguayensis.* Photo by Harald Schultz.

Paragoniates alburnus. Photo by Dr. A. Fernandez-Yepez.

Xenagoniates bondi, the glass characin. Photo by A. Norman.

ee. 48 or fewer lateral line scales . . . *B. opalinus*-group, with *B. coquenani* in Venezuela and *B. coxeyi* in Ecuador, *B. argenteus* in Panama, *B. stolzmani* in upper Amazon, and 3 more southern species: *B. bahiensis* (Bahia), *B. opalinus* (S.E. Brazil) and *B. travassosi* (Mato Grosso)

GENUS *CHALCEUS*

Chalceus is difficult to place. In several aspects it is very near *Brycon*, and maybe it does come from the same phylogenetic line. In other ways, such as the scales, the short anal fin, and the general aspect of the fish, it strangely resembles certain *Alestes*, which may indicate that a similar type existed before the drift of the continents.

Chalceus has the scales of the middle of the side very large (about 20-22 in longitudinal line); the lateral line is low and the perforated scales are much more numerous and smaller than those of the series above, owing to interposition of several additional ones (about 33-37 lateral line scales); the lower caudal lobe is longer than the upper one; the supraoccipital crest is rudimentary as in the erythrinids and alestids; the teeth are exactly like those of *Brycon*, with 3 rows on the upper jaw; there is a fontanel only in specimens less than 100-120 mm long, as in *Holobrycon*. The anal fin is short as in erythrinids, anostomids, hemiodids, etc.

Two species are known, differing in the width of the body, the length of the maxilla, the number of maxillary and mandibular teeth and chiefly in the coloration: *Chalceus macrolepidotus*, from the Guianas, has the fins pink to bright red (except the pectoral fins) and the body silvery; *Chalceus erythrurus* (type of the dubious genus *Plethodectes*) from the Amazon basin has the fins lemon colored (including the pectoral fins) with some red only on the middle dorsal rays, each caudal lobe and the adipose fin. Both reach about 200-250 mm in length and, like many *Brycon* species, have the most delicious flesh, better than the best trout.

Tribe Triportheini

In this tribe are included the *Brycon*-like species which have adopted a life near the surface. Thanks to the development of their breast into a sort of keel, like that of the hatchetfishes, associated with large, wing-like pectorals, they are able to jump several meters a few feet above the water surface, apparently to escape predators, without real flight.

The genus *Triportheus*, which was once called *Chalcinus*, with 9 species, and the genus *Moojenichthys*, with only one species (monotypic), constitute the tribe. There is a possibility that *Chalcinopsis*, which is somewhat intermediate between *Brycon* and *Triportheus*, could be included within the Triportheini. Finally a fourth genus, *Coscinoxyron*, has been proposed for *T. culter*. Its generic status is debatable, as it is no more than a *Triportheus* with an increased number of gill-rakers in a group where 40 or more rakers on the lower limb is not unusual.

The *Triportheus* species can be identified as follows:

a. Usually less than 40 lateral line scales; body depth less than 3.5 in standard length.
 b. 6 transverse scales from dorsal fin origin to lateral line (body depth about 2.66 in standard length; 28-33 anal rays).
  c. 32 or 33 lateral line scales ... *T. paranensis* (Parana basin)
  cc. 33-38 lateral line scales, exceptionally 39 or 40; series of spots on scale centers usually forming longitudinal lines ... *T. angulatus* (Amazon basin, very common)
 bb. 5 transverse scales from dorsal fin origin to lateral line.
  d. 23 or 24 anal rays (32 lateral line scales; body depth about 3 in standard length) ... *T. pictus* (Amazon basin)
  dd. 26-32 anal rays.
   e. Body depth 3.0-3.4 in standard length; anal fin just under, or somewhat behind, last dorsal ray.
    f. First anal ray behind level of last dorsal ray; usually 30-33 lateral line scales ... *T. albus* (Amazon basin, type of the genus)
    ff. First anal ray just under last dorsal ray; usually 33 or more lateral line scales ... *T. rotundatus*, synonym *Chalcinus brachypomus* (Guianas)
   ee. Body depth about 2.5 in standard length; beginning of anal fin said to be partly under dorsal fin (usually 31-34 lateral line scales) ... *T. guentheri* (Rio Sao Francisco)

aa. 40-48 lateral line scales; body depth usually more than 3.5 in standard length (6 or 7 transverse scales from dorsal origin to lateral line).
  g. 41 lateral line scales; beginning of anal fin in front of dorsal origin (35-41 anal rays) ... *T. magdalenae* (Colombia)
  gg. 43-48 lateral line scales; anal fin behind last ray or under middle of dorsal fin.
   h. 20-25 gill-rakers (anal fin behind dorsal fin, with 27-32 rays) ... *T. elongatus* (upper Amazon, very common)
   hh. 90-100 gill-rakers (anal fin under dorsal fin, with 30-36 rays) ... *T. culter*, type of *Coscinoxyron* (upper Amazon)

Most of the species, including *T. elongatus*, are plain silvery or golden, although certain species have a peduncular spot extending onto the middle

Aphyocharax (above) and *Prionobrama* (below) are usually associated in nature. Photo by Dr. Herbert R. Axelrod.

Aphyocharax anisitsi, the bloodfin, is often called *A. rubripinnis* in the hobby. Photo by Dr. Herbert R. Axelrod.

Aphyocharax anisitsi. Photo by Dr. Herbert R. Axelrod.

caudal rays, which are sometimes prolonged. *T. angulatus* may have longitudinal striae or a more or less dark back. *T. rotundatus* has characteristic black pectoral fins, at least when young. The three last-named species are by far the most common. They reach 200 mm or only slightly more in length.

Moojenichthys myersi, from the state of Bahia in Brazil, is a rather recently discovered species which is very similar externally to *Triportheus*, but with only two premaxillary and one mandibular rows of teeth, thus approaching the Tetragonopterinae. Perhaps this 125 mm long, plain species is a link with the said subfamily.

SUBFAMILY (5): CLUPEACHARACINAE

In many respects *Clupeacharax anchoveoides*, the single representative of the subfamily Clupeacharacinae, resembles an *Agoniates*. It has the same elongate, clupeoid body shape (the depth about 4.5 in the standard length), the same rather long anal fin, inserted before the posteriorly set dorsal fin (which is about at the same level in *Agoniates*), and the same compressed body with a ventral keel and developed pectorals.

The tooth structures, however, are different. The rather short maxillary is paddle-like (somewhat like that of the Alestidae), has only one tooth, or even none, and the multicuspid teeth are like those of the Tetragonopterinae.

Two specimens are known: one from Bolivia and one from the upper Amazon. They are silvery and measure only 50-60 mm in standard length.

Head of *Paragoniates alburnus*. Photo by Dr. A. Fernandez-Yepez.

SUBFAMILIES (6) and (7): PARAGONIATINAE and APHYOCHARACINAE

The Paragoniatinae are a small group of 6 monotypic genera (except *Prionobrama*, which may be composed of 2 species) which are characterized by a very compressed, usually elongate body with a long anal fin, the most extreme species resembling the African siluroids *Eutropius*, *Eutropiellus* or even the glass catfish, *Kryptopterus* (less the barbels). They usually have a single series of teeth on the upper jaw. At least one species has palatine teeth. The more generalized *Prionobrama* and *Rachoviscus* grade toward *Aphyocharax*, and the boundaries between the two groups (classically treated as subfamilies) are uncertain.

The species may be characterized as follows:

a. Adipose fin present.

b. Body relatively short; lateral line incomplete; less than 50 anal rays.

c. 29-37 anal rays (10-16 maxillary teeth; 35-41 scales; two genera which could well be included within the Aphyocharacinae).

d. Premaxillary series of teeth regular; anal fin falciform, with the first rays filamentous in males; caudal peduncle moderately deep . . . *Prionobrama filigera* (rather common in the Amazon basin; the form from the Paraguay and Uruguay Rivers is perhaps different; it is known under the name *P. paraguayensis*)

dd. Premaxillary teeth irregularly set, could pass for a tetragonopterine-like, two-seried dentition; anal fin rounded; caudal peduncle much deeper than long . . . *Rachoviscus crassiceps* (Rio de Janeiro)

cc. 43-48 anal rays (about 20 maxillary teeth; 39-46 scales) . . . *Paragoniates alburnus* (middle and upper Amazon basin; Venezuela)

bb. Body elongate; lateral line complete; 63-68 anal rays.

e. No teeth on palate (68 anal rays; 48 lateral line scales) . . . *Leptagoniates steindachneri* (a single specimen known, from Peruvian Amazon; not collected for almost a century)

ee. A few teeth on the ectopterygoid bone (63-66 anal rays; 50-51 lateral line scales; body translucent in life) . . . *Xenagoniates bondi* (Venezuela and Colombia)

aa. Adipose fin absent (body relatively short; lateral line incomplete; 52-56 anal rays; 38-44 scales) . . . *Phenagoniates macrolepis*, synonym *Ph. wilsoni* (Colombia and Venezuela)

A comparison of two similar looking characoids, *Hemibrycon* or *Bryconamericus* (above) and *Aphyocharax* cf *dentatus* (below). Photo by Dr. Herbert R. Axelrod.

Aphyocharax dentatus. Photo by Harald Schultz.

Aphyocharax erythrurus. Photo by Dr. Herbert R. Axelrod.

The most common species, or, more exactly, the least rare, is the Amazonian glass bloodfin, *Prionobrama filigera*, which has the body bluish and the caudal fin vermilion. The distinctive feature of this rather small fish (about 70 mm), which is sometimes seen in the aquarium, is its white first anal ray, prolonged into a long filament in the males.

The attractive *Rachoviscus crassiceps*, a little fish (30-35 mm) from the vicinity of Rio de Janeiro, seems to have disappeared since World War II, not only in aquaria, but in its habitat as well. It was described as colorful, having a blue-green body with brilliant small spots on its flanks and a broad reddish longitudinal band, together with yellowish fins, the dorsal and anal fins barred with black.*

The genus *Aphyocharax* (bloodfin tetras) is the only genus of the next subfamily (#7). It resembles *Prionobrama* and likewise has an incomplete lateral line, but has a shorter anal fin (17-27 rays) and the dorsal fin near the middle of the body. Its teeth, like those of the preceding group, are uniserial on the premaxilla and at most tricuspid. By comparison with the numerous other small fishes called tetras by aquarists, it is recognized at once by the development of its first postorbital bone (4th circumorbital), which covers, together with the large suborbital bone (or 3rd circumorbital), the whole cheek at the expense of the 2nd postorbital bone, which is much reduced. In the other tetras, the true Tetragonopterinae as well as certain Cheirodontinae in the broad sense (which were for a time classified near *Aphyocharax*), the postorbitals are often reduced and are always smaller than the large suborbital.

The following identification key to the genus *Aphyocharax* is compiled in part from Eigenmann's revision, which is quite dated (1915):

a. Maxilla not completely toothed (true *Aphyocharax*).
- b. Males with sexual hooks on all anal rays; black dorsal band and black marginal anal band present (mouth very small, maxilla very short, with 1-3 teeth) . . . *A. rathbuni* (Paraguay basin)

bb. Males with sexual hooks on anal lobe only; dorsal and anal fins usually plain (except in the poorly-known *A. avary* ?).
- c. Mouth small; only 2-4 maxillary teeth.
  - d. Body depth 3.0-3.75 in standard length; 30-35 scales; 18-23 anal rays . . . *A. anisitsi*, synonym *A. rubripinnis* (Argentina to Paraguay)
  - dd. Body depth 4.25 in standard length; 38 scales; 17 anal rays; color pattern said to resemble that of *A. rathbuni*, *A. agassizi* etc. . . . *A. avary* (known from a single specimen from Rio Madeira in Brazil)
- cc. Mouth moderate or large; more than 4 maxillary teeth.
  - e. Tip of maxilla not reaching large suborbital.

* The writer is informed that it has been rediscovered recently near Rio.

f. 5-8 maxillary teeth; 9 mandibular teeth; middle caudal rays black ... *A. pusillus* (upper Amazon basin, Rio Madeira)

ff. 9-20 maxillary teeth; 13-20 mandibular teeth; caudal red in life ... *A. alburnus* (Amazon and Orinoco basin) and *A. erythrurus* (Guianas), very difficult to tell apart, perhaps not distinct

ee. Tip of maxilla reaching the large suborbital (9-20 maxillary teeth; caudal plain) ... *A. dentatus* (Paraguay basin)

aa. Maxillary completely toothed; a black band from tip of last anal ray to anal lobe, then across the lobe to ventral fins (subgenus *Holoprion* ?, an available name)

g. Dorsal fin with black spot, caudal fin without caudal spot (27 anal rays) ... *A. agassizi* (Amazon basin)

gg. Dorsal fin plain; a caudal spot (22-23 anal rays).

h. Caudal spot extended onto middle caudal rays to their tips; 30 longitudinal scales ... *A. nattereri* (Amazon basin; it may be that *A. agassizi* and *A. nattereri* are the two sexes of a single species, in which case the name *agassizi* has priority)

hh. Caudal spot not extended up to end of middle caudal rays; 34 longitudinal scales ... *A. paraguayensis* (Paraguay basin)

The bloodfins are usually small (the largest, *A. dentatus*, is 70 mm long) and inhabit mostly the Upper Amazon, the southern part of the Middle Amazon basin and the basin of Rio de La Plata; the only exception is Guyana, whose fauna is very similar to that of the Upper Amazon.

The commonest species are perhaps *A. alburnus* in Upper Amazon (*erythrurus* may be the Guianan counterpart of it, possibly at the subspecific level) and the "false bloodfin", *A. dentatus* in the Rio Paraguay. The true bloodfin, *A. anisitsi*, is a well-known aquarium species; it is better known under the name *A. rubripinnis* which is likely to be a pure synonym (at least in the author's opinion). Unlike the two other species, it has not only the caudal fin bright red, but the other fins also (except pectoral and adipose fins).

Aphyocharax maxillaris and *Aphyocharax axelrodi* do not belong to the genus and will be treated later on.

SUBFAMILY (8): GLANDULOCAUDINAE

This important group is probably not natural, but is an artificial assemblage of, for practical purposes, a series of forms coming from different stems. The *Mimagoniates-Coelurichthys-Glandulocauda* complex, together with *Phenacobrycon*, *Argopleura*, *Planaltina*, *Acrobrycon*, *Landonia*

Unidentified *Gephyrocharax* from Camatagua, Venezuela. Photo by Dr. Herbert R. Axelrod.

Gephyrocharax caucanus from Colombia. Photo by Dr. Herbert R. Axelrod.

Gephyrocharax sp., female. Photo by Dr. Herbert R. Axelrod.

Gephyrocharax valenciae. Photo by Dr. Fernandez-Yepez.

(?) and *Pterobrycon*, issued from some *Hemibrycon* or *Bryconamericus*-like ancestor, whereas other genera such as *Pseudocorynopoma*, *Corynopoma* and *Hysteronotus* seem to be of a different origin. *Gephyrocharax* looks like a link between the two groups. Finally, *Tyttocharax* and allied genera (Xenurobryconini), here classified within (or near) the Glandulocaudinae, may be of a third origin which, we must confess, is presently unknown (although certain pigmy Aphyoditeina look rather similar to certain Xenurobryconini).

The Glandulocaudinae have in common certain male organs used to attract the females and to permit the internal fertilization of its eggs (at least in some species). All forms have a so-called caudal gland which is supposed to secrete an attractive odor (like the musk-gland of certain mammals), and certain of these forms have secondarily developed a lure, which they present to their female partner as if it was a prey or possibly another signal, thus inducing the female to come close-by (other examples of such a curious behavior are known in certain arthropods and birds, as well as in the most advanced animal, man, where, needless to say, gifts are used as lures).

At the same time, a few species were engaged in a way of life near the surface, thus having acquired a more or less trenchant breast, superior mouth, large pectoral fins, and a posterior dorsal fin in common with several other characoid fishes of different families or subfamilies.

The two major lines of Glandulocaudinae may be distinguished as follows:

a. Moderate size; mouth terminal or superior, with heavy teeth, usually tri-to pentacuspid, and biserial on the upper jaw . . . Tribe Glandulocaudini
aa. Very small size; mouth terminal or inferior, with thin, acute, conical teeth, often in bands and outside the mouth . . . Tribe Xenurobryconini

Caudal gland of *Glandulocauda terofali* and *Coelurichthys microlepis* (both after Gery, 1964).

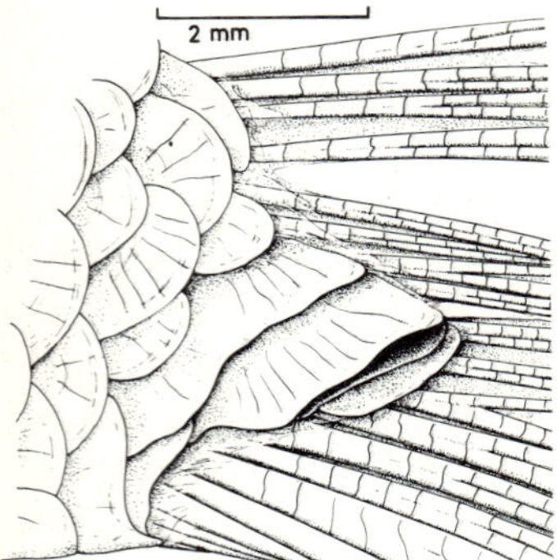

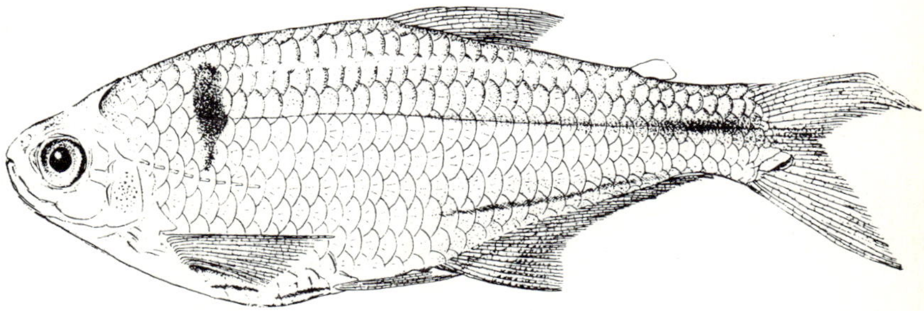

Glandulocauda melanogenys, G. melanopleura (both after Eigenmann, 1927), and type of *G. terofali* (after Gery, 1964).

Tribe Glandulocaudini

Croaking Tetras and others

The geographical distribution of the croaking tetras (so called because certain species, at least, are capable of emitting tiny noises or sounds) is extremely peculiar: *Glandulocauda* and allied forms, *Diapoma*, *Pseudocorynopoma* and allied forms are all confined to the most southeastern part of tropical South America (S.E. Brazil, Argentina, Uruguay and Paraguay), whereas the other genera are restricted to the most northwestern part, with the exception of *Acrobrycon*, of one *Hysteronotus* (Peru), and of *Planaltina* (state of Goiaz, near the modern town of Brasilia in Brazil).

The following key, based mostly on the sexual characters of the male, will aid in identifying the genera (in part after Géry, 1966):

Pseudocorynopoma doriae, male (above) and female (below). Photo by Dr. Herbert R. Axelrod.

Corynopoma riisei, the swordtail characin. Photo by R. Zukal.

Male *Corynopoma riisei.* Photo by Hilmar Hansen.

Coelurichthys tenuis is often improperly called *Mimagoniates barberi*, which is another species. Photo by Harald Schultz.

a. Opercle "normal", not notched or prolonged (adipose fin present).
  b. 6-10 maxillary teeth (species from the upper reaches of affluents of the Maranon in Ecuador, Peru and Bolivia).
    c. Breast rounded; characters of *Hemibrycon* except that males have a caudal "pouch"; about 11/58/7½ scales . . . *Acrobrycon* (2 species from Peruvian and Bolivian Andes)
    cc. Breast more or less keeled; number of inner premaxillary teeth variable but usually more than 4 which is typical for *Hemibrycon*; about 7/39-44/6 scales . . . *Hysteronotus hesperus* (Ecuador) and *Hysteronotus myersi* (Peru), two species apparently different from the type-species *megalostomus*, possibly at the generic level
  bb. Less than 6 maxillary teeth.
    d. Teeth similar to those of *Hemibrycon, Bryconamericus* or other generalized tetragonopterins, never flat crenulate incisors forming a continuous cutting edge.
      e. Lateral line complete.
        f. Breast and preventral region rounded or, if keeled, still narrowly rounded and not trenchant.
          g. Lower caudal rays separated from rest of fin in males (Bolivia to Central America, mostly Andean).
            h. Characters of *Bryconamericus*, breast rounded, dorsal fin at mid-body, mouth terminal, a caudal gland present . . . *Argopleura* (4 species from central and western Colombia)
            hh. Breast keeled, dorsal fin behind midbody, mouth superior; a modified triangular scale on base of lower caudal lobe, not twisted as in other genera . . . *Gephyrocharax* (6 or 7 species from the Bolivian Andes to Costa Rica)
          gg. Lower caudal rays not separated from rest of the fin.
            i. Breast rounded; characters of *Argopleura* except that males have no "spur" on inferior part of caudal fin; dorsal fin at midbody . . . *Phenacobrycon henni* (Pacific slope of Ecuador)
            ii. Breast keeled, ventral area narrowly rounded; dorsal fin more or less behind midbody (Rio Sao Francisco basin).
              j. Four inner premaxillary teeth, these teeth pentacuspid; dorsal fin just in front of anal fin, which is of moderate length (25 rays) . . . *Planaltina myersi* (Rio Preto)
              jj. Five or six inner premaxillary teeth, tricupid; dorsal fin behind anal origin, the fin long (35-36 anal rays) . . . *Hysteronotus megalostomus* (Rio das Velhas)
        ff. Breast and abdomen strongly keeled, trenchant (dorsal fin behind

midbody, the first rays filamentous in the males, as well as those of the anal lobe) . . . *Pseudocorynopoma* (2 species from La Plata basin to State of Sao Paulo)

ee. Lateral line incomplete.

k. No paddle-like expanded scale in humeral region of males; lower caudal rays not separated from rest of fin (southeastern tropical America) . . . *Glandulocauda-Mimagoniates-Coelurichthys* complex

kk. A paddle-like expanded scale in humeral region, which is probably used males by to attract females and permit internal fertilization through a spermatophore; a caudal "spur" protruding out of the lower rays, as in *Gephyrocharax* with which it has much affinities (northwestern tropical America) . . . *Pterobrycon* (2 species from Colombia (synonym *Microbrycon minutus*) and Costa Rica)

dd. Teeth incisiform crenulate, forming a continuous cutting edge somewhat like the tetragonopterine *Deuterodon* . . . *Landonia latidens* (Pacific slope of Ecuador, sympatric with *Phenacobrycon henni*)

ddd. Teeth conical (lateral line incomplete; no adipose fin; species of uncertain lineage, possibly close to the Aphyoditeina) . . . *Iotabrycon praecox* (Ecuador, sympatric with *Phenacobrycon henni*)

aa. Opercle of males forming a point or prolonged into a pen-like appendage.

l. Opercle lacking almost all of its upper half, the lower half thus forming a point; adipose fin present; lateral line interrupted in middle of body, continuing again on caudal peduncle; lower caudal rays not separated from rest of fin . . . *Diapoma speculiferum* (Rio Grande do Sul in Brazil, and Uruguay)

ll. Opercle with a pen-like appendage in males, provided with a paddle-like tip which, according to certain behavior specialists, is used as a lure; adipose fin absent; lateral line complete; lower caudal rays forming a "spur" in the males as in *Argopleura, Gephyrocharax* and *Pterobrycon* . . . *Corynopoma riisei* (Trinidad and upper Rio Meta in Colombia; the inland form may be different from that from Trinidad)

As seen above, certain genera are monotypic *(Phenacobrycon, Planaltina, Landonia, Diapoma, Iotabrycon)* or "almost" monotypic, the second species being dubious *(Acrobrycon, Pseudocorynopoma)*. These are rare, rather

Coelurichthys microlepis was formerly classified in the genus *Mimagoniates.* Photo by Dr. Herbert R. Axelrod.

Two croaking tetras, *Glandulocauda inequalis* (above) and *Coelurichthys microlepis* (below). Photo by Dr. Herbert R. Axelrod.

Tyttocharax madeirae. Photo by Harald Schultz.

Tyttocharax madeirae from Belem. Photo by Dr. Herbert R. Axelrod.

poorly known and what is said in the key seems to cover the scope of the present book. The same holds true for the largest genera, *Argopleura*, *Gephyrocharax* and *Hysteronotus*, whose needed revision has no place here. A few words will be said about the most common glandulocaudins, at least the better known among aquarists:

The swordtail tetra, *Corynopoma riisei* (the name *Stevardia albipinnis* has about the same "right" to be applied to this fish, but it seems that the "first revisor" rejected it, and this will last forever!), is a little fish (60 mm) whose interest lies not in the coloration (it is plain silvery) but in the finnage of the males which have very extended unpaired fins, as well as in its reproductive behavior (see for example W. Wickler, *Animal and Vegetal Mimicry*, 1968). The swordtail tetra is now considered as hardy in the aquarium, but the experienced aquarist would remember the times when it was so susceptible to ich (then incurable) that it was the first to suffer from it and was responsible for the loss of all other fishes when this disease occurred.

The croaking tetras have been rather recently reviewed (Géry, 1966). This group includes *Glandulocauda melanogenys*, *G. melanopleura* and *G. terofali* (respectively from near Sao Paulo and Rio Iguassu in Brazil, and Rio Lujan in Argentina), a fourth form of uncertain generic status, *G. inequalis* (from Rio Grande do Sul and Uruguay, like *Diapoma*), *Mimagoniates barberi* from Paraguay, and two *Coelurichthys* (a recently restored name), *C. tenuis* and *C. microlepis*, both from S.E. Brazil. Note that the old timer "*Mimagoniates barberi*" of aquarists, which reminds one of a danio, is now called *Coelurichthys tenuis* (the true *barberi* has never been imported, apparently). *Glandulocauda* (?) *inequalis* and *Coelurichthys microlepis* (the latter sympatric with *tenuis* near Santos in Brazil) are very difficult to tell apart if the caudal gland cannot be studied (its structure is much simpler in *inequalis*). However, it can be seen that *G. inequalis* has a submarginal anal band whereas that of *C. microlepis* (one of the "blue tetras") is very near the border of the fin, if not strictly marginal. The three species are highly desirable aquarium fishes. The other croaking tetras are quite rare and some even are not well known anatomically.

Tribe Xenurobryconini

Bristly-Mouthed Tetras

Like the preceding forms, the fishes allied to *Tyttocharax* (by far the commonest genus) have a caudal gland, and their shape shows some resemblance to the croaking tetras. However the minute, conical teeth are extremely different. In a review dating from 1963, the present author enumerated the known forms, as reviewed below.

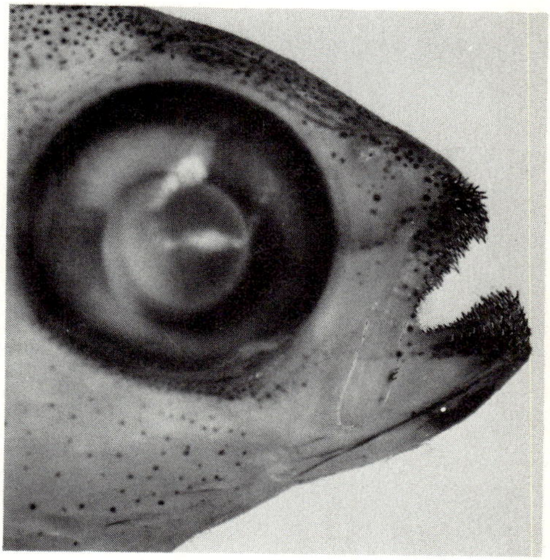

Head and caudal fin of *Tyttocharax madeirae,* head showing bristle-like teeth, caudal showing the characteristic "gland." Photos under the direction of the late Dr. O. Schindler.

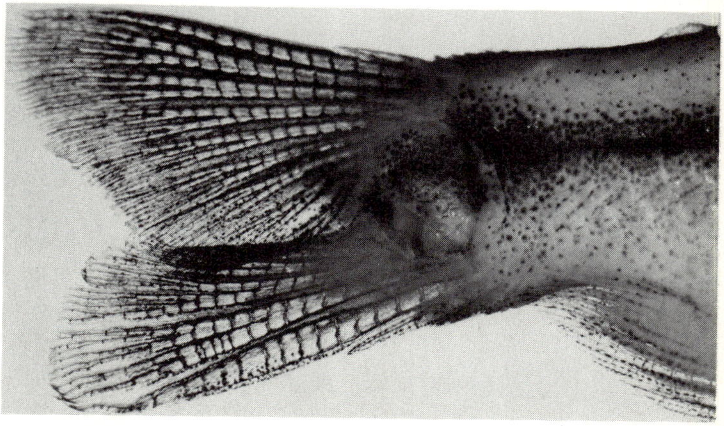

Tyttocharax madeirae, synonym *Microbrycon cochui*, from the upper and middle Amazon basin, is relatively common in certain localities. It is a very small (18 mm in standard length) blue fish with elegant black tipped finnage. Its muzzle and chin bristle with tiny conical teeth, a highly unusual situation, at least among characoids. Nothing is known of their use, if any. The rare *T. boehlkei*, also from the Upper Amazon, is close to it.

Tyttocharax rhinodus and *T. atopodus* are Andean species which inhabit more restricted basins (mostly Upper Rio Huallaga). The latter lacks an adipose fin and has a very short snout, in contrast with the former. Both have many fewer external teeth. *Xenurobrycon macropus*, from the Rio Paraguay, has no adipose fin either and probably no external teeth; it is not well known. Finally, *Tyttocharax odontocheilus* is even less known and may be synonymous with *T. madeirae*.

Stethaprion erythrops. Photo by Dr. Herbert R. Axelrod.

Poptella orbicularis. Photo by Dr. Herbert R. Axelrod.

Stethaprion erythrops. Photo by Dr. Herbert R. Axelroc

SUBFAMILY (9): STETHAPRIONINAE

Silver Dollars Tetras

We are arriving here at a very different group of fishes which have little similarity to preceding ones. The Stethaprioninae represent the disciform adaptation of some tetra (for example a *Tetragonopterus* or a *Gymnocorymbus*) which has acquired a predorsal spine, in a parallel evolution with that of the Serrasalmidae. All the species have a very compressed, almost round body, with the body depth contained 1.3 to 2 times in the standard length; there is a keeled preventral area, narrowly rounded or even trenchant; the caudal fin is scaled, the lateral line complete, and the mouth structures are exactly the same as those of a generalized tetra, with two series of multicuspid teeth on the upper jaw.

The form of the predorsal spine is used to separate the genera:

a. Predorsal spine blunt, saddle-shaped, concealed under the skin; no preanal spine . . . *Poptella*

aa. Predorsal spine pointed, triangular or spear-like, probably erectile; one or two triangular preanal spines.

b. Predorsal spine short (except in *franciscoensis* where it is intermediate between *Brachychalcinus* and *Stethaprion*); less than 40 lateral line scales . . . *Brachychalcinus*

bb. Predorsal spine long, spear-shaped; more than 60 lateral line scales . . . *Stethaprion*

Poptella orbicularis. Photo by G. Timmerman.

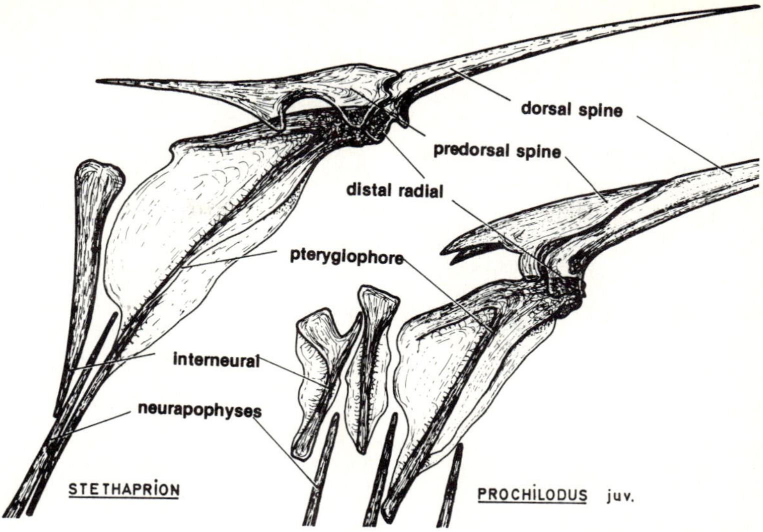

Dorsal spine of a *Stethaprion* (left) as compared with that of a *Prochilodus* (right) (after Gery, 1972).

GENUS *POPTELLA*
formerly called *Ephippicharax*

The single species, *Poptella orbicularis*, is still very close to a tetra, differing from *Gymnocorymbus thayeri* (which has also been called *Moenkhausia bondi* and *profunda*) only by its deeper body and, naturally, by the presence of the characteristic saddle-shaped predorsal "spine." The silver dollar tetra is rather popular among aquarists, thanks to its body shape more than to its coloration: it is just plain silvery, sometimes with black dorsal and anal filaments. The largest individuals seen were up to 115 mm in length.

This is a variable species. A study of meristic characters of different samples permits a separation of populations which probably reach subspecific level in Colombia, Guianas (including the Nickerie River, locality of *orbicularis longipinnis*) and perhaps Paraguay.

GENUS *BRACHYCHALCINUS*

The following key will give sufficient data, at least within the scope of the present book:

a. Body shape of a *Poptella*, body depth 1.7-2.3 in standard length; preanal spine (corresponding to first unbranched anal ray) concealed under scales and close to second anal ray; anterior branch of predorsal spine relatively long . . . *B. franciscoensis* (Rio Sao Francisco)

aa. Body almost round like a silver dollar, body depth 1.3-1.7 in standard length; preanal spine free and separated from second anal ray; predorsal spine relatively short.

Unidentified species of *Iguanodectes.* Photo by Dr. Herbert R. Axelrod.

Iguanodectes sp. Photo by Dr. Herbert R. Axelrod.

Iguanodectes spilurus from the vicinity of Humaita, Brazil. Photo by Dr. Herbert R. Axelrod.

Iguanodectes spilurus. Photo by Dr. Herbert R. Axelrod.

Iguanodectes geisleri. Photo by Dr. J. Gery.

b. Usually Iii, 9 dorsal rays (Iiii, 27-31 anal rays).

c. 18-22 transverse scales between dorsal and anal fins (body depth 1.4-1.7 in standard length) . . . *B. nummus* (Ecuador)

cc. 23-26 transverse scales between dorsal and anal fins (body depth 1.3-1.5 in standard length) . . . *B. guianensis* (Guianas)

bb. Usually Iii, 10 dorsal rays (predorsal spine usually very short, triangular; 24-26 transverse scales between dorsal and anal fins; body depth 1.3-1.7 in standard length) . . . (*B. copei*)

d. Iiii, 32-37 anal rays; depth of caudal peduncle 8.4-9.6 in standard length . . . *B. copei copei* (upper Amazon and Bolivia)

dd. Iiii, 29-32; depth of caudal peduncle more than 9.6 in standard length . . . *B. copei retrospina* (Rio Paraguay basin)

Brachychalcinus franciscoensis, paratype photographed at the California Academy of Sciences, San Francisco by the author.

GENUS *STETHAPRION*

The silver dollar tetras of the genus *Stethaprion* are like exaggerations of *Brachychalcinus*, being even deeper bodied, with a longer predorsal spine and a sharper preventral keel. They are very poorly known, being very rare, at least in collections (except for *S. erythrops*).

The type-species, *S. erythrops* (middle and upper Amazon basin), is distinguished from the two other species by its smooth scales (versus crenate, as in *Ctenobrycon*, a tetra) and its narrowly rounded ventral keel covered with alternating scales bent over the other side. *S. crenatum* (Rio Madeira basin) and *S. innesi* (lower Amazon) are said to have scale-like processes on either side of the trenchant ventral keel, forming a serrature somewhat similar to that of certain curimatins, with median scales every second "spine." They are said to differ from each other in having 45-46 anal rays and 66-67 lateral line scales in the former species, and 40 anal rays and 62 lateral line scales in the latter.

SUBFAMILY (10): TETRAGONOPTERINAE

Tetras

The American tetras (subfamily Tetragonopterinae) are without much doubt the most successful characoids, having invaded practically all Neotropical (and subtropical) biotopes. They differ from the other Characidae, as indicated in the key to the subfamilies, mostly in having two rows of teeth on the upper jaw (very rarely 3 or 1), the teeth similar in the young and in the adult, no predorsal spine, and exceptionally a ventral keel (in one case the keel is associated with a discoid body—see genus *Stichonodon*). Several groups, differing mostly in the body shape, may be recognized within the subfamily as follows:

Key to the tetragonopterine tribes and lower groups:

a. Body completely scaled or nearly so.

b. Elongate species somewhat resembling the Paragoniatinae; dorsal fin usually behind midbody and behind the long anal fin; outer premaxillary row of teeth sometimes weak or even absent ... Tribe Iguanodectini

bb. Body usually not very elongate, body depth rarely more than 4 in standard length except in *Bryconops*; dorsal fin usually at midbody or not much in front or behind; usually 2 premaxillary rows of teeth, occasionally 3.

c. Circumorbital bones, lateral line and eyes never completely absent, at least in young; anal fin short in only a few species (*Microgenys, Creagrutus melanzonus, Creagrudite, Hemigrammus analis, Bryconacidnus*, etc.) ... Tribe Tetragonopterini

d. Inner premaxillary row composed of only 4 teeth on each side; large suborbital usually entire; body never discoid.

e. Two premaxillary rows of teeth (the outer row quite irregular in certain *Bryconamericus*) ... *(A) Hemibrycon* and allied genera

ee. Outer premaxillary row with some backwardly displaced teeth clearly forming 3 premaxillary rows (teeth heavier than in the preceding section; anal fin usually rather short) ... *(B) Creagrutus* and allied genera

dd. Inner premaxillary row composed of 5 or more teeth on each side; large suborbital usually leaving a naked area on cheek; body discoid in a few species (2 premaxillary rows of teeth except in *Bryconops*, for example, where one or two displaced teeth may form the beginning of a third row).

f. No ventral keel ... *(C) Tetragonopterus* and allied genera

ff. A ventral keel (body strongly discoid as in *Stethaprion*) ... *(D) Stichonodon*

cc. Circumorbital bones, lateral line and eyes absent (the blindness apparently irreversible) ... Tribe Stygichthyini (a single cave species from Brazil)

aa. Body entirely naked ... Tribe Gymnocharacini (a single species from temperate region south of the Rio de la Plata)

Hemibrycon of the *polyodon-jelskii* group. Photo by Dr. Herbert R. Axelrod.

Carlastyanax aurocaudatus. Photo by Aaron Norman.

Carlastyanax aurocaudatus.
Photo courtesy Wardley Products Co.

Tribe Iguanodectini

With their elongate body and pluricuspid, flat incisiform teeth, the peculiar smelt-like Iguanodectini have been considered, for a time, as forming a subfamily in itself, different from the true tetras. However, the recently discovered *Iguanodectes geisleri* forms a perfect link with the tetras, and the whole group can no longer be treated separately. The two known genera, *Piabucus* and *Iguanodectes*, differ in the form of their breast, that of the former being trenchant, forming a real keel like in *Agoniates* or *Clupeacharax* for example, whereas the latter has its breast and ventral edge only narrowly rounded.

Piabucus dentatus is a rather large (about 175 mm long) silvery species which is among the oldest known South American fishes (it is the *Piabuca* of Marcgravius, 17th Century). It comes from the Guianas and Lower Amazon (two geographical forms), where it is quite rare except in certain restricted places. A second species, *P. melanostomus*, has been described from the Rio Paraguay. It is an exception in the Tetragonopterinae, having lost its outer premaxillary row of teeth. Otherwise it is much like *dentatus* in the meristic characters.

Iguanodectes spilurus (synonyms *Piabucus spilurus* and *Iguanodectes tenuis*) is by far the commonest species of the tribe. It differs from the other *Iguanodectes* in the following way (the key is taken from a recent review of the genus):

a. Anal fin relatively short, with 20-25 branched rays, beginning behind dorsal fin, which is inserted at midbody; no maxillary teeth (subgenus *Iguanobrycon*, a link with the Tetragonopterini) . . . *I. geisleri* (Rio Negro)

aa. Anal fin with 28-34 branched rays, beginning under dorsal fin, which is behind midbody; 1 or 2 maxillary teeth (nominate subgenus *Iguanodectes*).
- b. 56-64 lateral line scales.
  - c. Body depth usually 5 times in standard length . . . *I. spilurus* (Amazon basin and Guyana)
  - cc. Body depth 6.0-6.3 in standard length . . . *I. adujai* (Rio Negro)
- bb. 66-71 lateral line scales . . . *I. purusi* (Amazon basin)

In life the *Iguanodectes* species, which do not reach 60 mm in standard length, have a dark lateral band overlaid by a brilliant one, golden (or orange) on *spilurus* and cherry-red on *geisleri*.

Tribe Tetragonopterini

A. GENUS *HEMIBRYCON* AND ALLIED GENERA

For the layman, the number of 4 inner premaxillary teeth (versus 5 or more in other groups) seems to be a poor character to distinguish a group within the Tetragonopterini. However, this number is remarkably constant (except in a very small number of species where it is not genetically well fixed, somewhat in the same manner as the adipose fin in the genus *Lebiasina*). It is usually associated with an irregular outer tooth row and a strong armature of the cheek, the large suborbital (also called 3rd circum-orbital) in contact with the preopercular canal all around, without a "naked" zone on the cheek.

The number 4 seems indeed to be a basic one in characoids and may perhaps be called archetypal. It is to be found also in the Alestidae as well as in the Serrasalmidae, and even in the Anostomidae, but in the latter case the homology may be disputable. The "group" has apparently given rise to certain Glandulocaudinae and to the *Creagrutus*-group.

The genera may be identified as follows:

a. 6-20 maxillary teeth in full-grown specimens.
  b. Adipose fin present; maxilla long or rather long; caudal fin forked.
    c. Snout not very long, not produced; maxilla straight or convex.
      d. Lateral line complete; more than 4 scales under it; caudal lobes not scaled.
        e. Large suborbital always entire or nearly so; anterior nostril not tubiform; mandibular teeth all alike . . . (1) *Hemibrycon* (Guianas, Trinidad and N.W. South America)
        ee. Circumorbitals very weak; anterior nostril forming a short tube; 3rd mandibular tooth hooked . . . (2) *Carlastyanax aurocaudatus* (an aberrant species, endemic in the Rio Cauca)
      dd. Lateral line irregular; less than 4 scales under it; base of caudal lobes scaled . . . (3) *Boehlkea fredcochui* (upper Amazon) and *B. orcesi* (Ecuador) ?
    cc. Snout very long, produced; maxilla almost horizontal, with a downwardly directed concavity . . . (4) *Creagrudite maxillaris* (Rio Negro)
  bb. Adipose fin absent; maxilla rather short; caudal fin three-pronged (large suborbital slightly reduced; lateral line complete) . . . (5) *Nematobrycon* (Colombia)

aa. Less than 6 maxillary teeth, even in full-grown individuals.
  f. Adipose fin present; always more than 3 mandibular teeth; postorbitals developed.
    g. Snout not produced, rounded, mouth more or less terminal; eye rounded or horizontally elongate.
      h. More than 13 branched anal rays; large suborbital usually entire as in *Hemibrycon*.
        i. Lateral line usually complete (except in a few *Knodus*).
          j. Anal fin originating under middle or end of dorsal fin.
            k. Caudal fin not scaled . . . (6) *Bryconamericus* (mostly N.W. and S.E. South America)

Boehlkea fredcochui from Iquitos. Photo by Dr. Herbert R. Axelrod.

Boehlkea fredcochui. Photo by Stanislav Frank.

Boehlkea fredcochui. Photo by Dr. Herbert R. Axelrod.

kk. Caudal fin scaled, at least on its base
. . . (7) *Knodus* (mostly Amazon basin)

jj. Anal fin originating in front of first dorsal ray
. . . (8) *Piabarchus analis* (Rio Paraguay)

ii. Lateral line incomplete (anal fin originating below the first dorsal ray; caudal fin not scaled)
. . . (9) "*Hyphessobrycon*" *melanopleurus* (Rio Tiete, State of Sao Paulo; possibly close to *Piabarchus*)

hh. 13 or fewer branched anal rays; large suborbital usually not quite entire.

l. Lateral line complete; teeth not *Bryconamericus*-like.

m. Teeth narrow, tricuspid, those of inner premaxillary row larger than those of outer one; usually only 5 mandibular teeth; caudal naked
. . . (10) *Microgenys* (Colombia and Bolivia)

mm. Teeth broad, multicuspid; upper lip not covering outer premaxillary row, which is larger than and conceals the inner one; mandibular teeth numerous; base of caudal fin scaled
. . . (11) *Ceratobranchia* (Peru)

ll. Lateral line incomplete, at least lacking a few pores on the caudal peduncle; teeth *Bryconamericus*-like
. . . (12) *Bryconacidnus* (Peru, Ecuador and Bolivia)

gg. Snout produced, pointed, mouth subinferior; eye oval, somewhat vertically elongate (11-15 branched anal rays).

n. Lateral line complete; upper lip developed, covering the relatively weak outer row of teeth; anterior fontanel moderate; 11-12 branched anal rays
. . . (13) *Rhinobrycon negrensis* (Rio Negro)

nn. Lateral line interrupted on caudal peduncle; upper lip reduced, uncovering the prominent outer row of teeth; anterior fontanel almost closed; 15 branched anal rays
. . . (14) *Rhinopetitia myersi* (Rio Araguaia)

ff. Adipose fin lacking; only 3 multicuspid incisiform mandibular teeth; postorbitals weak
. . . (15) *Coptobrycon bilineatus* (Rio Tiete)

Genus *Hemibrycon*

The 16 or 18 species of the genus inhabit the periphery of the Amazon-Orinoco basins, where the rivers are fast running and well oxygenated. They are about 100 or 120 mm in standard length and resemble an ordinary

Astyanax, being of what is called "a generalized type," with nevertheless heavier head and jaws (usually). Most of the species are endemic to specific river systems as follows:

(*a*) the group composed of the poorly-known *H. tridens* from the Rio Apurimac in Peru, *H. tolimae* from the Rio Magdalena in Colombia, *H. beni* from the Rio Beni in Bolivia, *H. helleri* from the Rio Urubamba in Peru and *H. colombianus* from some part of the Colombian Andes. They all have a rather short anal fin (18-23 branched rays) and differ from each other in the number of scales and minor details;

(*b*) *H. dentatus* and *H. boquiae* from the completely isolated Rio Cauca in Colombia, with a long anal fin and numerous scales, differing from each other in body depth;

(*c*) *H. surinamensis* from Surinam, and *H. taeniurus* and *H. guppyi* from Trinidad. The former has the lowest number of scales under the lateral line (usually 4). The latter ones are very similar (sibling species ?);

(*d*) *H. dariensis* from Panama, which seems to have evolved from some generalized *Hemibrycon* such as *H. jabonero*;

(*e*) the other species, not endemic, namely *H. jabonero* from Venezuela and Ecuador, *H. polyodon* from Ecuador, Peru (?) and Rio Beni (?), *H. metae* from the Upper Orinoco basin, *H. decurrens* from the mouth of the Rio Magdalena and *H. jelskii* from Peru and Ecuador (?), constitute a group which is among the more difficult groups of the Tetragonopterinae (together with *Bryconamericus*, *Creagrutus*, which have a parallel geographical distribution, etc.).

The following key may aid in identifying these generalized species, provided that the exact locality is known, permitting the elimination of at least one of the already cited endemic species:

a. 23-26 branched anal rays, rarely 27.
  b. 38-42 lateral line scales . . . *H. jabonero*
  bb. 45-46 lateral line scales . . . *H. polyodon* (synonyms *H. coxeyi* and *H. huambonicus* ?)
aa. 27-32 branched anal rays, rarely 26 (42-44 lateral line scales).
  c. Maxilla not entirely toothed, said to have only 6 teeth . . . *H. decurrens* (a species close to a *Bryconamericus*)
  cc. Maxilla almost fully toothed.
    d. Lateral line low; body depth 3.0-3.5 in standard length . . . *H. metae*
    dd. Lateral line along midaxis of body or only slightly lower; body depth 2.33-3.0 in standard length . . . *H. jelskii*

Carlastyanax aurocaudatus

Like *Hemibrycon dentatus* and *H. boquiae*, the monotypic genus *Carlastyanax* is an isolated, very peculiar form of the Rio Cauca, the

Nematobrycon palmeri, the emperor tetra. This is the dark variety seen in aquaria. Photo by Dr. Herbert R. Axelrod.

Nematobrycon palmeri. Photo by Hilmar Hansen.

A color variety of *Nematobrycon* or a new form. Photo by Dr. Karl Knaack.

Nematobrycon lacortei (formerly known as *N. amphiloxus*), the rainbow tetra. Photo by Dr. Herbert R. Axelrod.

principal characters of which have been given in the key. In life, the fins are bright red, except the distal part of the caudal fin, which is hyaline, and the dorsal fin, which is yellow. The ground color of the body is also yellow, with a green tint anteriorly and a rosy tint posteriorly.

Boehlkea fredcochui and *Hemibrycon orcesi*

From the year 1956, certain aquarists had kept an imported fish from the vicinity of Leticia (upper Amazon basin), which had been called by someone *Microbrycon cochui* (this name is actually synonymous with *Tyttocharax madeirae*) or Cochu's blue tetra. Dr. James Böhlke showed that the fish was close to a *Hemibrycon*, differing from it nevertheless in having the caudal scaled (just like *Knodus* differs from *Bryconamericus*, *Moenkhausia* from *Astyanax*, *Hemigrammus* from *Hyphessobrycon*, etc.), as well as the lateral line somewhat irregular (certain individuals at least, raised in aquaria, have an incomplete lateral line). The present author named the genus in honor of Dr. Böhlke, and the species in honor of its importer. It is a small fish (about 40 mm in standard length), highly decorative and strongly resembling a croaking tetra (genus *Glandulocauda*). In life, its body is light metallic blue with a shimmer of purple; the tips of the caudal lobes as well as the adipose fin are white.

Hemibrycon orcesi, from Ecuador, may be a *Boehlkea*, having the caudal fin base scaled. As compared with *B. fredcochui*, it has the lateral line complete, has fewer anal rays and also more transverse scales.

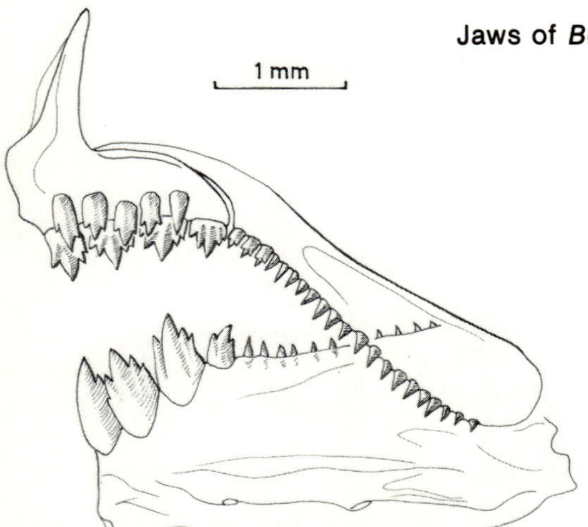

Jaws of *Boehlkea fredcochui* (after Gery, 1966).

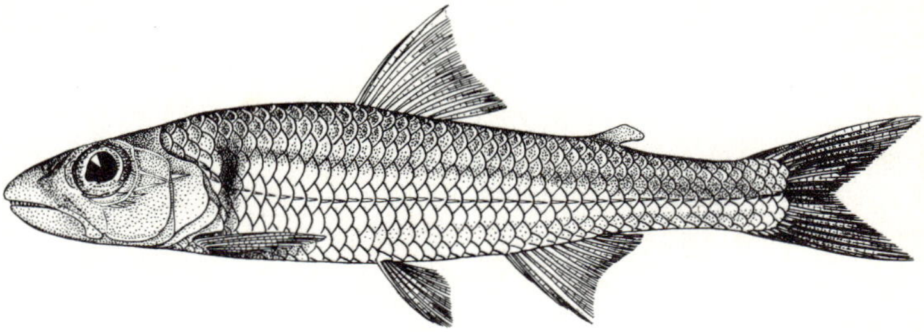

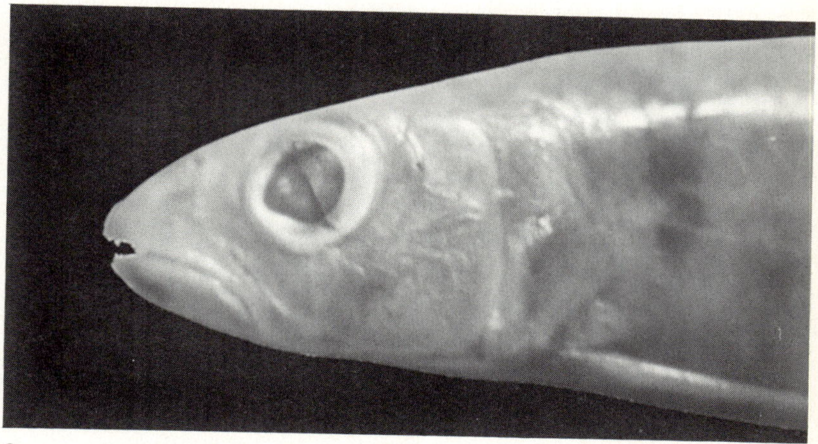

Creagrudite maxillaris. Original drawing of the type made by Mr. Pablo Bravo in 1930, published for the first time by permission of Dr. G.S. Myers, and head photo by the author.

Creagrudite maxillaris

This odd fish has been found in the Upper Rio Negro, near the Casiquiare Canal joining the Negro with the Orinoco. The largest known specimens measure less than 80 mm in standard length. They are elongate, spindle-shaped, and very little compressed. There is a general resemblance to a *Creagrutus*, but with a longer snout and a very peculiar concave maxillary. Such a form is otherwise to be found only in *Bramocharax*. Another feature distinguishing it from a *Creagrutus* is the teeth, which are not heavy and are only biserial on the premaxilla.* As concerns the evolution of the upper jaw, it is worth noting that *Creagrudite*, with its 10-12 maxillary teeth, is in the same position with respect to *Hemibrycon*, as is *Rhinobrycon* with respect to *Bryconamericus*.

*However, despite the noted differences, Boehlke and Saul (1975. Proc. Acad. Nat. Sci. Phila., 127 (3): 25-28) have just synonymized *Creagrudite* with *Creagrutus*.

Spawning sequence of *Nematobrycon palmeri*. All photos by R. Zukal.

Genus *Nematobrycon*

Emperor Tetras

The *Nematobrycon* species are two in number: the well-known emperor tetra, *Nematobrycon palmeri*, from the San Juan and Atrato Rivers in Colombia, and the rainbow tetra, known for a time under the name *N. amphiloxus*. S. Weitzman and W. Fink have recently named it *N. lacortei*, showing at the same time the real *M. amphiloxus* is a variety of *N. palmeri*.

The two species hybridize in aquaria, but are supposed to be geographically isolated in nature, the rainbow tetra coming from a different part of the basin of the Rio San Juan than the emperor tetra. They differ in the averages of certain counts of fin rays and teeth, as well as in the living color pattern. Emperor tetras coming from the typical locality are said to be lighter than the dark variety usually seen in the aquarium.

Genus *Bryconamericus*

The 30-40 *Bryconamericus* species, like the *Hemibrycon* species from which they probably evolved (most of the forms look like young *Hemibrycon*), are in need of revision. The few lines which can be devoted here to the genus are by far not in proportion to its importance. It is one of the less diversified genera, the species are very close to each other, diverging only in minor details, and the color patterns are especially invariable (there is usually a humeral spot and a lateral band which is silvery in life and which is often extended onto the middle caudal rays).

Within the genus, two artificial groups may be formed according to the length of the anal fin, which is more or less in correlation with the number of

Bryconamericus ternetzi. Original drawing of the type made by Mr. Pablo Bravo in 1930, published for the first time by permission of Dr. G.S. Myers.

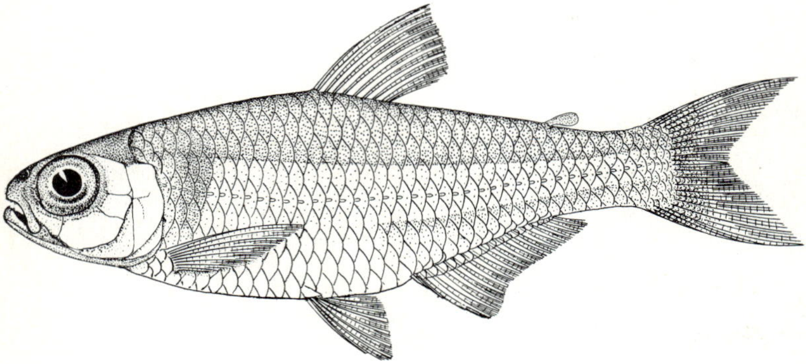

transverse scales. In the first division (*diaphanus*-group), there are usually 15-25 anal rays (total number) and 4-6 transverse scales above the lateral line. In the second one (*peruanus*-group), the anal rays are more numerous (usually 25-43), as are the transverse scales (6-9 above lateral line). However, the overlap is marked and certain species (*B. thomasi* for example) would belong to one group according to the anal rays and to the other according to the transverse scales.

(*a*) Key to the *Bryconamericus diaphanus*-group:

a. Teeth incisiform, multicuspid, resembling those of *Deuterodon* . . . *B. deuteronoides* (a sibling species of *B. cismontanus* in Colombia and Venezuela)*

aa. Teeth *Tetragonopterus*-like.
- b. Body elongate, the body depth 4.33-4.5 in standard length.
  - c. 15-16 anal rays . . . *B. bolivianus* (Rio Beni) and *B. hyphesson* (Guyana)
  - cc. 22 anal rays . . . *B. stramineus* (Rios Paraguay and Sao Francisco)
- bb. Body depth less than 4.33 in standard length.
  - d. Tips of caudal lobes black (outer premaxillary row of teeth extremely irregular) . . . *B. exodon*, the type species (Rio Paraguay)
  - dd. Tips of caudal lobes hyaline (outer premaxillary row of teeth wavy)
    - e. 30-32 lateral line scales . . . *B. megalepis* (State of Ceara, Brazil)
    - ee. More than 32 lateral line scales.
      - f. Body depth usually 2.6-2.8 in standard length . . . *B. iheringi* (Rio Grande do Sul and La Plata basin), and *B. thomasi* (Rio Paraguay—this species has 7 transverse scales above lateral line and may belong to the *B. peruanus*-group)
      - ff. Body depth usually more than 2.8 in standard length.
        - g. 4 or 4½ transverse scales above lateral line . . . *B. diaphanus* (upper Amazon), *B. novae* (Rio Tocantins) and *B. ternetzi* (upper Rio Negro)**
        - gg. 5 or 6 transverse scales above lateral line.
          - h. 2 broad maxillary teeth; head length about 3.8 in standard length . . . *B. cismontanus* (synonym *euryodous* (?)—Colombia and Venezuela)

* This species may be a *Knodus*.

** The latter species may be a *Knodus*.

Knodus moenkhausii. Photo by Dr. Herbert R. Axelrod.

Bryconamericus loisae (above) and *B. caucanus* (below). Photo by Dr. Herbert R. Axelrod.

Knodus breviceps. Photo by Dr. Herbert R. Axelrod.

Unidentified tetra of the *Hemibrycon-Creagrutus* section, possibly a species of *Microgenys* or *Ceratobranchia.* Photo by Aaron Norman.

hh. Usually more than 2 maxillary teeth; head length 4.0-4.5 in standard length.
  i. Body depth 2.9-3.2 in standard length (38 or 39 lateral line scales) . . . *B. eigenmanni* (western Argentina)
  ii. Body depth 3.25-3.75 in standard length (37-43 lateral line scales).
    j. 16-18 anal rays.
      k. Humeral spot faint . . . *B. grosvenori* (Peru)
      kk. Humeral spot large and conspicuous . . . *B. microcephalus* (southeastern Brazil)
    jj. 21-23 anal rays . . . *B. pachacuti* (Rio Urubamba) and *B. boops* (an abnormal (?) specimen from Rio Uruguay)

(*b*) Key to the *Bryconamericus peruanus*-group (except *B. alfredae* which may belong to *Acrobrycon* or even *Hemibrycon*):

a. Dorsal fin in front of midbody.
  b. 6/35 or 36/4 or 5 scales.
    c. 28-29 anal rays; body depth 2.5 in standard length . . . *B. baudoensis* (Colombia)
    cc. 36 anal rays; body depth 3.5 in standard length . . . *B. multiradiatus* (Rio Atrato in Colombia)
  bb. 7-8/37-43/6 scales.
    d. Body depth usually less than 3.2 in standard length . . . *B. ricae* (Costa Rica) and *B. scleroparius* (Central America and (?) Colombia) (sibling species in Costa Rica, differing from each other in the interorbital width as compared with the eye diameter)
    dd. Body depth 3.2-4.05 . . . *B. loisae* (Rio Meta in Colombia)
aa. Dorsal fin at midbody or behind.
  e. A caudal spot.
    f. Caudal spot not continued onto middle caudal rays.
      g. 6 or 7 transverse scales above lateral line . . . *B. ortholepis, B. scopiferus* and *B. guaitarae* (possibly a subspecies of *B. scopiferus*) (all from Colombia)
      gg. 8 transverse scales above lateral line . . . *B. cascajalensis, B. terrabensis, B. emperador*

and *B. zeteki* (all from Central America)

ff. Caudal spot continued onto middle caudal rays.

h. 25-32 anal rays; 36-40 lateral line scales.*

i. Body depth 2.5-2.8 in standard length; 11 dorsal rays (usually 36-37 lateral line scales) . . . *B. peruanus* (Peru and Ecuador)

ii. Body depth 2.9-3.3 in standard length; 9 or 10 dorsal rays.

j. 36 lateral line scales; 29 anal rays . . . *B. phoenicopterus* (known from a single specimen collected in the Peruvian Amazon basin)

jj. 37-40 lateral line scales; 25-28 anal rays . . . *B. caucanus* (synonym *B. alpha*, Colombia) and *B. osgoodi* (Peruvian Amazon)

hh. 36-43 anal rays; more than 40 lateral line scales . . . *B. brevirostris* (Peru and Ecuador)

ee. No caudal spot (6 or 7 transverse scales above lateral line; 36-40 lateral line scales; 26-30 anal rays) . . . *B. simus* (northern Ecuador)

Genus *Knodus*

According to certain authorities in fish systematics, the *Knodus* species (i.e. the 20 or 25 species which look like a *Bryconamericus* of the *diaphanus*-group, but which have the caudal scaled) do not deserve generic rank. They would have to be included within *Bryconamericus* proper, with or without a subgeneric name. In spite of the above mentioned criticism, the genus is here accepted for convenience, in order to split a group which is somewhat crowded. Actually it seems possible to tell if such and such species has the caudal scaled or not, provided that the biological material is abundant and well preserved. The genus is "workable," even if it does not seem to be natural (the same is true for a number of American tetras such as *Hyphessobrycon-Hemigrammus*, *Moenkhausia-Astyanax* etc.). Note also the geographic distribution of the genus is almost purely Amazonian, whereas that of *Bryconamericus* is more peripheral, and mostly from northwestern South America (including Panama and even north of it) and southeastern Brazil. Like the preceding group, *Knodus* is rather poorly known and the following key is tentative.

Key to the genus *Knodus* (see also *B. deuteronoides* and *B. ternetzi* under *Bryconamericus*):

*Fink (1976. Proc. Biol. Soc. Wash., 88 (30): 331-344) has just described a new genus and species from Panama, *Eretmobrycon bayano*, that may have derived from one of the forms which follow in the key.

Creagrutus of the *beni*-complex. Photo by Dr. Herbert R. Axelrod.

Another species of the *Creagrutus beni* complex. Photo by Aaron Norman.

Creagrutus cochui. Photo by Dr. Herbert R. Axelrod.

Pseudochalceus kyburzi. Photo by Dr. Herbert R. Axelrod.

a. Body elongate, body depth 3.5-4.25 in standard length.
  b. 4 transverse scales above lateral line (33 lateral line scales; 13-16 branched anal rays) . . . *K. meridae*, the type species (Venezuela)
  bb. 5 transverse scales above lateral line.
    c. 15-18 branched anal rays (body depth 3.75-4.25; 35-38 lateral line scales) . . . *K. heterestes* (Rio Tapajoz)
    cc. 19-21 branched anal rays (body depth 3.5-3.75; 39 lateral line scales) . . . *K. moenkhausii* (Rio Paraguay and upper Amazon) (*K. calliurus*, a poorly known species from "Capim," is said to differ from *K. moenkhausii* in having fewer scales and a black band on the upper caudal lobe)

aa. Body depth 2.5-3.5 in standard length ($4\frac{1}{2}$-6 transverse scales above lateral line).
  d. Head length 3.6-3.8 in standard length.
    e. 5 transverse scales above lateral line (body depth 2.7-3.15 in standard length; 34-37 lateral line scales).
      f. 16 branched anal rays . . . *K. orteguasae* (Colombia)
      ff. 20-24 branched anal rays . . . *K. megalops* (Peru)
    ee. 6 transverse scales above lateral line.
      g. 17 branched anal rays; 42 lateral line scales . . . *K. chapadae* (Mato Grosso)
      gg. 21 branched anal rays; 34 lateral line scales . . . *K. hypopterus* (Colombia)
  dd. Head length 3.8 (in *K. savannensis*) to 4.66 in standard length.

Knodus savannensis (after Gery, 1961).

h. 33-34 lateral line scales (20-23 branched anal rays) ... *K. caquetae* (Colombia)

hh. 35-40 lateral line scales.

i. 15-18 branched anal rays (35-37 lateral line scales) ... *K. victoriae* (Rio Parnahyba; subspecies *septentrionalis* from Ecuador)

ii. Usually more than 18 branched anal rays.

j. 19-21 branched anal rays (38-41 longitudinal scales).

k. Lateral line complete; head length 4.5-4.66 in standard length ... *K. breviceps* (Amazon basin)

kk. Lateral line somewhat shortened; head length 3.8-4.0 in standard length ... *K. savannensis* (northeastern Brazil)

jj. 23-30 branched anal rays.

l. 35-39 lateral line scales.

m. Dorsal fin in front of midbody ... *K. gamma* (Ecuador, Amazonian drainage)

mm. Dorsal fin behind midbody.

n. Lateral line complete; 50-55 mm in standard length ... *K. beta* (nominate form from the Rio Meta in Colombia, and one (or more) subspecies in Venezuela, Ecuador etc.)

nn. Lateral line incomplete; maximum size 35 mm in standard length ... *K. delta* (Ecuador, Amazonian drainage)

ll. 40 lateral line scales (dorsal fin behind midbody) ... *K. smithi* (synonym *K. jacunda* (?), Rio Madeira in Bolivia; possibly of the *K. beta*-group)

Piabarchus analis

This silvery fish from the upper Paraguay basin is known from a single specimen 37 mm long. It resembles a *Bryconamericus*, but with the anal fin inserted just in front of the first dorsal ray.

"Hyphessobrycon" melanopleurus

This species from southeastern Brazil looks somewhat like the preceding one owing to the position of the dorsal and anal fins, which are at the same level, but it has an incomplete lateral line and a conspicuous dark lateral band. It seems also to be quite a rare species.

Head study of *Pseudochalceus kyburzi.* Photo by Dr. Herbert R. Axelrod.

Astyanax (Astyanax) daguae. Photo by Dr. Herbert R. Axelrod.

Astyanax (Astyanax) metae may be a synonym of *A. maximus.* Photo of a specimen from Camatagua, Venezuela by Dr. Herbert R. Axelrod.

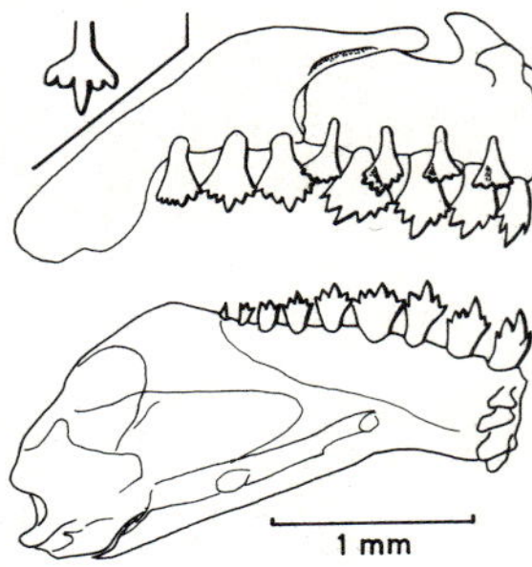

Ceratobranchia obtusirostris (after Eigenmann, 1927); head and jaws of type, the latter with dentition usually seen (above left, peculiar tooth structure seen only in the type of the species) (after Gery, 1966).

Genus *Microgenys*

Only two species, *M. minutus* and *M. lativirgatus*, are known from the upper Cauca and the Peruvian Amazon, respectively. With their very short anal fin (10-13 rays) and aquiline snout, they look like young *Creagrutus*, but with narrow, tricuspid teeth. Their color pattern is quite generalized: a lateral band, silvery in life and *(M. lativirgatus)* with a humeral spot.

Genus *Ceratobranchia*

The single (?) Andean species *C. obtusirostris* looks very much like a *Microgenys*, but with very different dentition. Its peculiar teeth have probably evolved along with a specialized diet, possibly a vegetarian one. Two other species have been described. One, *C. binghami*, is probably a synonym, and the other one, *C. elatior*, does not seem to belong to the genus.

Genus *Bryconacidnus*

The three Andean species, *B. ellisi* (possible synonym *Microgenys weyrauchi*), *B. hemigrammus* and *B. paipayensis*, have much in common with the *Microgenys* species except that their teeth are larger, *Bryconamericus*-like, and their lateral line is not complete. The type species *B. ellisi* has the body characteristically "peppered," at least in certain localities such as the Rio Beni in Bolivia.

Bryconacidnus ellisi, B. hemigrammus, B. paipayensis, and *Rhinopetitia myersi* (after Gery, 1964 and 1966); head of (a) *Rhinobrycon negrensis* and (b) of *Rhinopetitia myersi* (after Gery, 1964).

Two different forms of *Astyanax* (*Astyanax*) *fasciatus.* Photos by Harald Schultz.

The blind characin *"Anoptichthys jordani"* is actually an eyeless form of *Astyanax fasciatus mexicanus.* Photo by Dr. Herbert R. Axelrod.

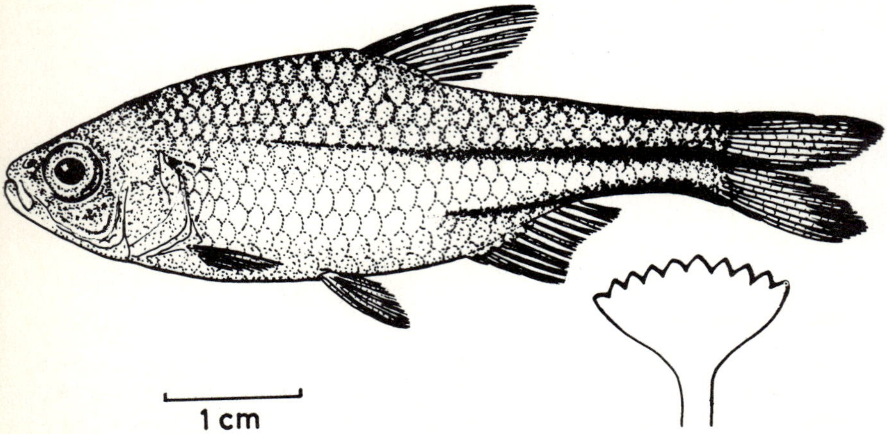

Coptobrycon bilineatus (after Gery, 1966).

Rhinobrycon negrensis and *Rhinopetitia myersi*

These two odd fishes, respectively from the Rio Negro and the Rio Araguaia in Brazil, have a *Creagrutus*-like snout and short anal fin, but a *Bryconamericus*-like dentition. In *R. myersi*, the tricuspid outer premaxillary teeth are very numerous for a tetragonopterine (6 or 7 teeth) and conspicuous, the upper lip being much reduced. There is no mark on the body of *R. myersi*, whereas the color pattern of *R. negrensis* is much like that of a generalized *Bryconamericus*.

Coptobrycon bilineatus

This is another odd fish which has been found only in the Rio Tiete basin in southeastern Brazil. Formerly known as a *Hasemania*, it is much regressed, having a very short lateral line, a short anal fin, very weak postorbitals, no adipose fin and only 3 mandibular teeth on each side (instead of 10 or more in the great majority of the Tetragonopterinae). The mandibular teeth are flat, expanded multicuspid incisors, like the premaxillary teeth of a *Parodon* for example.

B. GENUS *CREAGRUTUS* AND ALLIED GENERA

Certain tetras, probably coming from a *Hemibrycon* or *Bryconamericus*-like ancestor, have developed rather heavy upper teeth forming a sort of pavement in 3 irregular rows, in connection with a more or less overhanging snout. This specialization is apparently correlated with a diet composed of hard animals (snails, crustaceans ?) or plants (seeds or leaves ?), in part at least.

These forms can be identified as follows:

a. Lateral line complete.
  b. 18-21 branched anal rays ... (1) *Piabina*
  bb. 10-15 branched anal rays ... (2) *Creagrutus*
aa. Lateral line incomplete ... (3) *Creagrutops*

Genus *Piabina*

Piabina argentea, from the S. Francisco and Parana River basins, as well as *P. beni*, from the Rio Beni in Bolivia, are relatively small fishes (50-100 mm in total length) with a silvery lateral band and a faint humeral spot. They are like *Creagrutus* in having 3 rows of teeth on the upper jaw, a blunt snout, the maxilla short and the mandible included, but unlike it in the length of the anal fin, which has 20-23 rays. This gives to them a slightly different body shape.

The two known species are very close to each other, differing only in the disposition and number of the premaxillary teeth forming the intermediate row. In *P. argentea* the median row is not well separated from the outer one and the structure is somewhat intermediate between that of the last group (the fish is close to *Piabarchus* for example) and that of the next species, *P. beni*. This one has the median row well separated from the other rows, and it is close to the *Creagrutus beni*-group in this respect.

Genus *Creagrutus*

Like the genus *Bryconamericus* and a number of other tetragonopterine genera, the genus *Creagrutus* is so little diversified that its 15-20 species are difficult to identify and are in need of comprehensive revision. The poorly known type species, *C. muelleri*, together with *C. peruanus*, *C. phasma*, *C. bolivari*, *C. londoni* and *C. melanzonus*, have a premaxillary dentition composed of an outer row of 3 teeth (usually), a middle row of 2, apparently displaced from the outer row, and the usual inner row of 4 heavy teeth. These 3 rows are grossly parallel to each other, and the upper jaw makes a gentle curve when seen from below. This is the "typical" disposition of the teeth. A second structure is seen in the other species, at least in those where a good description of the dentition is available; is it called "atypical" despite its greater frequency. The number of teeth is about the same, but the rows are no longer parallel, the outer teeth following the border of the forward-pointing upper jaw and forming a V with the branches separated from each other.

Astyanax (*Astyanax*) *kennedyi* from the upper Amazon. Photo by Dr. Herbert R. Axelrod.

Astyanax (*Astyanax*) *fasciatus*, possibly the subspecies *mexicanus*. Photo by Dr. Herbert R. Axelrod.

Unidentified *Astyanax* from southeastern Brazil near Santos. Photo by Dr. Herbert R. Axelrod.

Astyanax (*Astyanax*) *kennedyi* from the Solimoes River. Photo by Dr. Herbert R. Axelrod.

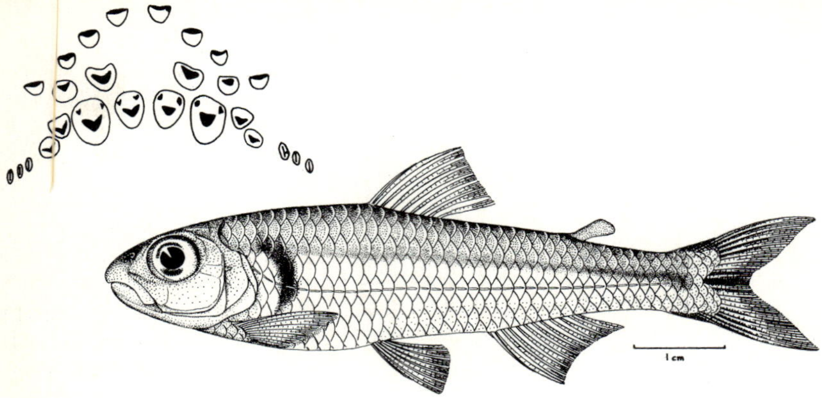

Type of *Creagrutus phasma* (original drawing by Mr. Pablo Bravo in 1930, published for the first time by permission of Dr. G.S. Myers), and upper dentition of a *Creagrutus* (*C. cochui*) (after Gery, 1964).

As concerns the other characters, they seem to be of little use to identify a species. The form of the humeral spot and of the lateral band, which are usually present, nevertheless seem rather constant within each population and can be used as a key character. This may also be the case for certain proportions such as the body depth at dorsal level and the head length (without membrane). The extent of the large suborbital has been said to sometimes be correlated with the anal rays and the lateral line scales. Unfortunately, it is not known for all species and will be mentioned in the following key (an "experimental" one!) for only a few forms.

Key to the genus *Creagrutus* (except *C. gephyrus*, recently described by Boehlke and Saul, 1975):

a. Dentition "typical" (*muelleri*-group).

b. Body depth 4.45-4.5 in standard length; 34-36 lateral line scales (8 or 9 branched anal rays) . . . *C. melanzonus* (Guyana and Rio Negro)

bb. Body depth 3.33-4.2 in standard length; 37-43 lateral line scales.

c. Humeral spot crescent-shaped; body depth more than 3.33 in standard length.

d. Body depth 4.2 in standard length; large suborbital well developed . . . *C. phasma* (Casiquiare Canal)

dd. Body depth usually 3.5-3.7 in standard length, rarely 4; large suborbital covering half cheek . . . *C. bolivari* (Venezuela)

cc. Humeral spot roundish; body depth 3.33 or less in standard length.

e. Head length 3.25-3.4 and body depth 3.0-3.1 in standard length; 6/36-37/4 scales . . . *C. londoni* (Colombia)

ee. Head length about 4 and body depth about 3.33 in standard length; 5/38-43/4 scales.

f. Snout blunt; 11-12 branched anal rays
. . . *C. muelleri* (type species; Eastern slope of the Ecuadorian Andes)

ff. Snout somewhat more pointed; 9-11 branched anal rays
. . . *C. peruanus* (eastern slope of the Peruvian Andes)

aa. Dentition "atypical" (*affinis*-group):

g. Head length 3.25-3.75 in standard length.

h. Body depth 3.25-3.75 in standard length (humeral spot vertically ovate; a caudal spot) . . . *C. anary* (Rio Madeira) and *C. hildebrandi* (Venezuela, a synonym (?))

hh. Body depth usually more than 3.75 in standard length.

i. No humeral spot; a series of dark spots along flanks; large suborbital almost entire; 5-6/34-35/3 scales
. . . *C. amoenus* (Colombia, Amazonian drainage)

ii. A humeral spot; no dark lateral spots along body; large suborbital incomplete; 4-4½/37-39/3-4 scales
. . . *C. boehlkei* (Ecuador, Amazonian drainage)

gg. Head length 3.9-4.33 in standard length.

j. Humeral spot horizontally elongate, followed by a lateral band to end of middle caudal rays.

k. Body depth 3.9-4.3 in standard length; 40-41 lateral line scales; large suborbital almost entire
. . . *C. cochui* (Upper Amazon)

kk. Body depth 3.5 in standard length; 35 lateral line scales; large suborbital moderately developed
. . . *C. atrisignum* (Upper Rio Tocantins)

jj. Humeral spot vertically elongate; usually no caudal pattern.

l. Last pores of lateral line not broad and not covered by flap resembling an accessory scale.

m. A black bar crossing dorsal fin; 6/4 transverse scales (body depth 3.0-3.45 in standard length)
. . . *C.* species (Lake Valencia in Venezuela; species called *C. beni* by authors)

mm. Dorsal fin usually plain; 4 or 5/3 or 4 transverse scales.

n. Body depth usually more than 3.25 in standard length.

o. 13 branched anal rays (average)
. . . *C. affinis* (north of Colombia), *C. notropoides* and *C. simus* (Panama), possibly a polytypic species

oo. Less than 13 branched anal rays (average)

Unidentified *Astyanax* (subgenus *Poecilurichthys*), rather close to *A. asymmetricus* but probably a new species. Photo by Harald Schultz.

Astyanax sp. from Lago Tefe. Photo by Harald Schultz.

Astyanax (*Poecilurichthys*) *zonatus*. Photo by Dr. Herbert R. Axelrod.

Astyanax (*Poecilurichthys*) *abramis*. Photo by Dr. Herbert R. Axelrod.

p. Usually 10-12 branched anal rays
. . . *C. beni*, possibly a polytypic species, from Rio Beni, in Bolivia (typical locality), Rio Meta in Colombia and Ecuador

pp. 9-10 branched anal rays
. . . *C. brevipinnis* (an endemic species from the upper Rio Cauca in Colombia)

nn. Body depth 2.75-3.25 in standard length (usually 10-11 branched anal rays and 33-37 lateral line scales) . . . *C. magdalenae* (middle Rio Magdalena)

ll. Last pores of lateral line enlarged and covered by flap resembling an accessory scale (12-14 branched anal rays, 39-41 lateral line scales)
. . . *C. caucanus* (an endemic species from the Rio Cauca in Colombia)

Creagrutops maracaiboensis

This Venezuelan species is a very small *Creagrutus* (22 mm in standard length) which has lost the major part of its lateral line, like *Hemigrammus* versus *Moenkhausia* for example. The upper teeth are atypically set. There is a very small round humeral spot as well as a similar caudal spot like that of *Creagrutus hildebrandi*. The latter species having similar meristics and proportions, its ancestors may have given rise to *Creagrutops maracaiboensis*.

Paratype of *Creagrutus atrisignum* (drawing by the author).

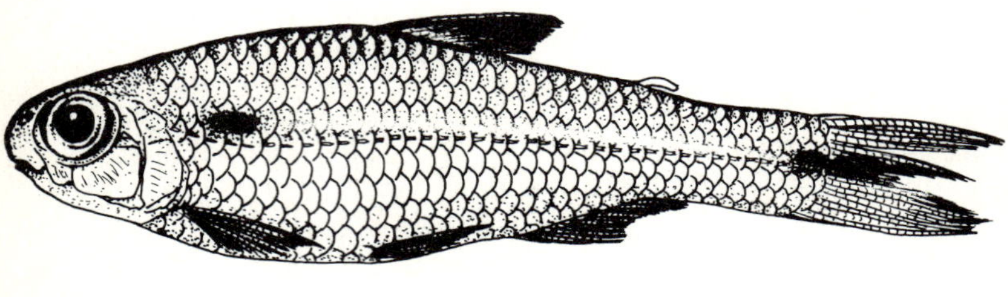

C. GENUS *TETRAGONOPTERUS* AND ALLIED GENERA
AMERICAN TETRAS

The tetragonopterines in the strict sense (true tetras), with the large genera *Astyanax, Moenkhausia, Hyphessobrycon* and *Hemigrammus*, plus almost 30 smaller genera, are the most successful of all Characidae. Except for a few species, they all have biserial premaxillary teeth, with at least 5 in the inner row, and usually a weaker armature of the cheek than that of *Hemibrycon* and allies. Owing to the great number of genera and species, it is necessary to split the true tetras into a few artificial groups, the principal characters of which are given in the following key:

a. Teeth usually not strictly conical, not backwardly recurved; mouth not clupeoid, maxilla not very broad and blade-like.
  b. Preventral area not very flat and not covered by elongate scales overlapping in the center and bent at the sides to form the lateral angle (premaxillary teeth not very numerous, without a gap in the outer series; maxilla fully toothed or not).
    c. Teeth not very compressed antero-posteriorly to form a more or less continuous, multicuspid cutting edge (except in a few *Hyphessobrycon*).
      d. Moderate to large size; lateral line incomplete in a few instances only (*Pseudochalceus, Psellogrammus* and *Brycochandus*).
        e. Caudal fin not scaled . . . Group (1): *Astyanax* and allied genera
      ee. Caudal fin scaled, at least onto 1/4 of its lobes . . . Group (2): *Moenkhausia* and allied genera (including *Tetragonopterus*)
    dd. Small species with incomplete lateral line, probably derived from some members of the preceding group (caudal fin scaled or not) . . . Group (3): *Hemigrammus* and allied genera
  cc. Teeth multicuspid, incisiform, very compressed antero-posteriorly and forming a more or less continuous cutting edge (herbivorous forms, otherwise much like *Astyanax*) . . . Group (4): Genus *Deuterodon*
bb. Preventral area flat, the elongate scales overlapping in the middle and bent at the sides to form the lateral angles; premaxillary teeth numerous, usually leaving a gap in the outer series; maxilla fully toothed, or at least with numerous teeth . . . Group (5): Genera *Phenacogaster* and *Vesicatrus*
aa. Teeth conical, backwardly recurved; mouth clupeoid, maxilla broad, blade-like (otherwise much like *Astyanax*) . . . Group (6): *Genycharax*

First group of American Tetras
ASTYANAX AND ALLIED GENERA

This important group comprises rather primitive tetras (at least looking rather old), namely *Pseudochalceus*, *Astyanacinus* and *Astyanax* itself, and

Astyanax (Poecilurichthys) anteroides. Photo by Harald Schultz.

Astyanax (Poecilurichthys) bimaculatus. Photo by Dr. Herbert R. Axelrod.

Ctenobrycon spilurus spilurus. Photo by Dr. Herbert R. Axelrod.

Astyanax (Poecilurichthys) bimaculatus. Photo by Dr. Herbert R. Axelrod.

a few others which have evolved toward a compressed body *(Ctenobrycon)* or, on the contrary, toward a spindle-shaped body (*Bryconops* and allies). The following key gives the principal characters of the genera:

a. Anterior maxillary margin not greatly arched above and not meeting premaxillary border at a right angle, usually not straight below; lateral line usually not very low on sides; body usually not spindle-shaped.

b. Maxilla completely toothed (lateral line incomplete) . . . (1) *Pseudochalceus*

bb. Maxilla with 0-6 teeth.

c. Maxilla long, forming simple curve with premaxilla, upper jaw almost half head length . . . (2) *Astyanacinus*

cc. Maxilla rather short, forming an angle with premaxilla, upper jaw never half as long as head.

d. Scales not ctenoid, even on breast; body usually not very elevated, not compressed; anal fin moderate in length.

e. Lower teeth composed of 4 or 5 large frontal ones, followed on sides by usually 5-10 much smaller ones . . . (3) *Astyanax* (3 subgenera, plus a blind form called *Anoptichthys*)

ee. Frontal mandibular teeth said to not be followed by series of smaller teeth (large suborbital very narrow) . . . (4) *Opisthanodus haerteli*

dd. Scales ctenoid, at least on breast; body compressed and elevated; anal fin long (39-47 rays).

f. Lateral line complete . . . (5) *Ctenobrycon*

ff. Lateral line incomplete . . . (6) *Psellogrammus*

aa. Anterior margin of maxilla greatly arched above and meeting premaxillary border at a right angle, straight or nearly so below, the bone partially concealed by antorbitals; lateral line usually low on flanks; body elongate, spindle-shaped.

g. Mandibular teeth as in *Astyanax* . . . (7) *Bryconops*, including subgenera *Brycochandus* and *Creatochanes*

gg. Mandibular teeth said to be lacking on sides, as in *Opisthanodus* (color pattern much like that of *B. (Creatochanes) caudomaculatus)* . . . (8) *Autanichthys giacopinii*

Genus *Pseudochalceus*

This interesting genus is probably quite old, dating possibly from the end of the Mesozoic. Its modern representatives inhabit only the southeastern and northwestern parts of South America, as if they had been pushed back, in the past, by competitors more successful in the newly formed Amazon basin.

P. lineatus, from Ecuador, and *P. multifasciatus*, from the States of S. Paulo and Sta. Catarina in Brazil, have a curious color pattern consisting of a longitudinal black line between each scale-row. The recently described *P. kyburzi* and *P. longianalis* from Colombia have the body more or less regularly spotted, again an unusual coloration. All have an incomplete lateral line and a quite long, fully toothed maxilla when adult; in the northwestern species (at least in 2 out of 3), the number of maxillary teeth increases with age. The species reach 85 mm in standard length. Following a recent revision of the genus, they may be identified as follows:

a. i, 7 or 8 ventral rays; usually i, 11 or 12 pectoral rays; maxilla not slipping under first suborbital, the number of its teeth increasing with age; teeth mostly pentacuspid; large suborbital almost covering cheek (nominate subgenus *Pseudochalceus*, Ecuador and Colombia).

b. 21-24 branched anal rays.

c. Body striated . . . *P. lineatus*

cc. Body spotted . . . *P. kyburzi*

bb. 27-30 branched anal rays (body spotted) . . . *P. longianalis*

aa. i, 6 ventral rays; usually i, 14 pectoral rays; maxilla partly slipping under first suborbital, the number of its teeth not increasing with age; teeth tricuspid; large suborbital leaving a broad naked zone on cheek (subgenus *Hollandichthys*, southeastern Brazil) . . . *P. multifasciatus*

Genus *Astyanacinus*

Astyanacinus shows a great resemblance to *Pseudochalceus*, mostly in the head profile (quite similar also to *Hemibrycon*), size (up to 140 mm) and body shape. A closer examination reveals that the lateral line is complete and that the maxilla has a few teeth only on its basal part. In these respects *Astyanacinus* is closer to the central genus of the group, namely *Astyanax*, and it can be described as an *Astyanax* with a lengthened upper jaw.

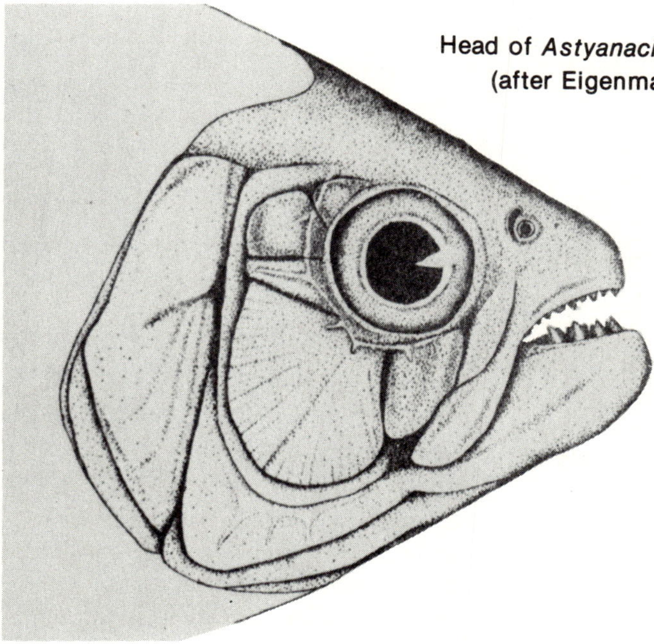

Head of *Astyanacinus moori* (after Eigenmann, 1927).

Ctenobrycon spilurus hauxwellianus, juveniles. Photo by Dr. Herbert R. Axelrod.

Ctenobrycon spilurus hauxwellianus, adult. Photo by Aaron Norman.

Bryconops (*Bryconops*) *caudomaculatus?* Photo by Dr. Herbert R. Axelrod.

Bryconops (*Bryconops*) *caudomaculatus?* Photo by Dr. Herbert R. Axelrod.

A. moori, from the upper Rio Paraguay and from the Rio Beni, as well as *A. multidens* from the Rio Beni only, differ mostly in the number of scales, more numerous in the latter species. They have the humeral spot and caudal band which are common among the tetras.

Genus *Astyanax*

The genus *Astyanax*, the largest unit of the Tetragonopterinae, has been split by Eigenmann into three subunits (usually treated as subgenera): *Astyanax* in the strict sense; *Poecilurichthys*, which has the dorsal line not covered by a regular series of scales (some scales at least, usually in front of the dorsal fin, not meeting in the middle); and *Zygogaster* from Colombia (and Para ?), which has the preventral area compressed, the scales of the two sides apposed in the midline.

All *Astyanax* spp., which are to be found almost everywhere in tropical and subtropical America and even in the USA, have a "generalized" body shape, moderately elongate with the dorsal and ventral profile usually rather symmetrical. Most of the species are about 50-100 mm long; certain ones reach 150 mm and one 200 mm (*A. maximus*).

(a) Nominate subgenus *Astyanax*

This very large unit is dominated (so to speak) by the ubiquitous *A. fasciatus* (from the Rio Negro in Patagonia to the south and to Mexico in the north), which is divided into several subspecies. All other species (40 or so) are much rarer, except in certain places. One species (possibly more) is the famous blind cave fish, *Anoptichthys*, derived from a subspecies of *A. fasciatus*.

Astyanax (*Astyanax*) *microlepis* (after Gery, 1966).

Astyanax (Poecilurichthys) bimaculatus. Photo by G. Timmerman.

Two subunits may be seen: (1) the majority of the species, which are usually more than 50-70 mm long and which have the large suborbital never entire, and (2) a few, usually rare, species grading toward *Bryconamericus*, owing to their small size and usually complete suborbital (*Astyanax hasemani*, from Rio Grande do Sul, with the dorsal fin behind midbody and without color pattern, may indeed be a real *Bryconamericus*). However, the Amazonian and Guianan species of this group, which will be called here the *A. paucidens*-group, have 5 teeth in the inner premaxillary row.

Finally, *A. albeolus*, a *fasciatus*-looking species from Panama and Costa Rica (synonym *A. regani* ?) also grades toward *Bryconamericus*, this time in having 4 inner premaxillary teeth (but the suborbital regressed). As its affinities are perhaps also with the *Hemibrycon*-group, it will not be included within the following key, together with *hasemani*.

Key to the species of the subgenus *Astyanax*:

a. More than 41 lateral line scales.
  b. 49-55 lateral line scales (Colombian species)
    c. 30 anal rays . . . *A. integer* (upper Rio Meta)
    cc. 21-25 anal rays . . . *A. microlepis* (upper Rio Cauca)
  bb. 43-45 lateral line scales . . . *A. cordovae* (Rio Primero in western Argentina)

Bryconops (Creatochanes) melanurus? Photo by Harald Schultz of specimens from Manaus.

Bryconops (Creatochanes) inpai. Photo by Dr. J. Gery.

Bryconops (*Creatochanes*) sp. from the Rio Negro. Photo by Dr. Herbert R. Axelrod.

Unidentified *Bryconops* from central Brazil, allied to *Bryconops* (*Creatochanes*) *affinis*. Photo by Harald Schultz.

aa. Usually 32-41 lateral line scales.

d. 23-33 anal rays.

e. No dark stripes along body.

f. Humeral spot horizontally elongate, bullet-shaped . . . *A. mucronatus* (Guyana)

ff. Humeral spot, when present, vertically elongate (sometimes faint or even absent).

g. Snout rounded, very blunt . . . *A. brevirhinus* (eastern Brazil)

gg. Snout "normal," more or less pointed.

h. No caudal spot . . . *A. daguae* (Colombia), *A. longior* (upper Amazon basin) and *A. marionae* (upper Rio Paraguay)

hh. A caudal spot or band.

i. Caudal spot not continued onto the middle caudal rays . . . *A. ruberrimus* (Colombia and Panama), *A. megaspilura* (Colombia) and *A. ribeirae* (southeastern Brazil)

ii. Caudal spot continued onto the middle caudal rays.

j. Dorsal fin in front of midbody . . . *A. maximus*, synonym *A. metae* (?) (Peru, Colombia and Venezuela), *A. venezuelae* (Venezuela) and *A. schubarti* (State of Sao Paulo, Brazil)

jj. Dorsal fin at midbody (several populations of what is apparently *A. fasciatus*, chiefly from northwestern, but also from southeastern, South America, may have the dorsal fin slightly in front of midbody).

k. Usually more than 2 maxillary teeth (up to 7) . . . *A. nicaraguensis* (Nicaragua) and *A. angustifrons* (Mexico)

kk. Usually 1 or 2 maxillary teeth . . . *A. fasciatus* (ubiquitous; several subspecies, including *A. f. mexicanus*, considered by some as a good species, *A. f. kompi* (Panama), *A. f. hanstroemi-gisleni* (Rio Magdalena) and *A. f. janeiroensis* (from Rio de Janeiro); this species has the color pattern of *A. bimaculatus*, but it is longer, with the predorsal line regular) ee. Dark stripes along body.

l. Stripes between rows of scales

. . . *A. lineatus* (Paraguay basin)

ll. Stripes throughout middle of scales

. . . *A. saltor* (State of Para in Brazil)

dd. 17-24 (rarely) anal rays (except in *A. eigenmanniorum*).

m. A caudal spot.

n. Caudal spot continued onto middle caudal rays (as in *A. fasciatus*).

o. Body depth 2.3-2.7 in standard length.

p. Snout blunt

. . . *A. giton* (eastern Brazil)

pp. Snout "normal," more or less pointed.

q. Usually 3 outer premaxillary teeth and 2 maxillary teeth

. . . *A. taeniatus* (eastern and southeastern Brazil)

qq. Usually 4 outer premaxillary teeth and 1 maxillary tooth

. . . *A. eigenmanniorum* (southeastern Brazil, Paraguay and Argentina)

oo. Body depth 2.6-3.6 in standard length

. . . *A. scabripinnis* (eastern and southeastern Brazil)

nn. Caudal spot not continued onto middle caudal rays.

r. 21-24 anal rays

. . . *A. mutator* (Guyana)

rr. 17-18 anal rays

. . . *A. rubropictus* (Argentina)

mm. No caudal spot.

s. Usually one maxillary tooth

. . . *A. gymnogenys* (southeastern Brazil)

ss. Usually 3-4 maxillary teeth

. . . *A. paranahyba* (Rio Parana basin)

One of the most famous tetras is the Mexico blind cavefish, *Anoptichthys jordani*, which has evolved from the surface form *Astyanax fasciatus mexicanus* quite recently, to the point that hybridization is still possible (even in nature). This fact renders the genus *Anoptichthys* somewhat dubious. Other species have been described from other caves in Mexico. Like other cave animals, they are without pigments and their blindness has caused the development of other senses such as taste.

Key to the *Astyanax paucidens*-group (mostly compiled from the literature):

Moenkhausia of the *lepidura*-group, possibly *M. copei.* Photo by Dr. Herbert R. Axelrod.

Moenkhausia collettii. Photo by Dr. Herbert R. Axelrod.

Unidentified *Moenkhausia,* close to *M. collettii.* Photo by Dr. Herbert R. Axelrod.

Unidentified *Moenkhausia,* which is more elongate than *M. collettii,* from the Rio Negro. Photo by Dr. Herbert R. Axelrod.

Moenkhausia collettii, race from Rio Aguaro, Colombia. Photo by Dr. Herbert R. Axelrod.

a. No caudal spot.

b. 18-19 anal rays; 1 maxillary tooth (5/4 transverse scales) . . . *A. scintillans* (Rio Orinoco)

bb. More than 24 anal rays; 4-7 maxillary teeth.

c. 6/5 transverse scales (27-31 anal rays) . . . *A. guaporensis* (Rio Guapore)

cc. 5/4 transverse scales (25-27 anal rays) . . . *A. guianensis* (Guyana and Venezuela)

aa. Usually a caudal spot (20-26 anal rays; 2-5 maxillary teeth).

d. No humeral spot; 3 transverse scales from lateral line to ventral fin . . . *A. paucidens* (Amazon basin; possibly a *Carlastyanax*)

dd. A humeral spot; 4 transverse scales from lateral line to ventral fin.

e. 36 or 37 lateral line scales; humeral spot faint, caudal spot conspicuous . . . *A. gracilior* (Amazon basin); *A. kennedyi* (upper Amazon at Iquitos)

ee. 32-35 lateral line scales; humeral spot(s) conspicuous, caudal spot faint.

f. Dorsal and anal fins plain; 24-26 anal rays; 5 maxillary teeth . . . *A. multidens* (Amazon)

ff. An oblique band on dorsal and anal fins; 20-22 anal rays; 3 maxillary teeth . . . *A. essequibensis* (Guyana)

(b) Subgenus *Poecilurichthys*

This unit is sometimes very difficult to tell apart from typical *Astyanax*. Its species usually have a greater number of scales than the "true" *Astyanax* species (except *A. microlepis* and *A. cordovae*), and most of them have the dorsal line naked, at least just in front of the fin. But some have an almost regular series of predorsal scales, with only a few of the scales overlapping in the middle. Certain populations of *A. bimaculatus* (the type species of *Poecilurichthys*) even have a predorsal series which is called "regular" in many other genera (such as *Moenkhausia*, *Hemigrammus* etc.), which often have one or two scales not quite in line.

We get thus a paradoxical situation: in most of the cases, while identifying a fish of this group, we have to deduce its subgeneric name (as given by the best worker on the genus, Eigenmann) from its species name. In the present case, if the structure of the dorsal line is not clearly apparent, it is recommended to run the specimen through both keys, that to typical *Astyanax* and the following one. Needless to say, most *Astyanax* species are unidentifiable from a photograph.

Key to the species of subgenus *Poecilurichthys*:

a. Usually more than 40 lateral line scales (*A. abramis*-group).

b. 45 anal rays (55 lateral line scales) . . . *A. erythropterus* (a poorly known species from the Rio Parana; concerning the other species keyed out by Eigenmann after *A. erythropterus*, see under *Ctenobrycon*)

bb. Less than 45 anal rays.

c. No humeral spot.

d. 54 or 55 lateral line scales; caudal spot mostly on lower part of caudal fin ... *A. asymmetricus* (Upper Amazon)

dd. 48 lateral line scales; caudal spot symmetrical ... *A. symmetricus* (Upper Amazon)

cc. A humeral spot, sometimes 2.

f. No caudal spot.

g. Humeral spot round, over 7th-9th lateral line scales ... *A. polylepis* (Guianas and Amazon basin)

gg. Humeral spot horizontally ovate, over first lateral line scales ... *A. poetzchkei* (Lower Amazon)

ff. A caudal spot (humeral spot(s) usually above first lateral line scales).

h. Caudal spot not continued onto middle caudal rays.

i. 33-40 anal rays; 2 conspicuous vertical humeral spots ... *A. festae* (western Ecuador)

ii. 28-34 anal rays; 1 horizontal humeral spot.

j. 34 anal rays; 53 lateral line scales ... *A. bourgeti* (Upper Amazon)

jj. 28-30 anal rays; 43-51 lateral line scales ... *A. abramoides* (Guianas; Venezuela)

hh. Caudal spot continued onto middle caudal rays.

k. Humeral spot(s) vertically elongate.

l. A faint humeral spot; caudal spot across entire caudal base ... *A. zonatus* (Upper Amazon)

ll. 2 humeral spots; caudal spot on middle caudal rays only.

m. 55 lateral line scales; 24 branched anal rays ... *A. myersi* (Venezuela)

mm. 52-53 lateral line scales; 25-26 branched anal rays ... *A. anteroides* (Upper Amazon)

kk. Humeral spot ovate or horizontally elongate.

n. 52-53 lateral line scales; humeral spot horizontally elongate (28-32 anal rays) ... *A. anterior* (Upper Amazon)

nn. 38-49 lateral line scales; humeral spot oval (30-34 anal rays) ... *A. abramis* (La Plata basin; Upper Amazon; Upper Rio Meta)

Above and below: Two species or subspecies of the *Moenkhausia lepidura* complex from the Rio Negro. Photos by Dr. Herbert R. Axelrod.

Moenkhausia dichroura. Photo by Dr. Herbert R. Axelrod.

Moenkhausia intermedia from Tefe. Photo by Dr. Herbert R. Axelrod.

aa. Usually 31-40 lateral line scales (humeral spot horizontally ovate; a caudal band usually to end of middle caudal rays) . . . *A. bimaculatus*-group

o. No dark stripes between rows of scales.

p. Inner premaxillary teeth concave in front, forming a crescent as seen from below.

q. Snout more or less pointed; large suborbital leaving a naked area on cheek; body depth 1.9-2.5 in standard length . . . *A. bimaculatus* (several subspecies from La Plata basin to Panama)

qq. Snout blunt; large suborbital entire; body depth about 2.7 in standard length (a modified *A. bimaculatus*) . . . *A. goyacensis* (State of Goyaz)

pp. Inner premaxillary teeth flat, the denticles forming a straight (or almost straight) line as seen from below.

r. 30-33 anal rays (a modified *A. bimaculatus*) . . . *A. orthodus* (western Colombia)

rr. 27-28 anal rays (humeral spot with vertical expansions) . . . *A. potaroensis* (Guiana)

oo. Dark stripes between rows of scales (5 or 6 maxillary teeth) . . . *A. superbus* (Venezuela)

(c) Subgenus *Zygogaster*

The few species belonging to the subgenus *Zygogaster* are modifications of some *bimaculatus*-like *Astyanax*. With the exception of *A. stilbe*, said to come from Para, Brazil, they inhabit Colombian rivers. They have a vertically oval humeral spot and a spot at the base of the caudal fin; the first dorsal and ventral rays are filamentous in the adult males of certain species; the preventral area is compressed, almost keeled, the scales of the two sides apposed at the midline, or sometimes with a median series of small scales; the predorsal line, as in the subgenus *Poecilurichthys*, is irregularly scaled and sometimes partly naked; there is only one maxillary tooth.

The species are close to each other. *A. stilbe* and *atratoensis* are sympatric in the Rio Atrato (as stated above, the types of *stilbe* are from the lower Amazon*), and *A. magdalenae*, *A. caucanus* and *A. filiferus* are sympatric in the Rio Magdalena.

They can be identified as follows (mostly compiled from the literature):

* A similar faunistic convergence exists in the case of *Hyphessobrycon inconstans*.

a. Preventral area narrowly rounded, with an incomplete median series of small scales (37-40 anal rays).
b. Body depth 1.9-2.3 in standard length . . . *A. atratoensis*
bb. Body depth 2.5-3.0 in standard length . . . *A. filiferus*
aa. Preventral area keeled, without a median series of scales, the ones of the two sides apposed in the middle.
c. 36-41 anal rays . . . *A. stilbe* and *A. caucanus*
cc. 33-34 anal rays . . . *A. magdalenae*

Opisthanodus haerteli

This 40 mm long tetra was imported into Germany in 1934 from the lower Amazon basin. It has a round humeral spot and a caudal band, and looks somewhat like a *Hyphessobrycon* as judged from Schreitmüller's drawing in the famous Arnold-Ahl's *Fremländische Süsswasserfische*. It has the body depth 2.75 in the standard length, 26-27 anal rays, 8/34-35/5 or 6 scales, the predorsal line regularly scaled, and the teeth tricuspid. The maxilla is not toothed. This is almost all that is known about this fish, which has a curious lower jaw (the tooth series of the sides is said to be replaced by an edge of the bone). A re-examination of the only 2 known specimens would be useful to trace its affinities.

Genera *Ctenobrycon* and *Psellogrammus*

The *Ctenobrycon* species are *Astyanax*-like tetras with pectinate scale borders (at least in the ventral region when young) and usually a compressed deep body with a long anal fin. *Psellogrammus* is simply a *Ctenobrycon* with an incomplete lateral line. It is currently undecided whether *C. spilurus*, *C. hauxwellianus* and *C. alleni* (formerly known as *Astyanax alleni*) are a single species or three good species. According to the overlap in their characters, the former hypothesis is to be favored.

C. spilurus spilurus comes from the Guianas and Venezuela; it is extremely abundant near Georgetown in Guyana, rarer elsewhere; it has the body depth 2.5 (on the average) in the standard length. *C. spilurus hauxwellianus* is an Amazonian form, and is also locally very abundant. It has the body depth about 2 times in the standard length but otherwise does not differ much from the former, having just a few more scales (44-51 in lateral line instead of 41-50), and about the same range of anal rays (39-47). *C. spilurus alleni* is a Paraguayan form with the same number of scales as *C. spilurus spilurus*, the same anal rays as both, the body depth of *C. hauxwellianus* (but the young more elongate) and the predorsal line naked. All are silvery in life, the rather faint humeral and caudal spots most apparent after preservation.

Two other species are not well known (at least by the writer): *C. multiradiatus*, an Amazonian form, which may be synonymous with *C.*

Moenkhausia lepidura icae? Photo by Dr. Herbert R. Axelrod.

Moenkhausia gracilima (formerly considered a subspecies of *M. lepidura*) from Tefe. Photo by Dr. Herbert R. Axelrod.

Moenkhausia takasei. Photo by Dr. Herbert R. Axelrod.

Moenkhausia takasei. Photo by Harald Schultz.

hauxwellianus (having only 2 or 3 lateral line scales less), and *C. correntinus* (usually called *Astyanax pellegrini*, which is very possibly a synonym), from Argentina and Paraguay. Again this species has the same number of scales and anal rays, but it is distinctly more elongate than *C. spilurus alleni* (the body depth 2.2-2.8 in the standard length).

Psellogrammus kennedyi, from the Rios Paraguay and S. Francisco basins, resembles *C. spilurus alleni*, but with the lateral line variously developed, usually incomplete.

Genus *Bryconops*

The species of this genus, like those of the preceding, could also have been derived from some *Astyanax*-like ancestor, but in a very different adaptive line than that of *Ctenobrycon*. Instead of being elevated, the body is becoming spindle-shaped, somewhat like that of *Iguanodectes*, and the maxilla is very arched at its meeting with the premaxilla, in an almost unique way among the Characidae.

The genus has been rather recently reviewed. It has been found that *Creatochanes* can no longer stand as a full genus and has to be included within *Bryconops* (the older name), as follows:

a. Maxilla relatively short, its tip not reaching suture between 2nd and 3rd circumorbitals; 28-35 branched anal rays.
 b. Lateral line complete; maxilla rarely toothed . . . (a) Subgenus *Bryconops* typical
 bb. Lateral line incomplete; maxillary teeth said to be numerous . . . (b) Subgenus *Brycochandus*

aa. Maxilla long, its tip usually reaching suture between 2nd and 3rd circumorbitals (with the exception of *B. inpai*, an intermediate form); 23-27 branched anal rays (lateral line complete; maxilla toothed) . . . (c) Subgenus *Creatochanes*

(a) Nominate subgenus *Bryconops*

Three or four species are known, all rather poorly except *B. caudomaculatus*: the rare species *B. alburnoides* and *B. lucidus* (very close to each other) from the Amazon basin, the rare *B. gracilis* from the Rio Tapajoz, and the rather common *B. caudomaculatus* mostly from the Guianas but also from the Amazon basin. The first two species have an incipient middle premaxillary row probably formed by 2 backwardly displaced outer teeth; the maxilla is never toothed. The last two species usually have two rows only on the premaxilla, and the maxilla may occasionally have one tooth. Both have the upper caudal lobe black or very dark. *B. gracilis* is very elongate, with 50 or 55 lateral line scales; *B. caudomaculatus*, less elongate, has only 43-45 lateral line scales.

(b) Subgenus *Brycochandus*

The single species, *B. durbini* from the Rio Tapajoz, looks much like *B.*

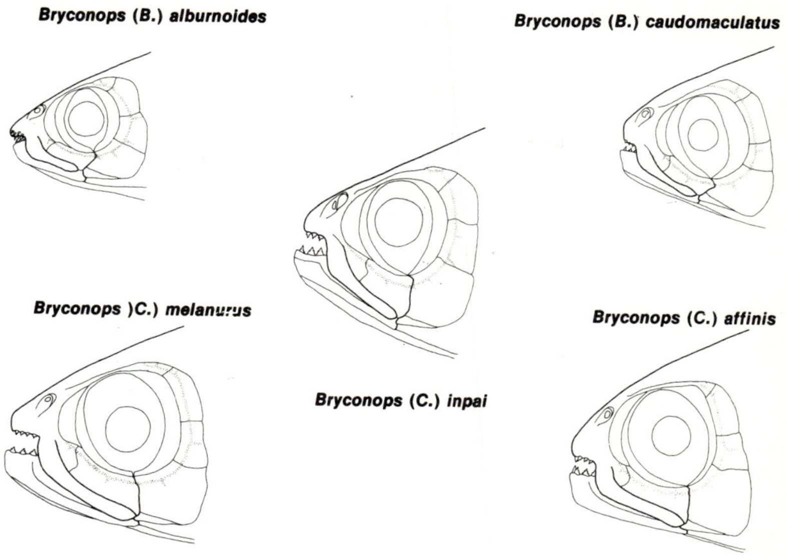

Heads of various species of the genus *Bryconops* (after Knoeppel, Junk and Gery, 1968).

caudomaculatus which has lost a part of its lateral line. It is admissible to doubt the validity of the subgenus and even of the species, if the known type specimens have no more maxillary teeth than "ordinary" *B. caudomaculatus*.

(c) Subgenus *Creatochanes*

The recently described *B. inpai*, from the lower Rio Negro, stands between *Bryconops* (nominal) and *Creatochanes* in having the maxilla short, not reaching the 2nd circumorbital suture.

The two other species, *B. melanurus* and *B. affinis*, are very close and difficult to tell apart. Adult *B. melanurus* have a caudal color pattern resembling that of *B. caudomaculatus*, i.e. the black color prominent on the upper caudal lobe and on the middle rays, with a light spot at the base of the upper lobe, whereas adult *B. affinis* have the color pattern of their caudal fin more symmetrical, both lobes being dark with a light spot on the base of both the lobes (these spots being yellow or orange in life). The form of the snout and of the premaxillary-maxillary junction is probably the best character to differentiate the two species. In *B. melanurus*, the type species of *Creatochanes*, common in the Amazon as well as in the Guianas, the profile of the snout is almost straight and the maxillary curve is very accentuated on the front part, forming with the premaxilla an angle of about 45°, usually situated at pupil level. In *B. affinis* (same geographic distribution less Colombia, and probably a little less abundant), the profile

Moenkhausia georgiae?, but could also be *Tetragonopterus chalceus*. Photo by Harald Schultz.

Moenkhausia grandisquamis from the upper Juruena basin in central Brazil. Photo by Harald Schultz.

Moenkhausia sanctaefilomenae. Photo by Dr. Herbert R. Axelrod.

Moenkhausia oligolepis. Photo by Harald Schultz.

of the snout is convex, the mouth lower, and the maxilla forms an angle of more than 45°, which is usually below pupil level.

It is likely that *B. cyrtogaster* (from the Oyapok river in French Guiana) is a subspecies of *B. melanurus*.

Autanichthys giacopinii

Like *Brycochandus*, *A. giacopinii* is a rare, *caudomaculatus*-like, elongate tetra which is said to differ from that species in the lack of a side series of mandibulary teeth (like in *Opisthanodus*). It comes from the Rio Autana, a southern tributary of the Orinoco in Venezuela.

Second group of American Tetras

MOENKHAUSIA AND ALLIED GENERA (including *TETRAGONOPTERUS)*

The genus *Moenkhausia*, together with 6 or so allied genera, is no less important than *Astyanax* and allies. It comprises also certain species with a disciform body, crenate scales, predorsal line irregularly scaled and maxilla fully toothed. All these species have the caudal fin more or less scaled (at least to an extent corresponding to the proximal fourth of the lower lobe, usually the more covered one). Like the preceding *Astyanax*-group, it is possibly unnatural, but in the absence of any other reliable character, the present arrangement is as good as any. The genera can be identified as follows:

a. Border of the scales not crenate.
  b. Predorsal line regularly scaled.
    c. Maxilla not completely toothed.
      d. Lateral line more or less straight; body compressed or not ... (1) *Moenkhausia*
      dd. Lateral line step-like at its beginning; body deep and compressed ... (2) *Tetragonopterus*
    cc. Maxilla completely toothed ... (3) *Schultzites axelrodi*
  bb. Predorsal line not scaled, at least partly.
      e. Body elongate (and colored like *Moenkhausia lepidura* for example) ... (4) *Gymnotichthys hildae*
      ee. Body deep and compressed (as in *Tetragonopterus* for example) ... (5) *Gymnocorymbus*
aa. Border of the scales crenate.
      f. Anal fin not scaled ... (6) *Bario steindachneri*
      ff. Anal fin scaled ... (7) *Markiana*

Genus *Moenkhausia*

The genus *Moenkhausia*, characterized by its scaled caudal, has been most successful in the Amazon basin and in the Guianas, in contrast to *Astyanax* which seems to be more diversified and more abundant at the

Type of *Moenkhausia eigenmanni* (after Gery, 1964).

periphery of the basin. This is likely to be the same concerning *Knodus* versus *Bryconamericus*, and *Hemigrammus* versus *Hyphessobrycon*, as if a scaled caudal fin would represent an advantage in certain types of waters. *Moenkhausia* also seems to be the stem from which most of the deep bodied species evolved, such as *Tetragonopterus*, *Stethaprion* and allies, and even the Serrasalmidae. Eigenmann had divided the species into two groups according to the shape of their body: one group of quite generalized, usually small forms with the back not much elevated (the body depth usually more than 2.75 times in the standard length), and a group of more specialized, usually large, deep bodied species (the body depth of adults 2-2.66 in the standard length).

The first group (or *lepidura*-group, according to the oldest name) is very close to a number of *Hemigrammus* species, differing only in the presence of a complete lateral line (which is even irregular in *M. cotinho* for example). The second group is closer to the above mentioned specialized forms. It has been split again into a subgroup of species with 3½ or 4 scales below the lateral line (*grandisquamis*-group) and a second one of species with 5 or more scales below the lateral line (*chrysargyrea*-group). This classification seems convenient and will be used here. However, it should be mentioned that certain forms are aberrant or intermediate: *M. newtoni*, which has the scales of the *chrysargyrea*-group and the body depth of the *lepidura*-group, for example, and *M. metae-eigenmanni*, which join the *chrysargyrea*-group with the *grandisquamis*-group.

(*a*) Key to the species of the *M. lepidura*-group (small species with the body depth usually more than 2.75 in the standard length—see also *M. eigenmanni* in group (*c*)):

Moenkhausia robertsi. Photo by Dr. Herbert R. Axelrod.

Moenkhausia agnesae from Igarape Preto. Photo by Harald Schultz.

Unidentified *Moenkhausia* close to *M. pittieri.* Photo by Dr. Herbert R. Axelrod.

Moenkhausia sp., also closely allied to *M. pittieri.* Photo by Dr. Herbert R. Axelrod.

a. 15 branched anal rays; 35-37 lateral line scales . . . *M. crisnejas* (Peru—a species close to *Knodus*, possibly really a *Knodus*)

aa. More than 15 branched anal rays; less than 35 lateral line scales.

b. A median caudal spot (caudal lobes not marked).

c. 18-21 anal rays.

d. Caudal spot very large, occupying all of caudal fin base (19-21 anal rays) . . . *M. cotinho* (an Amazonian and Guianan species, close to *M. sanctaefilomenae* but more elongate)

dd. Caudal spot occupying only middle of caudal fin base (18-19 anal rays) . . . *M. ceros* (Amazon basin)

cc. 28-31 anal rays (caudal spot faint; teeth broad, with seven cusps) . . . *M. newtoni* (State of Para, Amazon basin)

bb. No caudal spot.

e. A black line along anal base; caudal lobes not marked (see also *M. melogramma*, group (b)).

f. No humeral spot, or a very faint one; usually 18 or 19 anal rays . . . *M. copei* (Amazon basin and Guianas)

ff. A conspicuous humeral spot; usually 22-25 anal rays . . . *M. collettii* (sympatric with *copei* and usually more abundant than it)

ee. No black line along anal base; caudal lobes usually marked (except *M. lepidura hasemani*).

g. Distal part of both caudal lobes, except their tip, black (see also *Hemigrammus marginatus*).

h. Premaxilla very short; maxilla forming a strong anterior curve, its tip just reaching level of anterior margin of eye . . . *M. dichroura* (Amazon basin, Guyana, Rios Paraguay and Paranagua)

hh. Premaxilla and maxilla as in other *Moenkhausia* spp.; tip of maxilla extending beyond anterior margin of eye . . . *M. intermedia* (same geographical distribution, except in Guyana)

gg. Upper caudal lobe usually darker than the lower one, except in *M. lepidura hasemani* and *M. takasei* which have the caudal lobes hyaline (see also *M. costae* and *Bryconops* spp.) . . . *M. lepidura*-complex (Guianas and Amazon basin—several forms have been described; *M. hasemani*, *M. gracilima* and *M. takasei* possibly deserve full specific rank)

(*b*) Key to the *Moenkhausia grandisquamis*-group (large, deep bodied species with the body depth usually less than 2.66 in the standard length; usually 5 transverse scales above lateral line and 3-4 scales below (to ventral fins)—see also *M. metae* and *M. eigenmanni* in group (*c*)):

a. Eye moderate, its diameter 2.4 or more in head length.

b. Anal lobe black or barred by an oblique line (no caudal spot).

c. No dorsal spot; body depth 2.0-2.33 in standard length; about 36 lateral line scales; 33-34 anal rays ... *M. affinis* (Lower Rio Negro)

cc. A dorsal spot; body depth 2.45-2.75 in standard length; 31-33 lateral line scales; 25-28 anal rays ... *M. hemigrammoides* (Guianas; strongly resembling *Hemigrammus unilineatus*)

bb. Anal lobe usually plain.

d. Body depth about 2.0-2.2 in standard length (26-32 anal rays; 31-34 lateral line scales).

e. Longitudinal stripes through center of scales; 30-32 anal rays (humeral and caudal spots rather indistinct) ... *M. simulata* (Rio Pachitea, upper Amazon basin)

ee. No longitudinal stripes along body; 26-30 anal rays.

f. A conspicuous black caudal spot; radii of scales of flanks not diverging up and down.

g. 5/31/4 scales ... *M. ovalis* (upper Amazon)

gg. 6/32-34/4-5 scales (a species strongly resembling *Tetragonopterus chalceus*) ... *M. georgiae* (Guianas)

ff. No caudal spot; radii of scales of flanks diverging up and down (5/31-34/4 scales) ... *M. grandisquamis* (Guianas and Amazon basin)

dd. Body depth usually 2.2-2.66 in standard length, at least in adults.

h. A caudal spot (5 or 6 scales above lateral line; caudal lobes hyaline).

i. Broad black bar across entire caudal base as in *M. cotinho* (body depth 2.2-2.25 in standard length on average; 22-32 lateral line scales).

j. 29-32 lateral line scales ... *M. oligolepis* (Guianas and Amazon basin)

jj. 22-26 lateral line scales (lateral line often irregular or incomplete) ... *M. sanctaefilomenae* (South of the Amazon basin, Rio Paraguay and Parnahyba)

ii. Caudal spot on middle caudal rays only; body depth 2.6, on the average, in standard length; 34-36 lateral line scales ... *M. robertsi* (Iquitos, Upper Amazon)

hh. No caudal spot; usually 5 transverse scales above lateral line (except in *M. melogramma* and *M. orteguasae*).

k. Caudal lobes hyaline.

Male *Moenkhausia pittieri.* Photo by Dr. Herbert R. Axelrod.

Moenkhausia pittieri. Photo by Stanislav Frank.

Well developed male *Moenkhausia pittieri.* Photo by Van Raam.

l. A black line along anal base (as in *M. copei* and *M. collettii* for example) (about 34 lateral line scales and 26 anal rays) . . . *M. melogramma* (Upper Amazon)

ll. No black line along anal base.

m. 30-32 lateral line scales.

n. 26 anal rays . . . *M. xinguensis*, type species (Rio Xingu)

nn. 23-24 anal rays . . . *M. browni* (Guyana)

mm. 35-37 lateral line scales (26-29 anal rays) . . . *M. agnesae* (Upper Amazon)

kk. Caudal lobes marked, at least upper one.

o. Both caudal lobes marked (as in *M. dichroura* for example); 28-31 anal rays.

p. Body depth 2.6 in standard length; 35 lateral line scales . . . *M. orteguasae* (Colombia) (very similar to the next species)

pp. Body depth 2.4 in standard length; 37 lateral line scales . . . *M. barbouri* (Amazon basin)

oo. Upper caudal lobe with oblique band ending on anal base; 27-28 anal rays (32-33 lateral line scales) . . . *M. costae* (Sao Francisco and Itapicuru Rivers)

aa. Eye 2.0-2.2 in head length (34 lateral line scales).

g. No caudal spot; a vertical humeral spot; 28-30 anal rays . . . *M. megalops* (Amazon basin and Guianas)

gg. A small caudal spot; no humeral spot; 26 anal rays (see also *M. georgiae*) . . . *M. shideleri* (Guyana)

(*c*) Key to the *Moenkhausia chrysargyrea*-group (7 or more scales above lateral line and 5 or more below; usually large species with a relatively deep body, some of which probably gave rise to *Tetragonopterus*, *Gymnocorymbus*, etc. *Moenkhausia latissima* is probably a *Gymnocorymbus* and is treated as such in the present book):

a. Body depth 2.4 or more in standard length.

b. 20-22 anal rays; no caudal spot (6 or 7/4 or 5 transverse scales; species linking the former groups with the present one).

c. Body depth 2.8-3.1 in standard length; dorsal well in front of midbody . . . *M. eigenmanni* (Upper Rio Meta)

cc. Body depth 2.5 in standard length; dorsal fin at midbody . . . *M. metae* (Upper Rio Meta)

bb. 26-31 anal rays; a caudal spot.

d. Body depth about 2.6 in standard length; humeral spot conspicuous; a faint caudal spot (28-31 anal rays; scale formula 7-8/36-37/6) . . . *M. surinamensis* (Surinam)

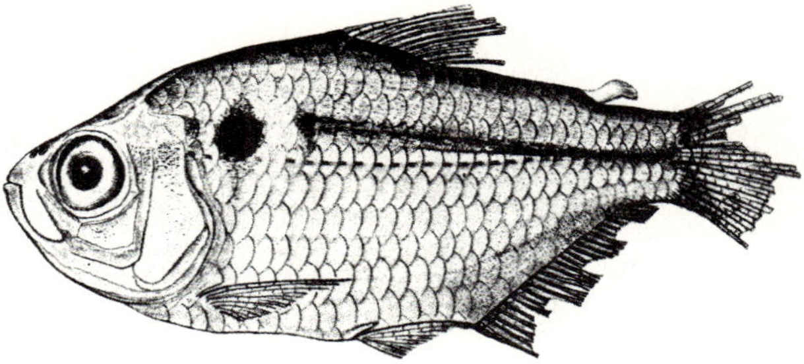

Unidentified *Moenkhausia* close to *M. comma* (after Gery, 1964).

dd. Body depth 2.4-2.5 in standard length; a faint humeral spot; caudal spot conspicuous (26-27 anal rays; scale formula 7/37-39/5) . . . *M. miangi* (Venezuela)

aa. Body depth usually 2.0-2.3 in standard length (in adults).

e. 28-35 branched anal rays.

f. No caudal spot; maxilla longer than snout, with 3-8 teeth (31-35 branched anal rays; scale formula 7-8/34-36/6-7; humeral spot horizontal) . . . *M. doceana* (eastern Brazil)

ff. A conspicuous caudal spot; maxilla not longer than snout, with only 0-1 tooth (28-32 branched anal rays; scale formula 7/35-37/5-6; humeral spot vertical) . . . *M. jamesi* and *M. justae* (Amazon basin)

ee. Less than 28 branched anal rays (no caudal spot; maxilla 3 or more in head length, with 1-4 teeth).

g. No humeral spot.

h. 5 inner premaxillary teeth; 8 gill-rakers on the lower arch (fins elongate in the adult male) . . . *M. pittieri* (Venezuela)

hh. Said to have only 4 inner premaxillary teeth and 12 gill-rakers on lower arch (fins not described) . . . *M. tridentata* (Amazon basin)

gg. One or two humeral spot(s).

i. Humeral spot single, not vertically elongate; no band over anal fin.

j. Humeral spot ovate, between opercle and dorsal fin level, which is at midbody or slightly before; usually 10 predorsal scales; 26-27 branched anal rays . . . *M. chrysargyrea* (Guianas and Upper Amazon basin)

jj. Humeral spot bullet-shaped, just behind opercle; dorsal fin clearly in front of midbody; usually 9 predorsal scales; 22-25 branched anal rays . . . *M. comma* (Amazon basin—very similar to the preceding, but apparently a distinct species)

ii. Two vertically elongate humeral spots; a conspicuous band over anal fin . . . *M. naponis* (Ecuador)

Bario steindachneri. Photo by Harald Schultz.

Tetragonopterus argenteus. Photo by Dr. Herbert R. Axelrod.

Tetragonopterus chalceus? Photo by Harald Schultz.

Tetragonopterus argenteus. Photo by Dr. Herbert R. Axelrod.

Genus *Tetragonopterus*

This genus differs from a deep, compressed *Moenkhausia* such as *M. georgiae*, for example, in the irregular disposition of the first scale rows. Instead of making a gentle curve after the opercle, the lateral line abruptly descends (step-like). The body is high, the preventral area flat and the predorsal scale series somewhat irregular but covering the predorsal area.

Two common species are known, *T. argenteus*, mostly from the Amazon and La Plata basin, which is considered as the type species, and *T. chalceus*, mostly from the Guianas and the Rio S. Francisco, but also quite abundant in certain parts of the Amazon. They differ in the number of lateral predorsal scales (12-16 on *argenteus* and 8-10 on *chalceus*) as well as in a few other minor details. They have two vertical black humeral spots (usually less prominent in *T. chalceus*) and a variable caudal spot.

Two other species (?), *T. huberi* and *T. gibbosus*, are too poorly known to deserve more than a mention.

Finally, it must be noted that a number of other species (130 or so) were first described in the genus *Tetragonopterus* and have been redistributed into other genera such as *Moenkhausia*, *Astyanax*, *Bryconamericus*, etc.

Schultzites axelrodi

This small species (35 mm), which has been found rather recently in the upper Rio Meta in Colombia, externally resembles *Moenkhausia dichroura* or *M. intermedia*. Technically it would belong to *Moenkhausia* except for its maxilla, which is toothed on its proximal two-thirds, having 7-10 teeth

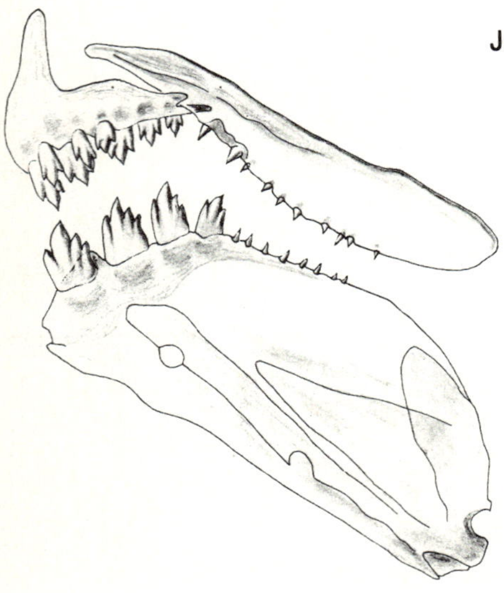

Jaws of *Schultzites axelrodi* (after Gery, 1964).

(usually 1-3, rarely more than 5 in all *Moenkhausia* species). In the *Moenkhausia-Tetragonopterus*-group it represents, in a way, the counterpart of *Pseudochalceus* in the *Astyanax*-group.

Gymnotichthys hildae

If not for the naked predorsal line which is mentioned in its description, this species (which the writer has never seen), from the Rio Autana in Venezuela,would easily fit within the definition of *Moenkhausia lepidura*.

Genus *Gymnocorymbus*

The *Gymnocorymbus* species (including the popular black tetra, *G. ternetzi*, the best known species among aquarists) represent the adaptation of some deep *Moenkhausia* toward a drisciform body (with the body depth 1.8-2.2 in the standard length in adults), accompanied by a naked predorsal line. The end of this evolutionary trend is represented by the Stethaprioninae, of which one member is exactly like *Gymnocorymbus thayeri*, with an even deeper body and a predorsal spine.

Spawning of the black tetra (*Gymnocorymbus ternetzi*). Photo by Ruda Zukal.

Schultzites axelrodi. Photo by Dr. Herbert R. Axelrod.

Photo of the holotype of *Gymnocorymbus socolofi*, a 52 mm specimen with adult coloration. Photo by Dr. Herbert R. Axelrod.

Gymnocorymbus socolofi, young of 27 mm. Photo by Dr. Herbert R. Axelrod.

Gymnocorymbus thayeri. Photo by Dr. Herbert R. Axelrod.

The species may be identified as follows:

a. About 6/6 transverse scales; coloration like that of *Bario*, i.e. with a horizontally elongate humeral spot and horizontal dark lines between the scale-series (body depth 2, on the average, in standard length; 31-36 anal rays, usually 33-35; 33-36 longitudinal scales) . . . *G. latissimus* (usually called *Moenkhausia latissima*—Upper Amazon near Tabatinga)

aa. 7-8/7-8 transverse scales; 2 vertical humeral bars; no dark stripes between the rows of scales (anal fin usually with more than 35 rays).

b. Body depth 2.2 and head length 3.66 in standard length in specimens 50 mm long (unpaired fins red in young) . . . *G. socolofi* (Upper Rio Meta)

bb. Body depth 1.8 and head length 3.85 (on average) in standard length in specimens 50 mm long and more.

c. Anal fin with 38 or 39 rays on the average, its margin straight, its base covered by 3 or 4 series of small scales; body light, the humeral bars faint as in *Tetragonopterus argenteus*, unpaired fins usually orange . . . *G. thayeri* (synonyms *Moenkhausia profunda* and *M. bondi*; strongly resembling *Poptella orbicularis* but less deep and without predorsal spine; Amazon and Orinoco basins)

cc. Anal fin with 40-42 rays on the average, its margin convex, its base covered by 5 or 6 series of small scales; body dark, humeral bars intense and unpaired fins also dark, even in young . . . *G. ternetzi*, the popular black tetra (Rio Paraguay basin; sympatric with the preceding in the Rio Guapore basin)

Bario steindachneri

This rather large Amazonian species, which was called for a time *Entomolepis steindachneri*, is a *Moenkhausia* with the scales strongly crenate when the fish reaches about 55 mm in standard length. Its proportions and meristics are those of a species of the intermediate group (*grandisquamis*-group), with the body depth about 2.25 in the standard length, 5/4 transverse scales and 25 or 26 anal rays; the preventral area is somewhat keeled. The color pattern, very characteristic, consists of a faint humeral spot, a conspicuous caudal spot and a series of longitudinal stripes along the sides between the scale rows as in *Pseudochalceus lineatus* and *P. multifasciatus*, *Astyanax lineatus* and *A. superbus*, *Moenkhausia agnesae* (very faint lines) and *Gymnocorymbus latissimus*, and not as in *Astyanax saltor* and *Moenkhausia simulata*, where the horizontal stripes are formed by series of dots on the middle of the scales.

Markiana nigripinnis. Photo by Dr. J. Gery.

Genus *Markiana*

Markiana nigripinnis, from La Plata basin, and *M. geayi,* from Venezuela, are deep bodied, *Moenkhausia*-like species with the scales crenate and smaller below the lateral line than above, with the short caudal lobes scaled and the rounded, convex anal fin also scaled.

Third group of American Tetras

HEMIGRAMMUS and ALLIED GENERA

In this group are placed, rather artificially, several small forms which have possibly evolved from some more conservative *Astyanax*- or *Moenkhausia*-like tetra by a process of paedomorphosis. This process, also called neoteny, is supposed to have played a large role in the evolution of several zoological groups, including man. The species would retain, it is believed, some juvenile characteristics of its ancestor, even at the adult stage. In the present case the small size and the incompleteness of the lateral line (which usually grows from front to back during ontogeny) are presumably retained.

The genera can be recognized as follows:

a. Adipose fin present.
  b. Caudal fin not scaled . . . (1) *Hyphessobrycon*
  bb. Caudal fin scaled, at least on its base to a certain extent.
    c. Maxilla long and slender, forming a simple curve with the premaxilla and armed with 6-12 conical or tricuspid teeth (caudal scaled on its base only—see also *Pseudochalceus*) . . . (2) *Parapristella*
    cc. Maxilla of moderate length, forming an angle with the premaxilla, usually with less than 6 teeth.
      d. Caudal lobes of equal length; swimming position horizontal.

Above and below: Two color varieties of *Gymnocorymbus thayeri*. Photos by Dr. Herbert R. Axelrod.

Above and below: Color variations in normal *Gymnocorymbus ternetzi.* Photos by Harald Schultz.

e. Maxilla strongly curved on its upper part, resembling that of a *Bryconops* . . . (3) *Ramirezella newboldi*

ee. Maxilla with a "normal", gentle curve.

f. Inner premaxillary row of teeth prevailing over outer one, as in the majority of tetras . . . (4) *Hemigrammus*

ff. Outer premaxillary row prevailing, owing to the regression of inner row (a link between the Tetragonopterinae and the *Pristella*-like genera ?) . . . (5) *Bryconella*

dd. Lower caudal lobe longer than the upper one; swimming position head up . . . (6) *Thayeria*

aa. Adipose fin absent (caudal fin usually not scaled) . . . (7) *Hasemania*

Genus *Hyphessobrycon*

American Tetras

As classically understood, the large genus *Hyphessobrycon* is a complex assemblage of several morphologic types, and it is probably not derived from a single stem. At least two species diverge from the bulk in having the scales quite numerous (45-48 instead of less than 37 in the other species), the predorsal area naked and the circumorbital bones very weak. These species, from Central America and Mexico, are northern offshoots of more generalized Amazonian forms such as certain members of the *callistus*-group, and will be treated at the end of the present account of the genus, just after the said group. However, for nomenclatural reasons, the name *Hyphessobrycon* applies properly to these two aberrant forms, and all others are not "true" *Hyphessobrycon.*

It is the author's opinion that at least certain of these "untrue" *Hyphessobrycon* deserve a new generic name, but also that this action would cause considerable trouble in the mind of most aquarists. This step has thus to be delayed until a complete revision of the American tetras has been achieved.

We may divide the 65 or so *Hyphessobrycon*-like species into the following quite artificial groups, according to their pattern:

(*a*) species without black markings on body;

(*b*) species with one or two humeral spot(s) (such as *Hy. flammeus*);

(*c*) species with a caudal spot (such as *Hy. minimus*);

(*d*) species with both humeral and caudal spots (such as *Hy. tropis*);

(*e*) species with a longitudinal pattern, usually a band uniting the humeral and caudal spots, which are no longer recognizable as such; sometimes the band is broken and the humeral spot is distinct from it. In this group, two subgroups can be formed according to the width of the the band: the so-called "*agulha*-group," where the band is usually not very dark but very broad, on the lower half of the body and most distinct only above the anal fin (example: *Hy. herbertaxelrodi*); and the so-called "*heterorhabdus*-group," where the intense band is relatively narrow and goes along the mid-axis of the body (example: *Hy. stegemanni*);

(*f*) species with a black spot on dorsal fin (such as *Hy. callistus*).

These groups will be keyed out in the following pages. The three last groups (*d*, *e*, and *f*), much discussed among zoologists as well as aquarists, will be described in the key with somewhat more detailed characters.

(*a*) Species without spots or bands on body. This group comprises only a few species, as follows:

a. Inner premaxillary teeth narrow and numerous (10-12); base of caudal fin with a "larval" pattern, as in certain *Microschemobrycon*, of two crescent-shaped brown spots . . . *Hy. riddlei* (certainly an aberrant member of the genus, close to certain Aphyoditeina, and known from a single Venezuelan specimen)

aa. Inner premaxillary teeth not very narrow and usually 5 in number.

b. 12-14 transverse scales . . . *Hy. panamensis* (Panama and Colombia)

bb. Less than 12 transverse scales . . . *Hy. gracilior* (Upper Amazon) (*Hy. gracilis* see under *Hemigrammus*)

Hemigrammus gracilis and *Hyphessobrycon riddlei* (after Eigenmann, 1918).

Aquarium developed strains of the black tetra, *Gymnocorymbus ternetzi.* Upper photo by Jiri Taborsky, photo below courtesy Wardley Products Co.

Long-finned black tetras, *Gymnocorymbus ternetzi,* obtained by selective breeding in aquaria. Photo by Dr. Herbert R. Axelrod.

Hyphessobrycon bifasciatus. Photo by Harald Schultz.

(*b*) Species with one or two humeral spots:

a. A single humeral spot (large suborbital never entire; maxillary teeth always narrow).

b. 22-26 anal rays (total count) (1-3 maxillary teeth).

c. Maxilla large, almost vertical, with 1 tooth . . . *Hy. maxillaris* (Rio Paraguay basin)

cc. Maxilla of moderate length, oblique (body depth 2.6-3.0; humeral spot vertically elongate; fins red; otherwise has many of the characters of *He. bellotti*, a species considered as a *Hemigrammus* in the present work) . . . *Hy. ecuadoriensis* (Pacific slope of Ecuador)

bb. 20 or fewer anal rays (total count).

d. 20 anal rays; 7 maxillary teeth; humeral spot vertically elongate . . . *Hy. catableptus* (a dubious species from Guyana, known from a single post-larval specimen)

dd. 13-14 anal rays; 2-4 maxillary teeth; humeral spot ovate . . . *Hy. albolineatum* (Venezuela) (possibly not a *Hyphessobrycon* in the usual sense)

aa. Humeral spot double, the marks vertically elongate.

e. 9-10 maxillary teeth; 14 transverse scales (29-31 anal rays, 35-36 longitudinal scales) . . . *Hy. tortuguerae* (Costa Rica)

ee. Less than 7 maxillary teeth; less than 14 transverse scales.

f. 29-32 anal rays (usually 1 maxillary tooth; maxilla broad and moderate in length; usually 34 longitudinal scales; anterior humeral bar prominent) . . . *Hy. bifasciatus* (S.E. Brazil)

ff. Less than 29 anal rays.

g. Usually more than 1 maxillary tooth; maxilla narrow and rather elongate.

h. Less than 5 maxillary teeth; 6 or even 7 scales above lateral line row.

i. 1-4 maxillary teeth, usually 3; humeral bars of equal intensity . . . *Hy. savagei* (Costa Rica)

ii. 1-2 maxillary teeth; posterior humeral bar prominent . . . *Hy. griemi* (S.E. Brazil)

hh. 5-6 maxillary teeth; 5 or 5½ scales above lateral line row (posterior humeral bar prominent) . . . *Hy. flammeus* (S.E. Brazil)

gg. 1 maxillary tooth; maxilla short (anterior humeral bar prominent; sometimes a darkening of the caudal peduncle and middle caudal rays; 20-23 anal rays; lateral line developed on 12-20 scales, sometimes complete or almost complete) . . . *Hy. balbus* (Goyaz)

Hyphessobrycon griemi, Ribeira River, South Sao Paulo, Brazil. Photo by Harald Schultz.

Hyphessobrycon bifasciatus. Photo by Harald Schultz.

Hyphessobrycon griemi. Photo by R. Zukal.

Hyphessobrycon minimus. Photo by Dr. Herbert R. Axelrod.

Hyphessobrycon flammeus. Photo by Klaus Paysan.

(*c*) Species with a caudal spot:

a. Caudal spot said to extend over fin to tip of middle caudal rays; no maxillary teeth . . . *Hy. erythrurus* (a poorly-known species; locality unknown)

aa. Caudal spot not extended onto the middle caudal rays.

b. 0-2 maxillary teeth.

c. Inner premaxillary teeth numerous (9) and narrow (tricuspid); maxilla without tooth; body depth 2.8 . . . *Hy. stramineus* (Upper Amazon)

cc. 5 or 6 inner premaxillary teeth.

d. Maxilla very short, with 2 broad teeth.

e. Body depth 2.15 . . . *Hy. latus* (Ceara)

ee. Body depth 2.8 . . . *Hy. iheringi* (Ceara, sympatric with the preceding)

dd. Maxilla not very short (body depth usually more than 2.8).

f. Premaxillary teeth 5-7 pointed; 1-2 broad maxillary teeth . . . *Hy. tukunai* (Upper Amazon)

ff. Premaxillary teeth tricuspid; 0 or 1 maxillary tooth.

g. Caudal peduncle not very deep; 20-21 anal rays . . . *Hy. parvellus* (S.E. Brazil)

gg. Caudal peduncle very deep; 23 anal rays . . . *Hy. piabinhas* (S.E. Brazil; restricted to the type)

bb. 2-3 very broad, 5-7 pointed teeth on maxillary . . . *Hy. minimus* (Guyana)

Hemigrammus iota and *Hyphessobrycon minimus* (after Eigenmann, 1912).

Hyphessobrycon reticulatus (Eigenmann, 1918).

(*d*) Species with a humeral and a caudal spot:

a. 14 transverse scales . . . *Hy. robustulus* (Peruvian Amazon)

aa. Less than 14 transverse scales.

b. Caudal peduncle symmetrically marked.

c. Caudal spot extended onto middle caudal rays.

d. Large suborbital leaving a naked area on cheek.

e. Dorsal fin behind midbody; humeral and caudal spots usually quite faint (20-22 anal rays; 2 maxillary teeth, each with 3 to 5 points) . . . *Hy. santae* (Paraguay and S.E. Brazil)

ee. Dorsal fin at midbody; humeral and caudal spots conspicuous.

f. Maxilla with only one narrow tooth (tricuspid) . . . *Hy. anisitsi* (southeastern Brazil, Paraguay and Uruguay basins)

ff. Maxilla with 2 very broad (6 to 9 pointed) teeth . . . *Hy. luetkeni* (S.E. Brazil and Paraguay basin)

dd. Large suborbital almost entire, in contact with preopercular canal.

g. 16-18 anal rays (see also *Hy. poecilioides* below); 36 longitudinal scales . . . *Hy. duragenys* (Rios Parahyba and Tieté)

gg. 18-21 anal rays; 31-44 longitudinal scales . . . *Hy. reticulatus* (S.E. Brazil)

Hyphessobrycon peruvianus from the upper Rio Meta. Photo by Dr. Herbert R. Axelrod.

Hyphessobrycon loretoensis. Photo by Karl Knaack.

Two aspects of *Hyphessobrycon metae.* Photos by Stanislav Frank.

cc. Caudal spot not extended onto middle caudal rays.

h. 1-3 maxillary teeth.

i. Maxilla rather small, with one broad tooth; body depth 2.5-2.75.

j. 24-26 anal rays . . . *Hy. inconstans* (Para) and *Hy. proteus* (Colombia)

jj. 31 anal rays . . . *Hy. maculicauda* ("Middle Brazil")

ii. Maxilla long and slender, with 1-3 conical teeth.

k. 21 anal rays; body depth 3.25 . . . *Hy. hildae* (Venezuela)

kk. 22-25 anal rays, body depth 3.7-4.5 . . . *Hy. tenuis* (Upper Amazon)

hh. 8 conical maxillary teeth (an anal keel in the male) . . . *Hy. tropis* (Rio Negro)

bb. Caudal peduncle not symmetrically marked, its upper part light (large suborbital broad).

l. 1-3 broad maxillary teeth; 22 anal rays . . . *Hy. melazonatus* (Amazon)

ll. 5-7 narrow maxillary teeth; 17-20 anal rays . . . *Hy. eos* (Guyana)

(*e*) Species with a longitudinal pattern.

The "*Hy. agulha*-group" (lower half of the body dark, especially above anal fin; usually a horizontally elongate humeral spot, more or less united with the asymmetrical, broad band):

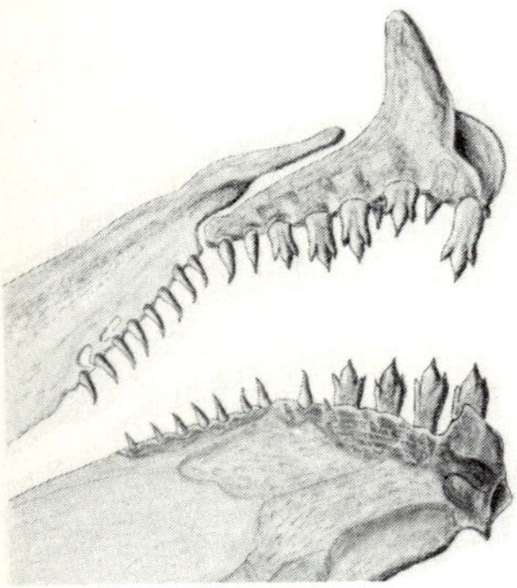

Hyphessobrycon tropis, type, and illustrations of its jaws (after Gery, 1963).

Hyphessobrycon herbertaxelrodi. Photo by Milan Chvojka.

a. Elongate, body depth 3.8-3.9 in standard length; caudal peduncle narrow, its depth about 1.8 in its length (head short, interorbital broad, snout short and round; dorsal fin before midbody; 9-10 predorsal scales; (8-10) 34-35 longitudinal scales; caudal fin red or brownish) . . . *Hy. peruvianus* (Peruvian Amazon; possibly a *Hemigrammus*)

aa. Body depth less than 3.6; caudal peduncle usually 1.2-1.75; snout of moderate length, more or less pointed.

b. Head length 3.4-3.5 in standard length; 10-11 predorsal scales (dorsal at midbody or scarcely before; (5-7) 33-34 longitudinal scales; caudal fin brick- or orange-red) . . . *Hy. loretoensis* (Peruvian Amazon)

bb. Head length 3.5-4.0; usually 9 or 10 predorsal scales.

c. Maxilla less than 3.3 times in head length; caudal peduncle usually 1.5-1.75.

d. Dorsal fin behind midbody; maxilla not reaching pupil level; (6-9) 30-31 longitudinal scales; 18-19 branched anal rays . . . *Hy. metae* (Rio Meta in Colombia; very close to *Hy. loretoensis*)

dd. Dorsal fin before midbody; maxilla reaching pupil level; (12-18) 33-35 longitudinal scales; 19-21 branched anal rays . . . *Hy. agulha* (Amazon basin; possibly a *Hemigrammus*, or a sibling species exists)

cc. Maxilla 3.4-3.5 in head length; caudal peduncle usually 1.2-1.4 ((7-8) 32-34 longitudinal scales) . . . *Hy. herbertaxelrodi*, the "black neon" (Mato Grosso)

Hyphessobrycon saizi. Possibly an abnormal form of *H. metae,* in the same relationship as *Hemigrammus armstrongi* is to *H. rodwayi.* Photo by Dr. Herbert R. Axelrod.

Hyphessobrycon agulha, or close to it. Photo by Harald Schultz.

The black neon tetra, *Hyphessobrycon herbertaxelrodi.* Photo by Dr. Herbert R. Axelrod.

Another color phase of the black neon, *Hyphessobrycon herbertaxelrodi.* Male above, female below. Photo by S. Frank.

The "*Hy. heterorhabdus*-group".

(*Hy. nigrifrons* Ahl may belong here; it is not included because of its incomplete description). The following is a transcription of the author's paper of 1966 (*Ichthyologica* 37 (2): 63-70) with suppression of *Hy. herbert-axelrodi* (now placed in the "*agulha*-group") and addition of *Hy. cachimbensis* and *Hy. simulans* (which is difficult to place, having however a faint broad longitudinal band after preservation in formalin).

a. Teeth narrow, the broadest tricupid; large suborbital very big or entire.

b. At least 4 maxillary teeth; dorsal fin at midbody or usually behind; large suborbital not quite covering cheek (body depth 2.5-3.0; maxilla about 2.7 in head; 14-17 branched anal rays; 6/4 transverse scales; 5 inner premaxillary teeth, the broadest tooth with 5 cuspids, all others with 3; vertically elongate humeral spot; band to tip of middle caudal rays) ... *Hy. coelestinus* (Goyaz) (see also "*Hy.*"*melanopleurus* in the key to *Hemibrycon* and allied genera)

bb. 1 maxillary tooth; dorsal fin at midbody; large suborbital covering entire cheek (body depth 2.8-2.9; maxilla more than 3 in head; 16-21 branched anal rays; 5/4 transverse scales; band equal in width, beginning behind head membrane) ... *Hy. sovichthys* (Venezuela)

aa. Teeth broad, at least quinquecuspid in inner premaxillary series and in front of mandible; large suborbital variable in size.

c. Dorsal fin at midbody or more or less behind (broadest teeth usually 7-pointed; 5 or 6/4 or 5 transverse scales).

d. Less than 18 branched anal rays; dorsal fin behind midbody; maxilla short, more than 4 times in head length.

e. Head heavy, interorbital width 2.2-2.5 in head length; large suborbital almost entire; body depth 2.6-3.0.

f. Head length 3.8-4.0; eye 3.0-3.25, interorbital 2.2 and maxilla almost 5 in head length; 12-14 branched anal rays; maxilla not toothed; band expanded on peduncle, forming a rather distinct caudal spot ... *Hy. taurocephalus* (Rio Iguassu)

Hyphessobrycon poecilioides (after Gery, 1966).

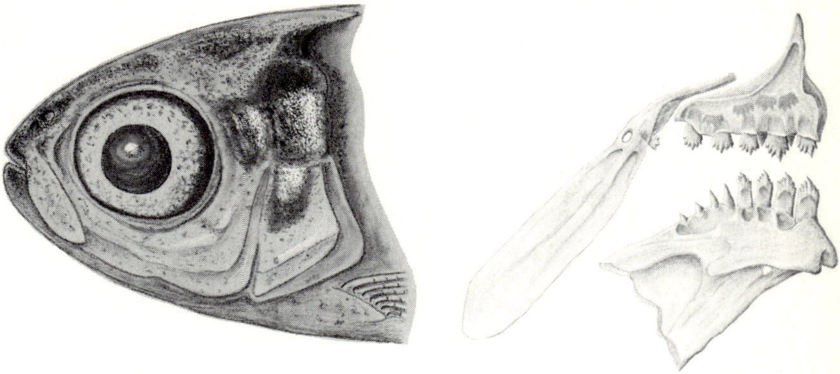

Head and jaws of *Hyphessobrycon simulans* (after Gery, 1963).

ff. Head length 3.66-3.75; eye 4, interorbital about 2.5 and maxilla more than 4 in head length; 14-16 branched anal rays; 0-2 maxillary teeth (usually 1); a vertically elongate humeral spot, distinct from lateral band ... *Hy. poecilioides* (Rio Cauca in Colombia)

ee. Head not heavy, interorbital width more than 3 in head length; large suborbital very reduced; body depth 3.4-3.8 (head length 3.65-3.85; eye 4.15-4.6 and maxilla 4.15-4.6 in head length; 16-17 branched anal rays; 1 maxillary tooth; coloration very similar to that of the neon tetra, *Paracheirodon innesi*) ... *Hy. simulans* (Rio Negro)

dd. More than 18 branched anal rays; dorsal fin at midbody or very slightly behind; maxilla not more than 4 in head length (interorbital width more than 2.75 in head length; body depth about 3).

g. Head length more than 4; maxilla about 4 in head length, with 0-2 teeth; large suborbital almost entire; 20-21 branched anal rays; lateral band expanded on caudal peduncle, forming a distinct caudal spot; fins yellowish ... *Hy. scholzei* (Para)

gg. Head length less than 4; maxilla 3.0-3.75 in head length, with 1-3 broad teeth (usually 2); large suborbital leaving naked area on cheek; 19-20 branched anal rays, lateral band not expanded on peduncle; fins bluish ... *Hy. vilmae* (Upper Rio Juruena, Brazil)

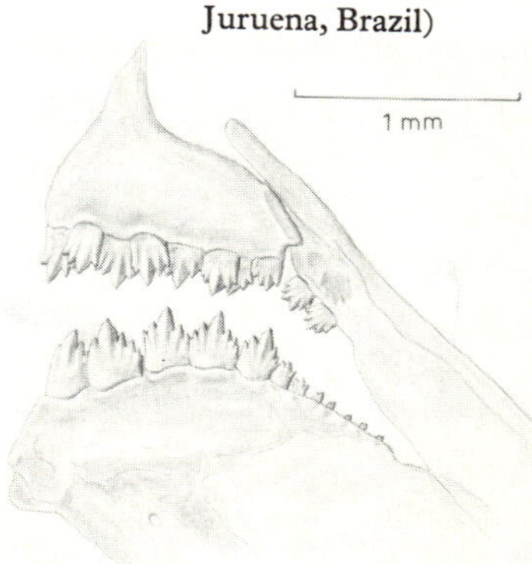

Jaws of *Hyphessobrycon vilmae* (after Gery, 1966).

Hyphessobrycon simulans. Photo by Dr. Herbert R. Axelrod.

Hyphessobrycon herbertaxelrodi. Photo by Dr. Herbert R. Axelrod.

Two photos of the false neon tetra, *Hyphessobrycon simulans.* Upper photo by Dr. Herbert R. Axelrod, lower photo by Hans Peter.

Hyphessobrycon heterorhabdus. Photo by Milan Chvojka.

cc. Dorsal fin in front of midbody (5/3 or 3½ transverse scales).

h. Broadest teeth 5-pointed; body depth 3.15-3.6; lateral band not expanded onto middle caudal rays (more than 18 branched anal rays; large suborbital leaving naked area on cheek; maxilla 2.8-3.1 in head length) ... *Hy. heterorhabdus* (middle and lower Amazon)

hh. Broadest teeth 7 or 9-pointed; body depth 2.5-3.1; lateral band expanded on peduncle and spreading over middle caudal rays (fins yellowish).

i. Maxilla reaching to pupil, 2.8 in head length; 20 branched anal rays ... *Hy. cachimbensis* (Para)

ii. Maxilla not reaching to pupil, 3.5-4.0 in head length; 17-18 branched anal rays ... *Hy. stegemanni* (lower Tocantins)

(*f*) The *Hy. callistus*-group.

The name "*callistus*-group" is extremely popular among aquarists. It probably covers an artificial assemblage of *Hyphessobrycon*-like forms (some rather close to the type, *Hy. compressus*) which share the possession of a "flag," a signal indicating (for meeting purposes, probably) their belonging to such and such species. That signal is a black, very conspicuous spot on the dorsal fin, usually underlined by a white or yellow zone. The species of the "*callistus*-group" frequently live in association with other species (belonging to other groups, i.e. having different teeth for example) such as *Pristella*, *Hemigrammus unilineatus*, and others.

Within the group, *Hy. bentosi* and the species close to it form a clear-cut subgroup which has been called the "*bentosi*-group." It is related to the type of the genus, *Hy. compressus*, and at the same time scarcely distinguishable from certain *Megalamphodus* species, except for the teeth.

Key to the *Hy. callistus*-group (see also *Cheirodon troemneri*):

a. Usually not more than 2-3 maxillary teeth and 6 inner premaxillary teeth; humeral spot usually conspicuous; pseudotympanum usually not very conpicuous in the adult (except in *Hy. serpae*) (so-called *callistus*-group in the strict sense).

b. Dorsal spot on two first rays only, forming band with yellow spot on tips; anal fin with similar marking on first rays, followed by a black border (meristics comparable to those of *H. serpae*) . . . *Hy. pulchripinnis*, the lemon tetra (Amazon basin)

bb. No such pattern, the dorsal spot on more than 2 first rays, first anal rays not black.

c. Anal fin with iv, 15-17 rays; no humeral spot; very small size (less than 20 mm in standard length) (4, more rarely 5, broad inner premaxillary teeth; 2-3 maxillary teeth) . . . *Hy. georgettae* (Surinam)

cc. Anal fin usually with iii or iv, 23-28 rays; humeral spot variable; size moderate (30-40 mm in standard length).

d. Humeral spot broad (much broader than eye diameter) and conspicuous, almost covering height of sides (meristics comparable to those of *Hy. serpae*) . . . *Hy. takasei*, the coffee bean tet (Amapa Territory in Brazil)

Hyphessobrycon pulchripinnis. Photo by Timmerman.

The black-band tetra, *Hyphessobrycon scholzei.* Photo by Klaus Paysan.

Hyphessobrycon scholzei? Photo by S. Frank.

Hyphessobrycon heterorhabdus, the flag tetra with "gold disease." Photo by Dr. Herbert R. Axelrod.

Normal *Hyphessobrycon heterorhabdus.* Photo by S. Frank.

Aquarium form of the *Hyphessobrycon callistus-serpae* group. Photo by Timmerman.

dd. Humeral spot variable in size, never broader than eye diameter, never covering almost height of sides.

e. 5 broad teeth in inner premaxillary series, usually 5-7 pointed; humeral spot very conspicuous, roundish, surrounded by a light area . . . *Hy. copelandi* (upper and middle Amazon basin)

ee. Usually six 3-5 pointed teeth in inner premaxillary series.

f. Humeral spot vertically elongate, lozenge-shaped, conspicuous; a broad black anal border; 6-7/4 transverse scales . . . *Hy. callistus* (Paraguay basin and (?) tributaries of the Amazon in the southern part of the basin)

ff. Humeral spot and anal border usually less conspicuous; 5 or 5½/3-3½ transverse scales . . . *Hy. serpae*-complex (Amazon basin and Guianas); 4 forms are known, which are difficult to tell apart: *Hy. serpae, Hy. minor, Hy. haraldschulzi* and the fish called "*red minor*" by aquarists

aa. Usually 4-11 maxillary teeth and 6-10 inner premaxillary teeth; humeral spot faint or absent; pseudotympanum conspicuous, even in adults (so-called *bentosi*-group).

g. 7-9/5-6 transverse scales ((8-11) 34-35 longitudinal scales; iii, 27-32 anal rays; body depth less than 2.6 in standard length (adults); eye 2.5-2.65 in head length; no black humeral spot, but a red one in life) . . . *Hy. erythrostigma* (synonym *rubrostigma*), the bleeding heart tetra (upper Amazon basin)

gg. 5-6/3-4 transverse scales ((5-7) 31-34 longitudinal scales and iii, 23-27 anal rays).

h. Humeral spot absent; 6-7 inner premaxillary teeth and 4-8 maxillary teeth; rarely more than 33 longitudinal scales *(Hy. bentosi)*.

i. Body depth 2.5-2.7; eye rather large in adults, about 2.3-2.5 in head length; 5/3-4 transverse scales; usually 4-5 maxillary teeth, rarely to 7 . . . *Hy. bentosi rosaceus* (Guianas and Rio Guaporé)

ii. Body depth 2.8-3.15; eye about 2.6 in head length in adults; 5-6/3-3½ transverse scales; 6-8 maxillary teeth . . . *Hy. bentosi bentosi* (synonym *Hy. ornatus* ?) (Amazon basin)

hh. Humeral spot present, faint, vertically elongate; 7-9 inner premaxillary teeth and 7-10 maxillary teeth; 34 longitudinal scales (apparently 6/4-4½ transverse scales; body depth 2.5-2.6) . . . "Tetra robertsi" of aquarists, an undescribed species (or hybrid ?) of the *bentosi*-group

Hyphessobrycon bentosi (after Eigenmann, 1918).

Hyphessobrycon vilmae. Photo by Harald Schultz.

Hyphessobrycon stegemanni from the savannah of northeastern Brazil. Photo by Dr. E. Schmidt.

Young *Hyphessobrycon stegemanni.* Photo by Harald Schultz.

Hyphessobrycon georgettae. Photo by S. Frank.

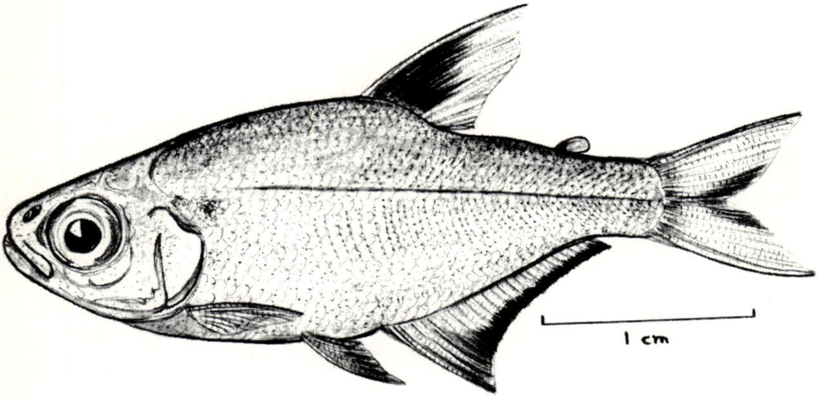

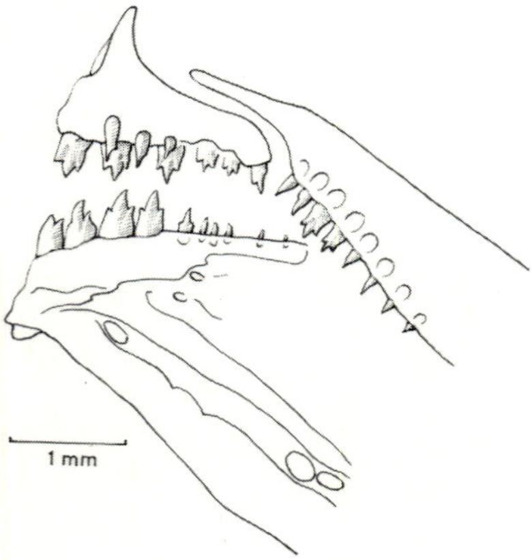

Hyphessobrycon sp. close to *compressus* from Guatemala (after Gery, 1971), and jaws of a paratype of *Hyphessobrycon compressus* (after Gery, 1966).

(*g*) The *Hy. compressus*-group ("true" *Hyphessobrycon*).

Hy. compressus, the type species, and *Hy. milleri*, respectively from Mexico and Guatemala, differ from the other species in having 45-49 longitudinal scales, 16-18 transverse scales, as well as the predorsal line naked. This character has a subgeneric value in *Poecilurichthys* (of the genus *Astyanax*), and a generic value in *Gymnocorymbus*, a *Moenkhausia*-like tetra. The dorsal fin is marked by a conspicuous black spot as in the *Hy. callistus*-group.

It has been suggested, following the study of a sample coming from near the locality of *Hy. milleri* (known from a single specimen, 40 mm long), that the latter could be the adult stage of *Hy. compressus* (largest known specimens 30 mm long), which is the older name, nomenclaturally speaking.

Genus *Parapristella*

American Tetras

Parapristella aubynei, from Guyana, has been separated from *Pristella maxillaris* on account of its tetragonopterine premāxilla (two well separated rows of teeth), straight premaxillary-maxillary border, and relatively feeble scalation of the lower caudal lobe. It may be considered as a *Hemigrammus* with an increased number of maxillary teeth, and it is certainly closer to that genus than to *Pristella*. It is a rather elongate fish with a humeral spot and a large spot on the base of the middle caudal rays, as well as a short anal fin. In life the caudal lobes are red or orange on their base.

A second species, *P. georgiae*, from the upper Rio Meta, differs in having even fewer anal rays and the maxillary teeth (as well as the lateral mandibular ones) tricuspid instead of conical.

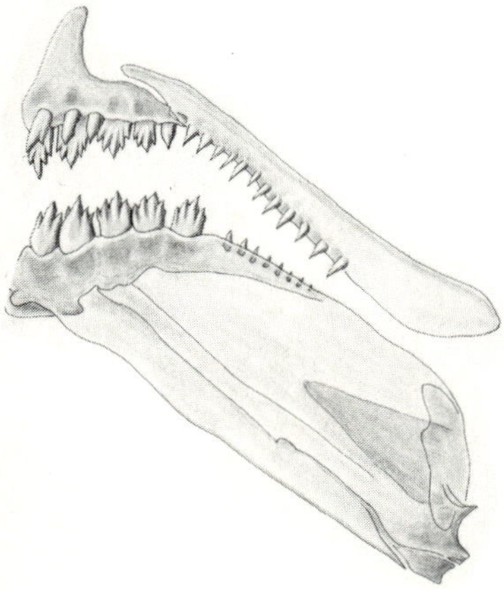

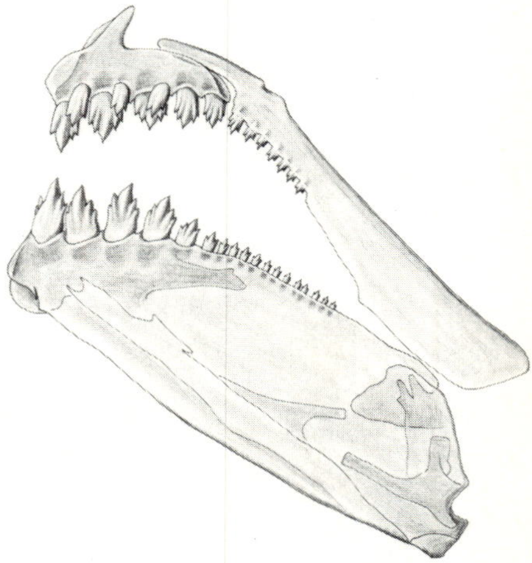

Jaws of *Parapristella aubynei* and *Parapristella georgiae* (after Gery, 1964).

Ramirezella newboldi

This Venezuelan tetra is said to have a toothless maxilla forming a strong curve at its meeting with the premaxilla, like a *Bryconops*. Otherwise it resembles a *Hemigrammus* of group d, with a rather faint, round humeral spot and a very large, conspicuous caudal spot (like *He. megaceps* for example, but with fewer anal rays).

Hyphessobrycon pulchripinnis, the lemon tetra. Photo by R. Zukal.

A color phase of *Hyphessobrycon pulchripinnis.* Photo by Stanislav Frank.

Hyphessobrycon copelandi. Photo by Dr. Herbert R. Axelrod.

Genus *Hemigrammus*

American Tetras

The *Hemigrammus* species, which are somewhat less numerous than the *Hyphessobrycon* species (about 40 instead of 65), are like the latter but with the caudal fin scaled. A few species, *He. stictus* and *He. bellottii* for example, are intermediate, having a few scales only on the basal part (about one fourth) of the caudal lobes; they are here classified within *Hemigrammus*.

Hemigrammus parallels the geographic distribution of *Moenkhausia*, being mostly Guianan and Amazonian, with only a few species from the La Plata basin. Certain species, such as the "head and tail light" (of uncertain origin) have iridescent spots, chiefly on the caudal peduncle but also on the eye and sometimes on the caudal base, which probably act as recognition signals.

As in the genus *Hyphessobrycon*, a number of artificial groups are to be found, depending on the color pattern:

(*a*) Species without black markings (for example *He. gracilis*);

(*b*) Species with one (or two) humeral spot(s) and no caudal spot, such as *He. bellottii*;

(*c*) Species with a caudal spot and no humeral spot (for example the Costello tetra);

(*d*) Species with both a humeral (eventually two) and a caudal spot (for example the false "head and tail," *He. ocellifer*);

(*e*) Species with a black spot on the dorsal fin such as the type species, *He. unilineatus*.

In the genus *Hemigrammus* there is no group corresponding to the one with a longitudinal band recognized in the genus *Hyphessobrycon*. Almost all species have a faint line along the body. The few species which have a more or less distinct "band" have been distributed throughout the other groups (*He. erythrozonus*, the glowlight tetra, in group (*a*), *He. levis* in group (*c*), *He. barrigonae* in group (*d*), and *He. ulreyi* and *He. coeruleus* in group (*e*)). However *Hy. agulha* or its sibling (?), if it is a *Hemigrammus*, could be the exception to the rule.

(*a*) The *He. gracilis*-group: a few rather elongate *Hemigrammus* having very faint or no black markings on body, namely *He. erythrozonus*, the glowlight tetra of aquarists (incorrectly called *gracilis*), *He. gracilis* (the true *gracilis*, for a time classified within *Hyphessobrycon*) and *He. mimus*. They can be identified as follows:

a. Inner premaxillary teeth narrow and numerous, numbering 10 or 11 on each side (3 or 4 tricuspid maxillary teeth; 17 anal rays) (an aberrant member of the genus, homologous with *Hy. riddlei*) . . . *He. mimus* (a mimic of *Microschemobrycon* in the Rio Negro)

aa. Inner premaxillary teeth 3- to 5-pointed, numbering 5 on each side as in most *Hemigrammus* (17-24 anal rays).

Hemigrammus erythrozonus. Photo by G. Timmerman.

b. A lateral band, brilliant red in life (light after preservation); dorsal fin also red; 2 or 3 tri-to pentacuspid maxillary teeth . . . *He. erythrozonus*, the glowlight tetra (an endemic species from the Essequibo river basin in Guyana and a popular aquarium species)

bb. No brilliant red band but some bluish iridescence; fins not colored; occasionally a very faint caudal spot (after preservation); 1 to 3 very small maxillary teeth . . . *He. gracilis* (Amazon basin)

(*b*) Key to the *He. bellottii* group: a humeral spot; no caudal spot; usually a narrow longitudinal line along body.

a. 6 4 transverse scales; dorsal fin behind midbody; humeral spot round, very conspicuous, surrounded by light area and situated over the 7-8th lateral line scales; largest teeth very broad, with 7-9 cusps (body depth 2.75-3.25 and head length 3.5-3.8 in standard length; 26-31 anal rays; 33-35 longitudinal scales; caudal fin not much scaled; caudal peduncle red in life) . . . *He. stictus* (formerly within *Hyphessobrycon*; possibly a transition form; Amazon basin, Guianas and Rio Meta)

aa. 5 3 or $3\frac{1}{2}$ transverse scales; dorsal fin at midbody or slightly in front of it; humeral spot variable, situated over the first lateral line scales, to the fourth or fifth; largest teeth with 3-7 cusps.

Hyphessobrycon of the *serpae*-complex which has been called *H. haraldschultzi.* Photo by Dr. E. Schmidt.

Hyphessobrycon callistus. Photo by Dr. Herbert R. Axelrod.

Hyphessobrycon serpae. Photo by Dr. Herbert R. Axelrod.

The coffee bean tetra, *Hyphessobrycon takasei.* Photo by Harald Schultz.

b. Anal fin very short, with 12-14 rays (humeral spot faint; a longitudinal band, red in life, and a short black band over anal fin; largest teeth with 5-7 cuspids; body depth 3.25-3.50 and head length 3.5-3.75 in standard length; 30-32 longitudinal scales; caudal fin half scaled) . . . *He. analis* (Amazon and Guianas)

bb. 17-26 anal rays.

c. Usually 5 inner premaxillary teeth, the largest ones with 5-7 cusps.

d. 0 or 1 conical maxillary tooth; base of anal fin underlined by a characteristic double line (humeral spot vertically ovate, conspicuous; dorsal fin usually in front of midbody; 20-24 anal rays; body depth 3.3-3.9 and head length 3.6-4.1 in standard length; one of the commonest tetras, formerly considered a *Hyphessobrycon*) . . . *He. bellottii*, synonym *He. orthus* ? (Guianas and Amazon)

dd. 2-4 maxillary teeth; line at anal base, when present, single.

e. Body depth 2.75-3.5 in standard length; dorsal fin at midbody; body silvery.

f. 19-22 anal rays; head length 3.6-4.0 in standard length; large suborbital narrow; mouth small; largest teeth with 4-7 cusps; humeral spot circular . . . *He. microstomus* (Amazon basin)

ff. 23-26 anal rays; head length 3.0-3.5 in standard length; large suborbital broad; mouth large; largest teeth with 3-5 cusps; humeral spot semicircular (truncate below) . . . *He. lunatus* (Amazon basin)

ee. Body depth 4 in standard length; dorsal fin in front of midbody; body brassy (one of the so-called "copper tetras") (25 or 26 anal rays; head length 3.55 in standard length; largest teeth with 5 cusps; 4 conical maxillary teeth; humeral spot round, faint) . . . *He. aereus* (French Guiana)

Hemigrammus bellottii, the color pattern after preservation (after Gery, 1963).

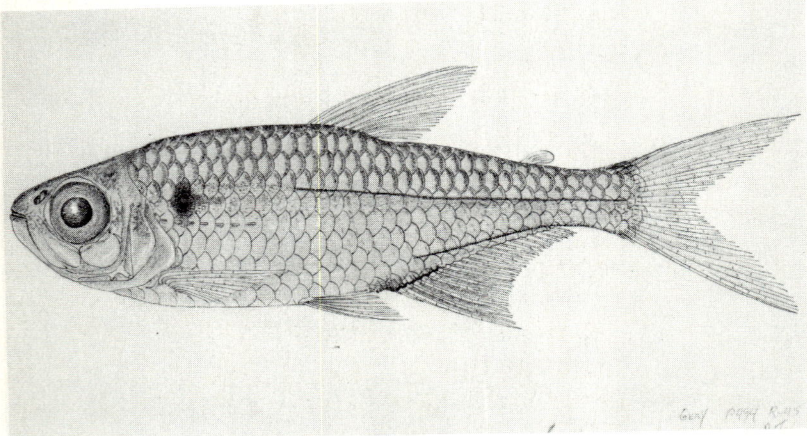

Hemigrammus cupreus and *H. cylindricus* (after Eigenmann, 1918).

cc. Usually 6 or 7 inner premaxillary teeth, largest ones with 3 cusps (dorsal fin at midbody; maxilla elongate, with 3-6 conical or tricuspid teeth; 17-22 anal rays).

g. Body compressed, shining copper (another one of the "copper tetras"), body depth 3.0-3.75 and head length 4 in standard length; large suborbital broad; humeral spot vertically elongate . . . *He. cupreus* (Amazon)

gg. Body subcylindrical, silvery; body depth 3.66-4.66 and head length 3.3-3.66 in standard length; large suborbital narrow; humeral spot round or triangular . . . *He. cylindricus* (Guyana)

(*c*) Key to the *He. tridens*-group: a caudal spot; no humeral spot; usually a longitudinal band, or line, along body (for *He. nanus*, see under *Hasemania*):

a. Caudal lobes not conspicuously barred.

b. Caudal spot on middle caudal rays only, not extended onto caudal peduncle; maxilla toothless; largest inner premaxillary teeth very broad, with up to 9 cusps (body depth 3.2-3.6 in standard length; 17-20 anal rays; a brilliant silvery line along body under the dark line; no iridescent spot on caudal peduncle) . . . *He. levis* (a common species in the middle Amazon basin)

bb. Caudal spot on peduncle as well as on middle rays; maxilla with 1-4 teeth; largest inner premaxillary teeth usually with 5 cusps (7 in the single known specimen of *He. micropterus*).

Unidentified tetra, either a *Hyphessobrycon* of the *callistus*-group or *Cheirodon troemneri.* Female above. Photo by J. Elias:

Aquarium form of the *Hyphessobrycon callistus-serpae* group. Female above. Photo by S. Frank.

The bleeding heart tetra, *Hyphessobrycon erythrostigma,* which was formerly *H. rubrostigma.* Photo by S. Frank.

Hemigrammus hyanuary. Photo by Milan Chvojka.

c. Body depth 2.6 in standard length; caudal spot elongate, not as wide as the peduncle, extending from last anal rays to tip of middle caudal rays (dorsal fin at midbody; 20-22 anal rays; large suborbital broad; mouth large) . . . *He. brevis* (Rio Sao Francisco basin)

cc. Body depth 2.75-3.8 in standard length.

d. An iridescent spot on upper part of caudal peduncle.

e. 14-15 anal rays; dorsal fin in front of midbody; body depth 3.3-3.8 in standard length; caudal spot elongate, occupying the whole peduncle except its upper part and extending onto middle caudal rays to their tips . . . *He. hyanuary*, the costello tetra or green neon (Middle and Upper Amazon)

ee. 23 anal rays; dorsal fin behind midbody; body depth 3 in standard length; caudal spot arrow shaped, not reaching tips of middle caudal rays . . . *He. mattei*, possibly the head and tail light, which is not the true *He. ocellifer* (exact locality unknown, said to be Argentina)

dd. No iridescent spot on upper part of caudal peduncle.

f. 16-19 anal rays; dorsal fin behind midbody; inner premaxillary teeth with a prominent middle point and two smaller ones on each side; caudal spot very large, characteristically diamond shaped with longitudinal and lateral points (sometimes a few chromatophores scattered in the humeral region) . . . *He. tridens* (Rio Paraguay)

ff. Usually 22-24 anal rays; dorsal fin at midbody; inner premaxillary teeth with median cusp not much longer than lateral ones; caudal spot hastate, without expansions above and below.

g. Largest teeth with 7 cusps; one very broad maxillary tooth (caudal spot not continuous with the lateral band; 32 longitudinal scales; body depth 3.2 in standard length)

. . . *He. micropterus* (Venezuela, a poorly known species)

gg. Largest teeth with 5 cusps; usually more than one maxillary tooth, quinquecuspid at the most.

h. Caudal spot not continuous with lateral band; base of anal fin with a series of parallel transverse dashes; caudal scaled up to proximal third of the lobes ((6) 33-35 longitudinal scales; body depth 3.33-3.5 in standard length)

. . . *He. boesemani* (rather common in Guianas and Upper Amazon; formerly considered a subspecies of *micropterus*)

hh. Caudal spot continuous with lateral band; base of anal fin with only a few chromatophores; caudal more than half scaled ((9-15) 32-34 longitudinal scales; body depth 2.75-3.2 in standard length) . . . *He. rodwayi* (sympatric with the preceding; *He. armstrongi*, one of the brass or copper tetras, is probably a pathological variety of *He. rodwayi*)

Hemigrammus armstrongi (currently thought to be only a parasitized *H. rodwayi*). Photo by Dr. Herbert R. Axelrod.

Two color phases of the "rosaceus" form, *Hyphessobrycon bentosi rosaceus.* Photo above by Dr. Karl Knaack, photo below by Harald Schultz.

Two color phases of the "ornatus" form, *Hyphessobrycon bentosi bentosi.* Photos by S. Frank.

aa. Both caudal lobes conspicuously barred (see also *Moenkhausia dichroura* and *M. intermedia, Petitella georgiae*).

i. 14 or 15 anal rays; dorsal fin in front of midbody; no maxillary tooth . . . *He. rhodostomus* (Amazon basin; rare)

ii. 20-24 anal rays; dorsal fin at midbody; 2-4 maxillary teeth . . . *He. marginatus* (ubiquitous; Paraguay, Sao Francisco, Amazon basin, Upper Rio Meta and Venezuela (?); rather common in certain places)

(*d*) The *He. ocellifer*-group: one or two humeral spots; a caudal spot; usually a longitudinal line, this color pattern very conspicuous in *He. barrigonae*:

a. Humeral spot single.

b. 6 or 7/4-6 transverse scales (body depth 2.5-3.25 in standard length).

c. 26-28 anal rays; 34-37 longitudinal scales; humeral and caudal spots distinct; no lateral band uniting the two spots; no black line along anal base.

Hemigrammus barrigonae. Photo by Dr. Herbert R. Axelrod.

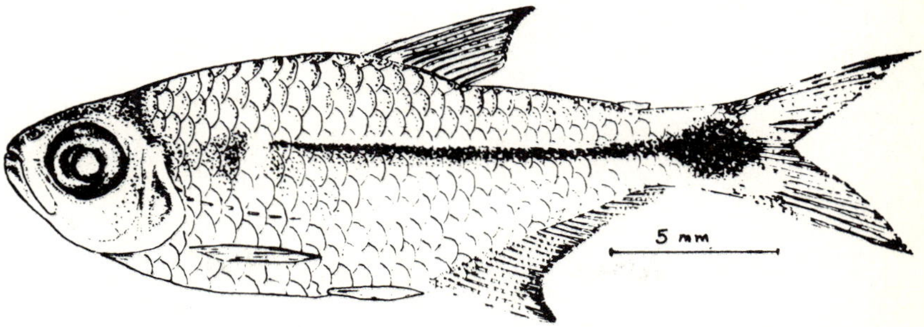

Hemigrammus guyanensis, type (after Gery, 1959).

d. Humeral spot large and conspicuous, roundish; caudal spot very broad and conspicuous, not reaching to tip of middle caudal rays; preventral area roundish ... *He. megaceps* (Peru)

dd. Humeral spot faint, vertically elongate; caudal spot hastate, reaching to tip of middle caudal rays; preventral area flat ... *He. caudovittatus*, the popular Buenos Aires tetra (said to come from Argentina)

cc. 24-25 anal rays, 33-35 longitudinal scales (lateral line irregular, sometimes almost complete); a lateral band continuous with the humeral and caudal spots; black line along anal base ... *He. barrigonae* (Rio Meta in Colombia)

bb. 5/3 transverse scales.

e. An iridescent spot on the upper half of the caudal peduncle; 24 anal rays on average (humeral spot faint, triangular; caudal spot lozenge-shaped, not reaching to tip of middle caudal rays; 3-5 narrow maxillary teeth) ... *He. guyanensis* (French Guiana; not on the coast, where it is replaced by *He. ocellifer*)

ee. No iridescent spot in upper half of caudal peduncle; 15-22 anal rays.

f. Caudal spot vertically elongate, I-shaped; 15-17 anal rays (humeral spot vertical, conspicuous; size small; large suborbital in contact with preopercle; dorsal fin behind midbody) ... *He. iota* (Guyana)

ff. Caudal spot horizontally elongate or more or less lozenge-shaped, usually not quite reaching to tip of middle caudal rays; 18-23 anal rays.

Parapristella georgiae. Photo by Dr. Herbert R. Axelrod.

The "tetra robertsi" of aquarists, not yet named. Photo by Dr. Herbert R. Axelrod.

Hemigrammus stictus. Photo by Dr. Herbert R. Axelrod.

The glowlight tetra, *Hemigrammus erythrozonus.* Female above. Photo by R. Zukal.

g. Lower caudal lobe dark; body depth 2.66 in standard length; humeral spot vertical, very distinct; a longitudinal band, not continuous with the caudal spot (dorsal fin clearly behind midbody; large suborbital in contact with preopercle; 22 anal rays) . . . *He. boulengeri* (a single specimen known, from Rio Grande do Sul)

gg. Lower caudal lobe hyaline; body depth 2.75-3.85 in standard length; humeral spot usually faint; lateral, narrow band usually continuous with caudal spot.

h. 18 anal rays (average); dorsal fin slightly behind midbody; 2 or 3 maxillary teeth; humeral spot vertically elongate (29 or 30 lateral scales) . . . *He. vorderwinkleri* (Rio Negro)

hh. 20-23 anal rays; dorsal fin at midbody or in front of it; one, rarely 2 maxillary teeth; humeral spot roundish (30-34 lateral scales).

i. Body depth 2.8-3.6 and head length about 3.5 in standard length; caudal spot very deep; dorsal fin usually not much in front of midbody . . . *He. schmardae*, synonym *proneki* (Amazon basin)

Hemigrammus ocellifer. Photo by Ruda Zukal.

Type of *Hemigrammus vorderwinkleri,* a 33 mm female. Photo by the author.

ii. Body depth 3.85 and head length 4.2 in standard length; caudal spot not covering more than half of peduncle; dorsal fin distinctly in front of midbody . . . *He. melanochrous* (Rio Madeira)

aa. Humeral spot double (always an iridescent spot on upper part of caudal peduncle).

j. Caudal spot covering only end of caudal peduncle (24-27 anal rays; 30-33 lateral scales).

k. Caudal peduncle as deep as long; caudal fin scaled on more than its basal half; dorsal fin in front of midbody (body depth 2.55 in standard length; great suborbital almost entire) . . . *He. pretoensis* (Upper Amazon basin)

kk. Caudal peduncle longer than deep; caudal fin scaled on its basal third; dorsal fin at midbody or behind.

l. First humeral spot roundish; caudal spot small; body depth 3.25-3.85 in standard length; large suborbital almost entire (32-33 lateral scales) . . . *He. luelingi* (lower Ucayali)

ll. First humeral spot vertically elongate; caudal spot large; body depth 2.4-2.8 in standard length; large suborbital leaving a naked zone on cheek (usually 30 or 31 lateral scales) . . . *He. ocellifer* (Amazon and Guianas, very common)

jj. A black band from above anal fin to end of caudal peduncle (21-24 anal rays; 28-31 lateral scales; body depth 2.25-2.6 in standard length) . . . *He. pulcher*, the garnet tetra (upper Amazon basin; two subspecies)

Hemigrammus stictus. Photo by Dr. Herbert R. Axelrod.

Another aspect of *Hemigrammus stictus.* Photo by Dr. Herbert R. Axelrod.

A young example of *Hemigrammus bellottii.* Photo by Dr. Herbert R. Axelrod.

Hemigrammus bellottii from the Rio Negro. Photo by Dr. Herbert R. Axelrod.

(*e*) The *He. unilineatus*-group: only a few *Hemigrammus* have a dorsal spot (including the type species, *He. unilineatus*), and the group is much less complicated than the homologous *callistus*-group in the genus *Hyphessobrycon*.

a. A conspicuous black lateral band and a black line over anal fin, which is plain; dorsal spot only on base of first rays (5 or 6 tricuspid teeth in inner premaxillary row; 2 or 3 similar teeth on maxilla) . . . *He. ulreyi* (Paraguay basin)

aa. Lateral band, if any, light or very faint or narrow; no black line over anal fin, which has some black markings.

b. Inner premaxillary teeth 5- to 7-pointed; maxilla with fewer than 6 teeth; dorsal and anal markings not very conspicuous.

c. A vertically elongate humeral spot and a light lateral band; 20-22 anal rays; premaxillary teeth pentacuspid; 2-5 tricuspid maxillary teeth . . . *He. coeruleus* (Amazon)

cc. Humeral spot faint or absent; no light lateral band; 23-24 anal rays; premaxillary teeth with up to 7 cusps; 3 pentacuspid maxillary teeth . . . *He. elegans* (Amazon)

bb. Inner premaxillary teeth tricuspid; maxilla usually with 6 narrow teeth; a very conspicuous, large dorsal spot and a conspicuous, oblique band on anal lobe (2 humeral spots, the first one vertically elongate, usually conspicuous; usually 23-29 anal rays) . . . *He. unilineatus*, the featherfin tetra (Trinidad, Guianas and Amazon; a subspecies from French Guiana has a few more scales and anal rays than the nominate form)

Hemigrammus unilineatus **from the upper Amazon region near Benjamin Constant, Brazil. Photo by O.C. Beldt.**

Bryconella pallidifrons. Photo by Dr. J. Gery

Bryconella pallidifrons

Ten years ago, the author found in a collection from the Igarape Preto (Upper Amazon basin at the boundaries between Brazil, Peru and Colombia), made by the late Harald Schultz, a *Hemigrammus*-like tetra which differed in having no postorbital bones and especially in its very irregular premaxillary teeth. It was named *Bryconella haraldi*. Instead of having, as usual, 1, 2 or 3 outer teeth (sometimes more) and 5 inner teeth (rarely more), this tetra had 4 or 5 outer teeth and 3 or 2 inner teeth, the two rows very close together. In a few specimens the two rows were even confounded in a unique, irregular row, as in several *Pristella*-like genera usually classified within the Cheirodontinae such as *Pristella*, *Pseudopristella*, certain *Megalamphodus*, *Paracheirodon* and *Petitella*. Later on it was found that the fish had been described previously under the names *Cheirodon pallidifrons* and *Hyphessobrycon thompsoni*, and that its correct name is *Bryconella pallidifrons*.

It is a nice little fish which has some resemblance in its body shape and general structure to the neon tetra, yet without a neon-like color pattern. Instead it has a yellow color with bluish tints on the abdomen, a conspicuous brown humeral spot followed by a longitudinal band to the end of the middle caudal rays and some red on the caudal and dorsal fins.

Hemigrammus levis. Photo by Dr. Herbert R. Axelrod.

Hemigrammus levis with a more prominent caudal spot. Photo by Aaron Norman.

The Costello tetra, *Hemigrammus hyanuary.* Photo by S. Frank.

A different color phase of *Hemigrammus hyanuary.* Photo by J. Elias.

Thayeria obliqua. Photo by Klaus Paysan.

Genus *Thayeria*

Penguin Tetras

The three known "penguins" swim head up at an angle of about 30° with the horizontal; correlatively, there is an elongation of the lower caudal lobe, which is 10% (in the black-line) to 25% (in the true penguin) longer than the upper one and marked with a very contrasting black band, more or less prolonged onto the body (this band, associated with the oblique position, breaks up the contours of the animal, which is thus no longer recognizable as a fish; this is called a disruptive color pattern). Except for their interesting protective adaptation, these tetras have nothing differentiating them from *Hemigrammus* species.

They can be recognized as follows:

a. 20 or 21 anal rays (average); large suborbital leaving a naked area on the posterior part of the cheek (black band very oblique, running from just in front of and below the dorsal fin to the end of the lower caudal lobe, which is only 10-15% longer than the upper one) . . . *T. ifati* (Maroni basin in French Guiana and Surinam; probably the most primitive species, still close to *Hemigrammus*; unknown in the aquarium)

aa. 16-19 anal rays (average); large suborbital usually almost entire.

b. A conspicuous, almost longitudinal black band from upper angle of gill-cover to tip of lower caudal lobe, which is 10-25% longer than the upper one . . . *T. boehlkei*, the popular blackline thayeria (first imports probably from the Peruvian Amazon or the Rio Araguaia; now bred in quantity)

bb. Black band very oblique and short, running from just behind dorsal fin (sometimes farther behind) to tip of lower caudal lobe, which is 25-30% longer than the upper one . . . *T. obliqua*, probable synonym *T. sanctaemariae*, the penguin tetra (Amazon basin; sympatric with *T. boehlkei* in upper Amazon and possibly in the Rio Tocantins basin; not bred in captivity)

Thayeria boehlkei. Photo by G. Timmerman.

Hemigrammus ocellifer. Photo by S. Frank.

The head-and-tail light or "false ocellifer," possibly *Hemigrammus mattei.* Female above. Photo by S. Frank.

Hemigrammus boesemani. Photo by Dr. Herbert R. Axelrod.

Hemigrammus rodwayi. Photo by Dr. Herbert R. Axelrod.

A brass tetra, possibly *Hemigrammus rodwayi.* Such a parasitized fish was once considered a full species. Photo by Klaus Paysan.

Hasemania melanura (after Eigenmann, 1921).

Genus *Hasemania*

American Tetras

The genus *Hasemania* is an assemblage, possibly unnatural, of forms which resemble *Hyphessobrycon* or even more closely, *Hemigrammus* (most have some scales on the caudal fin base), but with the adipose fin usually completely lacking (it may be present, but reduced, in certain populations of *Hasemania nana*, the silver-tipped tetra). They all come from the southeastern part of Brazil, from Rio S. Francisco to Rio Parana basins. Their color pattern is quite plain. A recent key can be summarized as follows:

a. Usually 18-21 anal rays.

  b. Body depth 2.65-3.2 in standard length; 7/5 transverse scales; 3 maxillary teeth . . . *H. maxillaris* (Rio Iguassu, S.E. Brazil)

  bb. Body depth 3.3-3.5 in standard length; 5/4 transverse scales; usually no maxillary teeth (?) . . . *H. hanseni* ("State of Goiaz" in Brazil)

aa. Usually 16-17 anal rays (usually no maxillary tooth).

  c. Body depth 2.5-3.0 in standard length; 6/6 transverse scales; head wide . . . *H. melanura* (same locality as *H. maxillaris*)

  cc. Body depth usually more than 3 in standard length; 4 or 5/3 or 3½ transverse scales; head narrow . . . *H. nana*, the silver-tipped tetra (Rio Sao Francisco basin; formerly known as *Hemigrammus nanus* and *Hasemania marginata*, which are synonyms; a red strain has become established in the aquarium)

Fourth group of American Tetras

Genus *Deuterodon*

A *Deuterodon* externally resembles an *Astyanax*, and the only differences are in the teeth, which tell us that these fishes have chosen a different diet than that of the omnivorous *Astyanax* (the stomach contents reveal that the diet is a purely vegetarian one). *Deuterodon* teeth are flat, multicuspid incisors like those of the earlier mentioned Iguanodectini, or of the Cheirodontinae (where they are uniserial on the upper jaw). In certain species the upper lip is weak, the teeth exposed and the snout long, giving to the profile of the head a kind of "smile" which is distinctive. The other characters, such as incomplete suborbitals, complete lateral line and naked caudal fin, are also those of *Astyanax*.

The species of this genus are found in several places in tropical America, and its distribution is clearly discontinuous: Nicaragua, Guianas (and Rio Negro according to unpublished data), Amazon basin, upper Paraguay basin and especially southeastern Brazil. It is likely that in each locality the group has arisen from a different *Astyanax*-like ancestor and that the genus is polyphyletic (which is also the case in several other tetragonopterine genera).

Head of a species of *Deuterodon* (after Eigenmann, 1927).

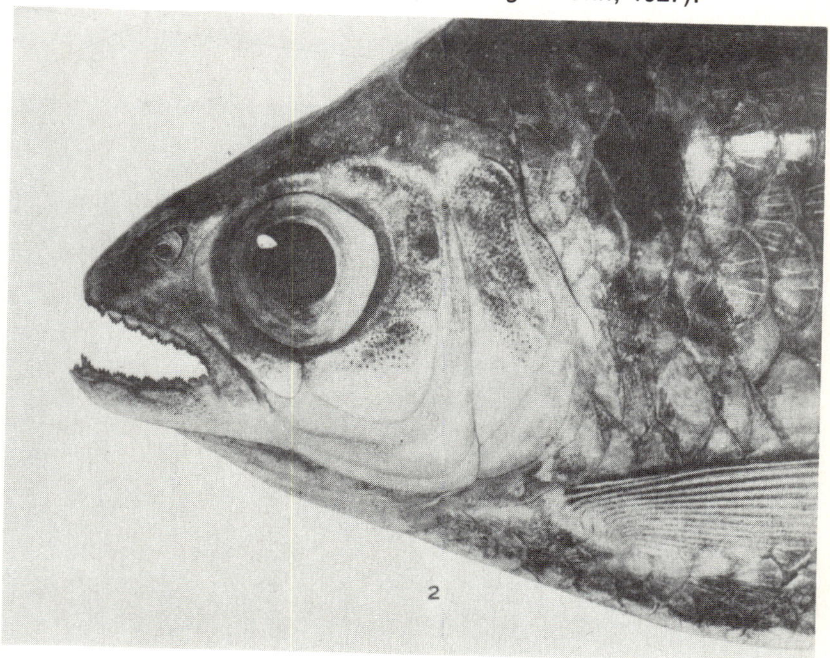

The rummy-nose tetra, *Hemigrammus rhodostomus.* Photo by Dr. Herbert R. Axelrod.

Another aspect of *Hemigrammus rhodostomus.* Photo by Klaus Paysan.

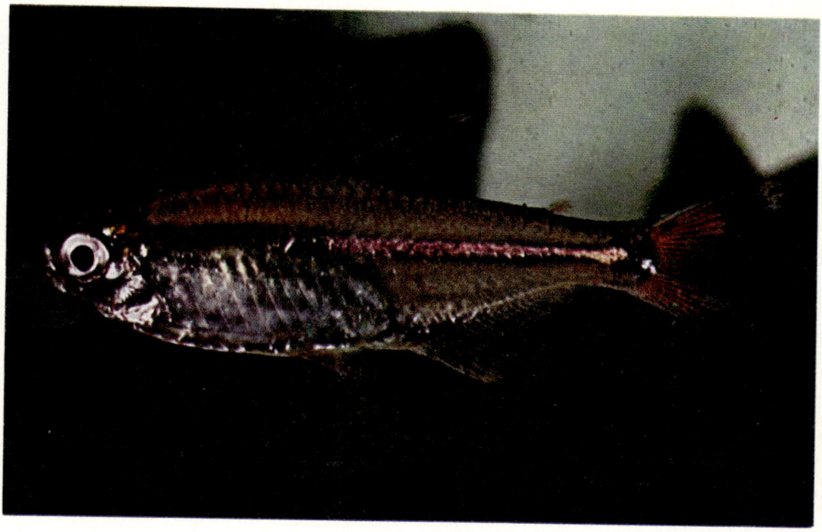

Hemigrammus marginatus. Photo by Dr. Herbert R. Axelrod.

Hemigrammus marginatus? Photo by S. Frank.

In the following key to the species, *Deuterodon euspilurus*, from the upper Amazon basin in Colombia, has been omitted: its teeth are too poorly described to raise any doubt concerning its referral to a different genus, namely *Odontostilbe* in the Cheirodontinae.

Key to the species of *Deuterodon*:

a. Broadest teeth with 5-7 cusps; preventral area narrowly rounded or compressed (small, 50 or 75 mm, rather generalized species from the Guianas; 24-25 anal rays; 3-5 maxillary teeth, the broadest with five cusps).

b. Body depth 2.5-2.7 in standard length; 36-37 lateral line scales; 2 vertical humeral bars; longitudinal band with lateral expansions, chevron-like; caudal spot extended onto caudal base . . . *D. pinnatus* (Guianas)

bb. Body depth 3.2-3.5 in standard length; 37-40 lateral line scales; a comma-shaped humeral spot; series of dots on scales forming longitudinal stripes, the ones corresponding to the lateral line broader; caudal spot extended onto lower caudal margin . . . *D. potaroensis* (Guyana)

aa. Broadest teeth with 7-11 cusps; preventral area rounded or flattened (usually larger, 55 to 115 mm, specialized species from Nicaragua, southeastern Brazil and Upper Paraguay basin; one Amazonian species, *D. minor*, known from specimens only 28 mm in standard length).

c. Preventral area rounded, not flattened owing to development of spinous bones from pelvic girdle; broadest teeth usually with 7 cusps (again except in *D. minor*).

d. Teeth not very compressed, rather heavy at least on lower jaw (25-28 anal rays; 7-8/36-37/6-7 scales; large suborbital leaving very wide naked area on cheek; 4-6 maxillary teeth; a diffuse humeral spot, a dark lateral band, and a rather large caudal spot extended onto middle caudal rays to their end) . . . *D. nasutus* (Nicaragua)

dd. Teeth very compressed, incisor-like.

e. Broadest teeth with 7-8 cusps; size up to 115 mm; a triangular humeral spot and a lateral band, ending in a conspicuous lozenge-shaped caudal spot (*D. iguape*-group, S.E. Brazil and State of Minas Gerais).

Deuterodon nasutus (after Eigenmann, 1921).

Deuterodon pinnatus and *D. potaroensis* (after Eigenmann, 1912).

f. Snout normal, *Astyanax*-like, shorter than eye, which is 2.6-3.75 in head length; free portion of maxilla longer than horizontal extent of premaxilla, with 2 or 3 teeth.

g. Large suborbital very narrow; head length 4.0-4.33 in standard length; 3 or 4 outer premaxillary teeth; 2 or 3 maxillary teeth; 10 mandibular teeth . . . *D. pedri* (Minas Gerais, Brazil)

gg. Large suborbital almost complete; head length 3.5-4.0 in the standard length; 2 or 3 outer premaxillary teeth; 2 maxillary teeth; 7-9 mandibular teeth . . . *D. parahybae* (Rio Itapemirim)

ff. Snout rather elongate, about equal to eye, which is 3.3 in head length; profile of head distinctive; free portion of maxilla equalling horizontal extent of premaxillary, usually with more than 3 teeth.

h. Branched anal rays 21-22; eye 2.6-3.2 in head length; 6/4 transverse scales . . . *D. langei* (Rio Parana)

hh. Branched anal rays usually 19-20; eye 3.1-3.6 in head length; 6-7/5 transverse scales.

Hemigrammus schmardae. Photo by Dr. Herbert R. Axelrod.

Hemigrammus schmardae (above) and the very similar *H. vorderwinkleri* (below). Photo by Dr. Herbert R. Axelrod.

Hemigrammus caudovittatus. Note its great similarity in form and color pattern to *Astyanax fasciatus.* Photo by Dr. Herbert R. Axelrod.

i. 2 or 3 outer premaxillary teeth; 3 or 4 maxillary teeth (8-12 mandibular teeth) . . . *D. iguape*, the type species, synonym *D. depressirostris* (State of Sao Paulo)

ii. 2 outer premaxillary teeth, rarely 3; 5-6 maxillary teeth, rarely 4 (10-12 mandibular teeth) . . . *D. rosae*, synonym *D. stigmaturus* (States Sta. Catharina and Rio Grande do Sul)

ee. Broadest teeth with 10 cusps; size small; humeral spot inconspicuous; caudal spot large, round, not extended onto middle caudal rays (24-26 anal rays; 6/35-37/5 scales; 3 maxillary teeth; body depth 3.1-3.8 in standard length) . . . *D. minor* (Rio Tapajoz basin)

cc. Preventral area flattened, owing to pelvic spines protruding anteriorly; broadest teeth with 9-11 cusps (23-26 anal rays; 7/35/5 scales; 2 maxillary teeth; body depth 2.3 in standard length; a humeral bar and an oval caudal spot, somewhat extended onto caudal base) . . . *D. acanthogaster* (Upper Rio Paraguay)

Deuterodon acanthogaster (after Eigenmann, 1911).

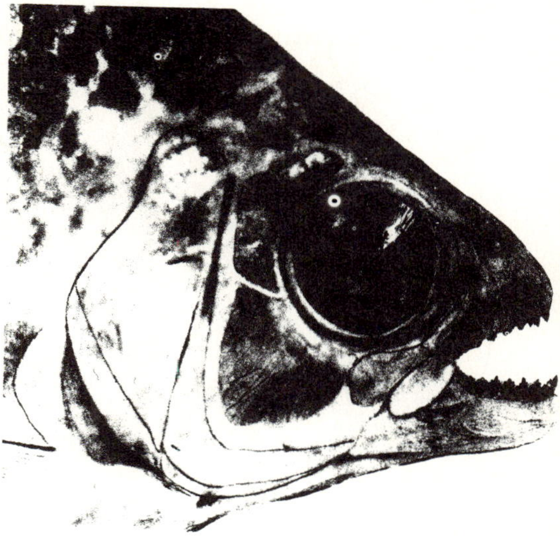

Head of *Deuterodon iguape* (after Eigenmann, 1917).

Fifth group of American Tetras

Genera *Phenacogaster* and *Vesicatrus*
Glass Tetras

These two genera are odd tetras, approaching in body shape some characins such as *Roeboides* and the lesser-known *Eucynopotamus* or even *Asiphonichthys*. According to the general organization of the fishes, it may be that *Phenacogaster* is the link with the Characinae and even that it might belong to that subfamily.

The *Phenacogaster* species are compressed fishes of small size (up to 85 mm), with a rather long anal fin (31-45 rays) and peculiar ventral area as well as an unusual dentition. The preventral region is flattened, covered by specialized elongate scales overlapping in the center and bent at the sides. Certain species have a few median interpolated small scales. Their teeth are more numerous than usual in the tetras. The outer premaxillary row, composed of tricuspid to conical teeth, often leaves a gap between the mesial part and the distal part (the one near the maxilla) of the bone; the inner row, also composed of tricuspid to conical teeth, usually has more than 5 teeth and possibly up to 10. The maxilla is armed with similar teeth (with three points basally and one point near its tip) on its entire or almost entire length. The mandible is less unusual, with a front series of 5 or so large tricuspid teeth and several (up to 15) narrow conical teeth.

Hemigrammus barrigonae. Photo by Dr. Herbert R. Axelrod.

Hemigrammus luelingi. Photo by Dr. Herbert R. Axelrod.

Hemigrammus coeruleus. Photo by Dr. Herbert R. Axelrod.

Hemigrammus coeruleus. Photo by Dr. Herbert R. Axelrod.

Hemigrammus coeruleus showing red color of formalin preserved specimen. Photo by Dr. J. Gery.

Other peculiarities worth mentioning are the conspicuous triangular loss of the muscles above the anterior part of the gas-bladder (called pseudotympanum or humeral hiatus), the series of minute sensory pores on the head (called pit-lines) and the fact that some species, if not all, are translucent in life (glass tetras).

According to a recent review of *Phenacogaster*, the species can be identified as follows:

a. Less specialized species with a median series of small preventral scales, a not fully toothed maxilla and a complete outer maxillary row (a large, conspicuous humeral spot) . . . *Ph. megalostictus* (Guyana and Rio Negro)

aa. Specialized species with few, if any, median preventral scales and a fully toothed maxilla (up to 34 teeth).

b. Outer premaxillary row complete, with 7 teeth (elongate form, body depth about 3.45 in standard length; a large, conspicuous humeral spot) . . . *Ph. franciscoensis* (Rio Sao Francisco)

bb. Outer premaxillary teeth not forming a continuous series, but with a gap between mesial and distal teeth (formula 2 or 3 + 4 or 5)

c. 28-30 branched anal rays; body depth 3.2 in standard length; maxilla short, more than 4 in head length; humeral spot conspicuous . . . *Ph. calverti* (Upper Rio Tocantins)

cc. 31-39 branched anal rays; body depth less than 3.2 in standard length; maxilla long, less than 4 in head length; humeral spot usually small or inconspicuous (*Ph. pectinatus*-group).

d. 31-34 branched anal rays; body depth 2.6-2.66 in standard length (humeral spot small) . . . *Ph. beni* (Rio Guapore basin) and *Ph. suborbitalis* (Eastern coast of Brazil)

dd. 33-39 branched anal rays; body depth about 2.75 in standard length.

e. Dorsal rays ii, 7i; 3 + 5/9 premaxillary teeth; a small humeral spot . . . *Ph. microstictus* (Guianas)

ee. Dorsal rays ii, 8i; 2 + 3 or 4/5 premaxillary teeth; usually no humeral spot . . . *Ph. pectinatus*, the type species (Middle and Upper Amazon basin; rather common)

The single species of *Vesicatrus*, *Vesicatrus tegatus*, from the Upper Paraguay basin, resembles *Ph. megalostictus* in the unspecialized structure of its premaxilla (outer series of teeth complete), and maxilla (not fully toothed), as well as in the presence of a few median preventral scales. It differs in having only 8 perforated scales in the lateral line, which is considered a generic character (of not much weight in the present case), as well as in the coloration: the conspicuous lateral spot is clearly behind the humeral region, just in front of the dorsal fin level; there is a large caudal spot prolonged (after preservation) onto the middle caudal rays.

Sixth group of American Tetras

Genycharax tarpon

As far as known, this strange fish inhabits exclusively the upper Rio Cauca in Colombia. It is of moderate length (about 100 mm), and looks like an *Astyanax* but with a very different clupeoid mouth armed with backwardly recurved, pointed teeth and a blade-like, broad maxilla. Its position is uncertain. It may be an adaptation of some *Astyanax* toward a specialized diet, but its affinities with a characin are not to be excluded.

Genycharax tarpon and close-up of its head (after Gery, 1966).

Hemigrammus pulcher pulcher. Photo by Dr. Herbert R. Axelrod.

Hemigrammus pulcher haraldi. Photo by Dr. Herbert R. Axelrod.

Hemigrammus ulreyi from the Rio Japura. Photo by Harald Schultz.

Hemigrammus unilineatus, the garnet tetra. Photo by S. Frank.

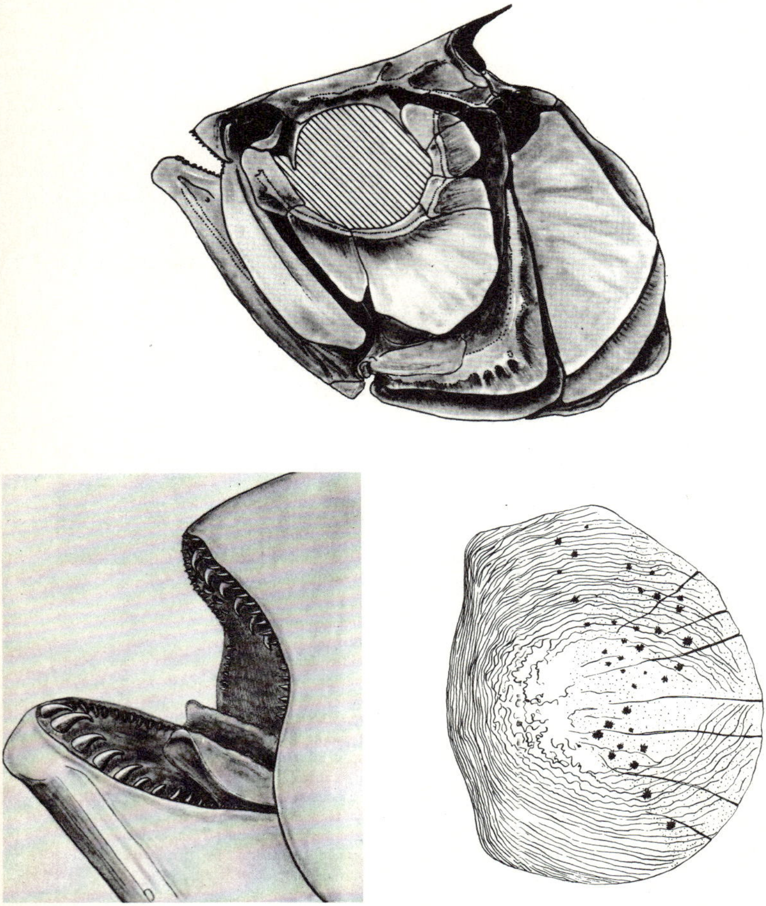

Lateral view of skull, dentition, and scale of *Genycharax* (after Gery, 1966).

D. GENUS *STICHONODON*

There is but little doubt that the 80 mm long Amazonian *Stichonodon*, with the single species *insignis*, belongs to the Tetragonopterinae and that it is close to a *Gymnocorymbus* or a *Poptella*. In the course of its evolution, it has acquired a strong ventral keel associated with a disciform body (but no predorsal spine), powerful pectoral fins and scaled caudal and anal fins. The dentition is similar to that of a *Moenkhausia*, but without maxillary teeth; the large suborbital is entire, the lateral line complete. There is a large black humeral spot well above the lateral line.

Tribe Stygichthyini

In the 1960's, a well-drilling operation in the State of Minas Gerais, Brazil, obtained from a subterranean sheet of water a single specimen of a very odd tetra. This specimen was named *Stygichthys typhlops* by Drs. M. Brittan and J. Böhlke. The fish, which is only 24 mm long in standard length, is entirely white without any pigment, and is entirely devoid of a lateral line system on its body, of circumorbital bones and of eyes. Its only visible sensory organs are some small papillae on the head and a leaf-like adipose fin whose function is unknown.

Other peculiarities are the great number of inner premaxillary teeth (7 or 8) and of maxillary teeth (at least 10), as well as the small number of anal rays (only 8 branched rays). These characters are quite aberrant among the modern Tetragonopterinae, and it may be that the eyed ancestor of *Stygichthys* is now extinct, the blind fish being what is called a relict.

Tribe Gymnocharacini

Another odd tetra is *Gymnocharacinus bergi*, a medium-sized fish from a small river south of the Rio de la Plata in Argentina (the most southern characoid, clearly not in the subtropical but in the temperate zone). It is a rather elongate fish, entirely naked, without an adipose fin and with the anal fin short. The obtuse profile of its head as well as its color pattern (a broad dark lateral band) suggest that its ancestor could have been some generalized *Hemibrycon*-like tetragonopterine.

Gymnocharacinus bergi. Photo by Dr. J. Gery.

Bryconella pallidifrons. Photo by Harald Schultz.

Thayeria boehlkei. Photo by R. Zukal.

Thayeria boehlkei, the black line thayeria. Photo by Harald Schultz.

Thayeria obliqua, the penguin thayeria. Photo by Dr. Herbert R. Axelrod.

SUBFAMILY (11): RHOADSIINAE

This small family, with only 2 genera from western Ecuador, western Colombia, and Costa Rica (perhaps three genera if the Costa Rican form is generically different), shows great differences between the young and the adults. They usually have a single premaxillary row of teeth (like the Cheirodontinae) when young, and usually a double one (like the Tetragonopterinae) when adult, usually composed of 2 conical outer teeth and 5 inner multicuspid incisors. The maxilla and the mandible also change considerably during growth. The former is round and short, with only 2 cuspidate teeth in the young, and grows out of proportion with the general increase in size, along with the appearance of 15-20 conical teeth along the whole length of the bone (somewhat as in the male *Poecilocharax weitzmani*). On the mandible the frontal multicuspid incisors lose some of their cusps and the lateral teeth become canine-like. Finally the dorsal fin lengthens, so that its longer rays reach the caudal fin in the adults.

The genus *Parastremma* is restricted to western Colombia. Its type species, *P. sadina*, is characterized by numerous scales (55-64), a complete lateral line, a large size (up to 180 mm) and a moderately deep body (body depth 2.5 in standard length). It has a double humeral spot and, when young, a caudal spot.

G. Dahl described, in 1960, two other forms which would differ by their smaller size (95-100 mm) and their larger eye (its diameter 1.4 or less in the interorbital width, instead of 1.8 or more in *P. sadina*), as well as by an increased number of gill-rakers and fewer lateral mandibular teeth: *P. pulchra*, from the Rio S. Juan, which has no spot when adult, and *P. alba*, from the Rio Sando, which has about the same color pattern as *P. sadina*.

The genus *Rhoadsia*, from western Ecuador and Costa Rica, differs from *Parastremma* in having fewer lateral scales (about 37-41) and the lateral line perforating only the first 10-20 scales. The type species, *R. altipinna*, grows up to 170 mm and is restricted to the lowlands. A dwarf mountain form, *R. minor*, living in an altitude of over 1000 meters, has been distinguished from it, being more elongate. Both (which may be geographical subspecies instead of good species) have a very characteristic large black spot on the sides at the level of the dorsal fin. The unpaired fins of *R. minor* are red and yellow, and the fish seems to be very suitable for the aquarium.

The Costa Rican form, called *Rhoadsia eigenmanni* (sometimes *Carlana eigenmanni*), has about the same meristics as those of the preceding forms, but a different color pattern; the spot on the sides is in the humeral region at the level of the 8th-9th lateral line scales and there is a conspicuous horizontally elongate caudal spot extending to the end of the middle

Parastremma sadina (from top to bottom: male, female, young) (after Eigenmann and Myers, 1929).

Unidentified tetra, possibly the silver-tipped tetra, *Hasemania nana* (in spite of the adipose fin). Photo by Harald Schultz.

A new species of *Hasemania* close to *H. melanura.* Photo by Harald Schultz.

A red strain of *Hasemania nana* developed by Mr. Hearin. Photo by Harald Schultz.

Hasemania nana. Photo by Dr. Herbert R. Axelrod.

The three species of *Rhoadsia: Rhoadsia altipinna* (top), *R. minor* (center), and *R.* (or *Carlana*) *eigenmanni* (bottom) (after Eigenmann and Myers, 1929).

caudal rays. In this form there seems to be no outer premaxillary row of teeth, even in the largest individuals known.

Fink and Weitzman, who have recently (1974) studied several specimens (including the type) from Costa Rica and Nicaragua, believe this form (which they call *Carlana*) to belong to their "so-called cheirodontin fishes" rather than to the Rhoadsiinae. However, the general organization of the adult is exactly that of a *Rhoadsia* that has lost its outer premaxillary teeth, and the resemblance could hardly be coincidental.

SUBFAMILY (12): CHEIRODONTINAE AND ALLIED GENERA

This composite group concludes the lengthy enumeration of the characid subfamilies. Most of the authors include within it not only the Cheirodontinae in the strict sense, but also the genera which share with them the character of a single premaxillary row of teeth (excluding a siluroid body as in the Paragoniatinae and *Piabucus melanostomus*, or the species with very large postorbital bones as in the Aphyocharacinae, as well as the recently redescribed *Rhoadsia eigenmanni*).

A single row of teeth is probably not a primitive condition in most characoid groups. The ancestor of the Cheirodontinae probably had two premaxillary rows and the outer row was lost during evolution (more often than the inner one, except in the *Henochilus-Psalidodon* line). Incidentally, this so-called regressive evolution also concerned other characters, some already discussed, such as the lateral line. A regression of this type is usually polyphyletic: it probably occurred several times, in different characid lines, most of them altogether tetragonopterine-like. This is why the present arrangement, which is followed here for convenience, cannot be natural. The Cheirodontinae proper may be of a different origin from that of *Megalamphodus*, *Pristella*, *Grundulus*, *Aphyodite* and several allied genera. Likewise *Henochilus* and *Psalidodon* are of a third line, whereas the scale-eater *Probolodus* may be of a fourth one (still unknown).

In a completely arbitrary way, therefore, the genera have been arranged here according to the structure of their teeth. The order will be first the genera with usually at least 5 cusps on the broadest teeth (or Cheirodontinae proper) and, in a second part, all other genera with conical or tricuspid teeth, some of which are actually impossible to properly classify by their phylogenetic affinities: *Psalidodon* et al., *Probolodus*, *Megalamphodus* and *Pristella* et al., *Grundulus* et al., and the Aphyoditeina.

(A) The Cheirodontinae in the strict sense

This group is fairly homogeneous, and most of the genera probably have a common origin. There are a few exceptions, for example the cardinal tetra,

The glass tetra, *Phenacogaster pectinatus*. Photo by Dr. Herbert R. Axelrod.

A different color phase of the glass tetra, *Phenacogaster pectinatus* from Tefe. Photo by Dr. Herbert R. Axelrod.

Phenacogaster pectinatus, or close to it. Photo by Harald Schultz.

Phenacogaster sp., probably another form of *P. pectinatus.* Photo by Dr. Herbert R. Axelrod.

Cheirodon axelrodi, the origins of which are unknown (it is rather remote from the neon tetra, despite the common color pattern, according to the study of its chromosomes by J. J. Scheel), or *Cheirodon troemneri*, which may be a *Hyphessobrycon* instead of a *Cheirodon*, depending on the point of view. However, as compared with those of the second cheirodontine group, these exceptions are not a real burden.

Of the 13 known genera, 7 are monotypic (i.e. with only a single species). This is a very high percentage, equalled only by the next groups. Most of these monotypic genera are rare, very specialized forms which will be characterized in a short way in the following key. The more detailed description of the larger genera will follow. Among them are two genera, *Compsura* and *Saccoderma*, which have evolved parallel to the Glandulocaudinae, the males having acquired a similar caudal gland. They have not been classified within the last named subfamily because they have clearly evolved from a *Cheirodon* or an *Odontostilbe* (respectively), their sexual adaptation being only convergent with that of the Glandulocaudini and the Xenurobryconini (which may be convergent, too).

Key to the Cheirodontinae in the strict sense (broadest teeth with 5-9 cusps, except in a species from southern Chile):

a. Lateral line complete.
  b. Adipose fin present; large suborbital usually broad; maxilla usually with few teeth, if any.
    c. Caudal fin scaled; teeth not very compressed, of the *Hemigrammus-Hyphessobrycon* type (maxilla small, without teeth; resembling an *Odontostilbe* with rather elongate body, the body depth 3.5-4.0 in standard length, which is 40-45 mm; 37-40 lateral line scales; 19-22 anal rays) . . . *Aulixidens eugeniae* (Upper Rio Negro)
    cc. Caudal fin not scaled; teeth compressed antero-posteriorly.
      d. Upper and lower teeth not similar: cutting edge of the former rounded, with 5-7 cusps; cutting edge of the latter straight, with 3 median points of the same size and a small notch on each side.
        e. Maxilla with a few teeth . . . *Holoshestes* (2 species)
        ee. Maxilla without teeth (body depth 3.0-3.4 in standard length, which is about 40 mm; 35-37 lateral line scales; 23-25 anal rays; faint humeral and caudal spots) . . . *Cheirodontops geayi* (Venezuela)
      dd. Upper and lower teeth similar, lower ones only slightly less rounded than upper ones, never forming continuous straight cutting edge.
        f. Lips well developed; mouth small, oblique; 2-7 maxillary teeth; a single series of gill-rakers.
          g. No caudal gland on adult males . . . *Odontostilbe* (7 species)
          gg. A caudal gland on adult males . . . *Saccoderma* (3 species)
        ff. Upper lip lacking, mouth horizontal; 2 rows of gill-rakers . . . *Othonocheirodus*, synonym *Odontostoechus* (2 species)

bb. Adipose fin absent (otherwise much like the preceding; both lips lacking; mouth horizontal; a single series of gill-rakers; body depth 3.75-4.25 in standard length, which is about 45 mm; 34-37 lateral line scales; 11-13 anal rays; a small humeral spot and a lateral band; upper caudal lobe longer than lower)

. . . *Monotocheirodon pearsoni* (Rio Beni basin in Bolivia)

aa. Lateral line incomplete.

h. Teeth not very compressed, *Hyphessobrycon*-like.

i. Caudal fin scaled on its base as in *Hemigrammus*; premaxillary teeth in a somewhat irregular row; an almost perfect mimic of *Hemigrammus rhodostomus*, although no real mimicry may be involved, the two species being allopatric; similarities include scaled caudal fin, proportions, meristic characters and caudal color pattern (the posterior anal band is nevertheless broader on *Petitella*, which has no dark spot on lower part of caudal peduncle)

. . . *Petitella georgiae*, the false rummy-nose tetra (Upper Amazon basin)

Skull and jaws of *Petitella georgiae* (a) as compared with those of *Hemigrammus rhodostomus* (b) (after Gery and Boutiere, 1964).

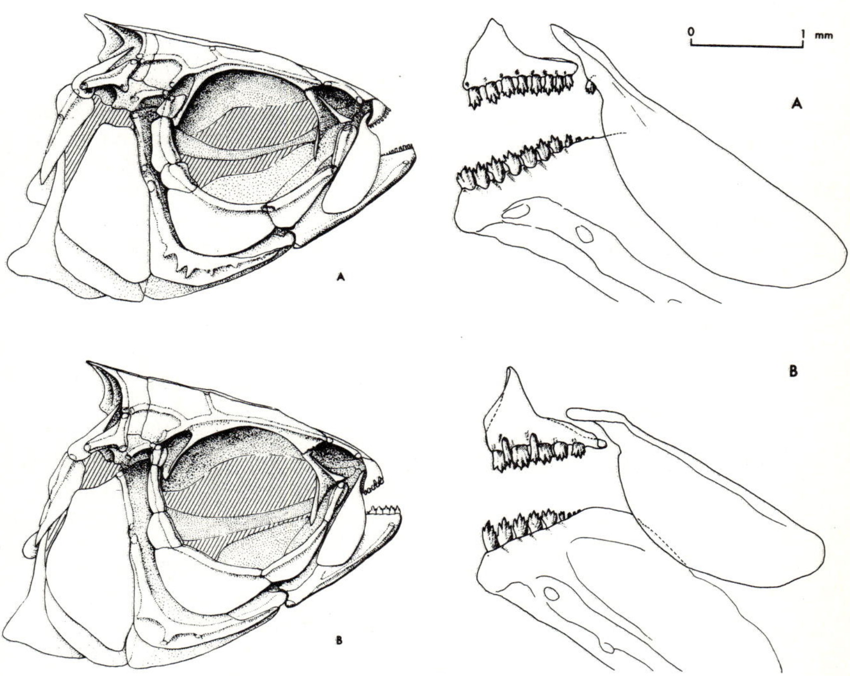

Many of the glass-tetras look alike. Compare this *Asiphonichthys* (possibly *condei*) with the *Phenacogaster* sp. shown here. Photo by Dr. Herbert R. Axelrod.

Phenacogaster megalostictus? Photo by Aaron Norman.

Vesicatrus tegatus, the spotted glass tetra. Photo by Harald Schultz.

Stichonodon insignis from Tefe. Photo by Dr. Herbert R. Axelrod.

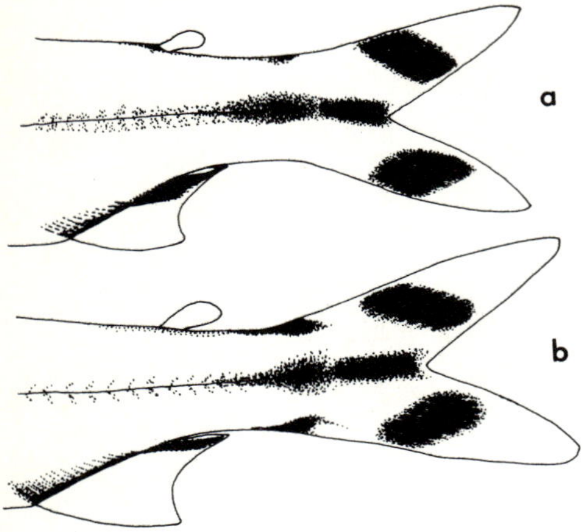

Color pattern differences of caudal area of *Petitella georgiae* (above) as compared with those of *Hemigrammus rhodostomus* (after Gery and Boutiere, 1964).

ii. Caudal fin naked as in *Hyphessobrycon*; 4 premaxillary teeth (body depth 3 in standard length, which is 35 mm; 33 lateral scales; 26 anal rays; no maxillary teeth; 4 premaxillary and mandibular teeth; a large humeral spot, a lateral stripe and a caudal spot up to end of middle caudal rays)
. . . *Mixobrycon ribeiroi* (a single specimen known from Rio Paraguay)

hh. Teeth compressed antero-posteriorly.

j. Upper and lower teeth not similar, much like those of *Holoshestes* (body depth about 3 in standard length, which is 50 mm or so; 34-37 lateral scales; 23-27 anal rays; maxilla long, with 4 teeth; a conspicuous caudal spot)
. . . *Aphyocheirodon hemigrammus* (Rio Tiete basin)

jj. Upper and lower teeth similar, usually like those of *Odontostilbe*, with 5-12 points (only three in the regressed *Cheirodon australe* from Chile).

k. No caudal gland or modified scale on adult males (genus *Cheirodon*).

l. No prominent interhaemal spines on adult males
. . . Subgenera *Lamprocheirodon* (1 species) and *Pseudocheirodon* (3-4 species ?)

ll. A series of prominent interhaemal spines on adult males, forming a keel on lower edge of caudal peduncle
. . . Subgenus nominate *Cheirodon* (16 species)

kk. A caudal gland or modified scale on adult males
. . . *Compsura* (2 species)

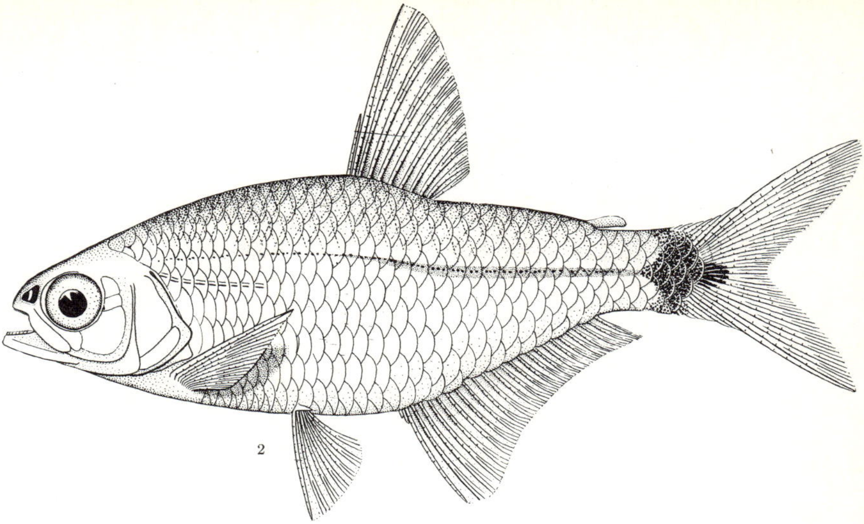

Aphyocheirodon hemigrammus (after Eigenmann, 1915).

GENUS *HOLOSHESTES*

Holoshestes pequira, the type species, from Rio Guapore basin and Rio Paraguay, differs from *H. heterodon* (synonym *Odontostilbe iheringi* ?), from the Rio S. Francisco, mostly in the color pattern. The former has 3 conspicuous black spots, one on the distal part of the first dorsal rays, one on the distal part of the first anal rays and one astride the peduncle and caudal fin base. The second has a narrower caudal spot and the fins are only a little dusky. Both species measure 50-55 mm, and usually have 20-25 anal rays and 34-39 lateral line scales. It is worth mentioning that *H. heterodon*, save for the characteristic lower teeth, is not markedly different from *Odontostilbe microcephala*.

Holoshestes pequira (after Eigenmann, 1915).

Top left: False rummy-nose, *Petitella georgiae.* Photo by S. Frank. *Bottom left and top and bottom right:* The fishes in these photographs are usually considered by German aquarists as *Petitella.* Actually they are likely to be a mixture of *Petitella* (the more slender fish) and a new species close to the rummy-nose, but more colorful. Photo below left by H.J. Richter; photos on the right by S. Frank.

GENUS *ODONTOSTILBE*

The *Odontostilbe* species* are very close to each other, and the following key will show a number of overlaps; they actually seem unavoidable. Owing to the scope of the present book, the suggested synonymies cannot be demonstrated. The interhaemal spines were said to be not projecting in the adult males as contrasted with certain *Cheirodon* species. However, a conspicuous peduncular keel, formed by developed interhaemals, has been observed in at least 2 *Odontostilbe*. The color pattern usually consists of a common caudal spot and nothing else.

Fink and Weitzman, in a revision of the Central American cheirodontine characids, have synonymized *Odontostilbe* (together with *Compsura* and *Pseudocheirodon*) with *Cheirodon*. They argue, among other things, that

Odontostilbe paraguayensis **(after Eigenmann, 1915) and** ***O. caquetae*** **(after Fowler, 1943).**

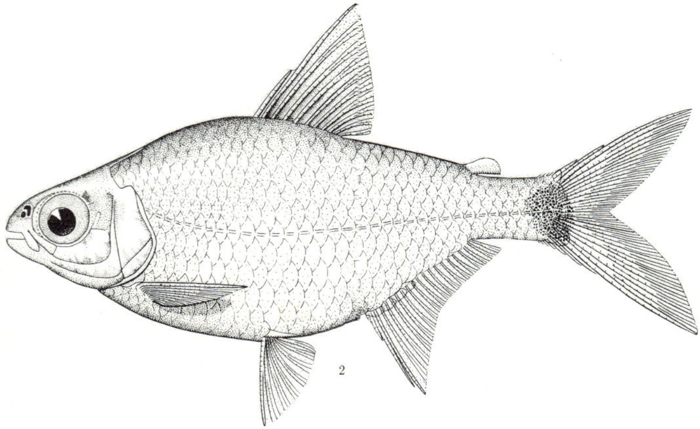

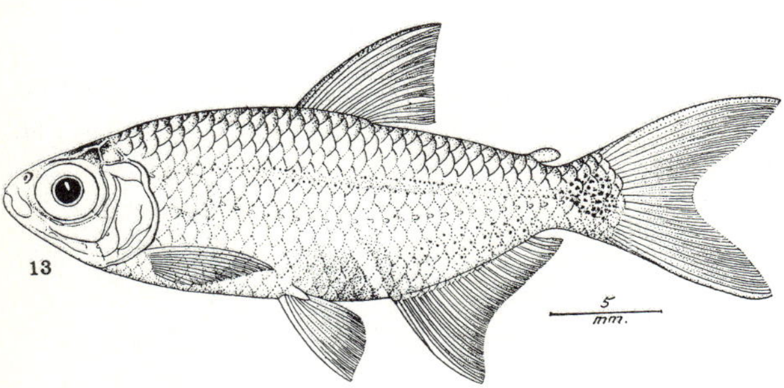

* See also *Deuterodon euspilurus*.

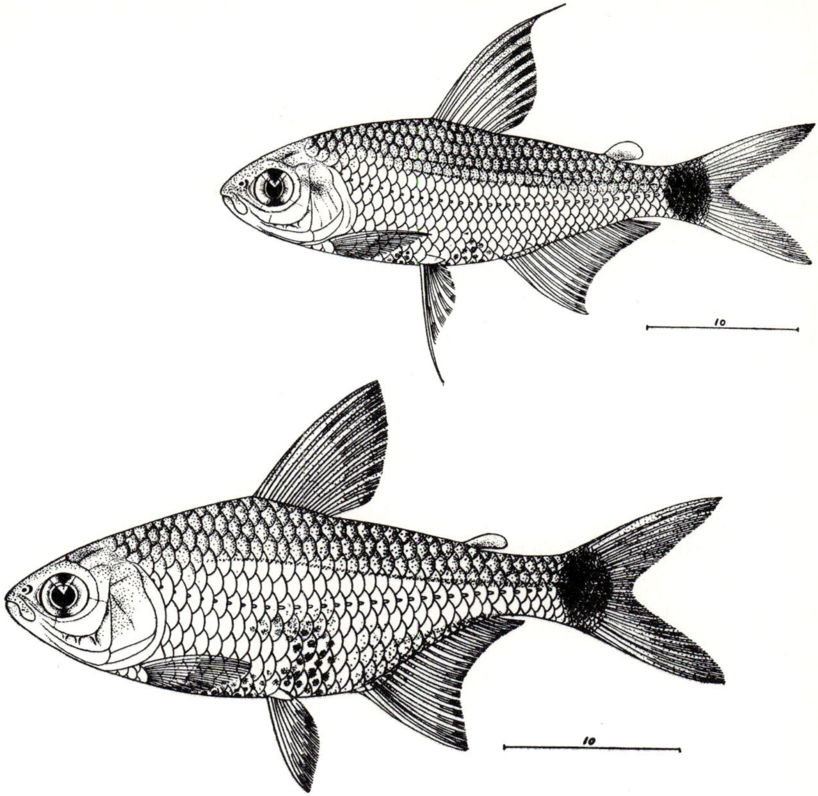

***Odontostilbe drepanon* and *O. madeirae* (after Fowler, 1913).**

certain *Cheirodon* may have a complete or almost complete lateral line in certain populations. It is evident that such an action can only be accepted following a complete revision of the genera and especially of their respective type species (*Cheirodon pisciculus* and *Odontostilbe fugitiva* for the two principal taxa). Therefore, and despite certain arguments that sound logical, *Odontostilbe* is here provisionally preserved.

Needless to say, as it was pointed out several times in the present book, the same reasoning about lateral line variability would apply, for example, to *Moenkhausia* versus *Hemigrammus* and *Astyanax* versus *Hyphessobrycon*, which also have "transitory" forms (as well as to several other genera, including certain African tetras). The same dubious cases may be found also in other critical characters such as scaled caudal fin, "fully" toothed maxilla and even in the number of inner premaxillary teeth or the setting of the anteriormost perforated scales. At the limit, one could imagine some "modern" worker lumping all Tetragonopterine genera, for example, into the single old genus *Tetragonopterus*, thus returning to the good old days of the "nineties."

Cheirodon kriegi. Photo by Harald Schultz.

A cheirodontin from Camatagua, Venezuela, probably *Cheirodontops geayi.* Photo by Dr. Herbert R. Axelrod.

Cheirodon notomelas. Photo by Harald Schultz.

Cheirodon insignis from Rio Calima, Colombia. Photo by Dr. Herbert R. Axelrod.

Key to the genus *Odontostilbe*:

a. Body depth 2.5-3.0 in standard length; 32-35 lateral line scales.

b. Dorsal fin at midbody; maxilla extending to level of front margin of eye; caudal spot small, blackish (body depth 2.5-2.75 and head 4.5-4.66 in standard length; iii, 20-22 anal rays; scale formula $5\frac{1}{2}/32\text{-}34/4\frac{1}{2}$) . . . *O. pulchra* (Trinidad and Venezuela)

bb. Dorsal fin in front of midbody; maxilla not extending to level of front margin of eye; caudal spot large, deep black (body depth 2.6-3.0 and head 3.75-4.5 in standard length; iii, 18-21 anal rays; scale formula 6/32-35/4) . . . *O. paraguayensis* (Rio Paraguay)

aa. Body depth 2.9-4.0 in standard length; 34-38 lateral line scales.

c. 6 or $6\frac{1}{2}/4$ or 5 transverse scales; maxilla longer than radius of eye.

d. iii, 16-19 anal rays (rarely 15); head length 4.0-4.6 in standard length; mouth slightly inferior . . . *O. microcephala* (Rios Paraguay and Parana)

dd. iii, 19-20 anal rays; head length 3.95-4.2 in standard length; mouth terminal . . . *O. dierythrura*, synonym *O. hasemani* (Bolivia, Amazon drainage)

cc. Usually 5 or $5\frac{1}{2}/3\frac{1}{2}$ or $4\frac{1}{2}$ transverse scales; maxilla about equal to radius of eye.

e. iii, 17-18 anal rays; body depth 3.35-3.45 and head length 4.15-4.3 in standard length.

f. Mouth apparently terminal . . . *O. mitopterus* (endemic in central Panama, described as a *Cheirodon*)

ff. Mouth slightly inferior . . . *O. roloffi* (Ecuador)

ee. iii, 21-22 anal rays on average; body depth 2.9-3.33 and head length 3.75-4.2 in standard length . . . *O. fugitiva*, the type species and the most common, synonyms *O. madeirae*, *O. drepanon* and *O. caquetae* (Amazon basin, Colombia)

GENUS *SACCODERMA*

According to L. P. Schultz, who proposed the genus in 1944, the species are like *Odontostilbe* but with a caudal sack or gland on both sexes. They are rather small (30-40 mm) and are restricted to the N.W. part of South America: *S. hastata* in the Rios Atrato and S. Juan in Colombia, *S. melanostigma* around the Lago de Maracaibo in Venezuela, and the endemic *S. robusta* only in the Rio Sinu in Colombia. They have 20-22 anal rays, 33-35 lateral line scales and a color pattern resembling that of *Holoshestes pequira*, i.e. with more or less conspicuous dark spots on the peduncle and distal part of the first dorsal and anal rays. They differ in minor details of coloration and position of the fins.

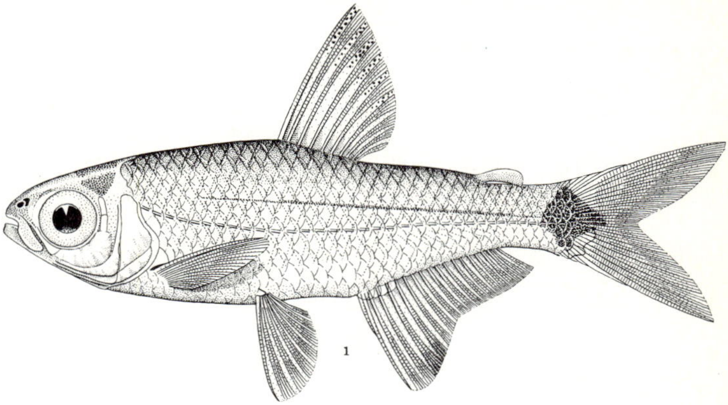

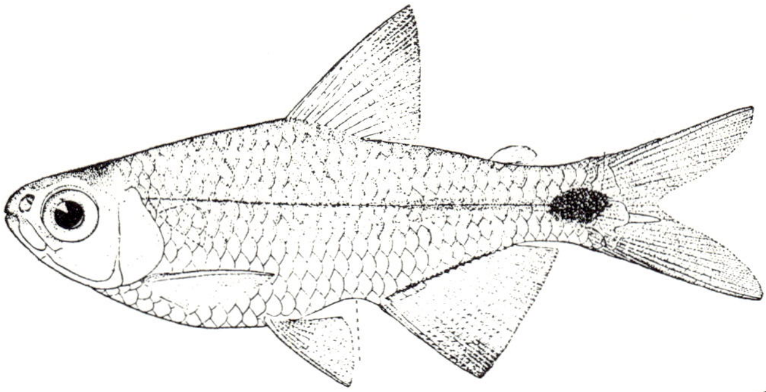

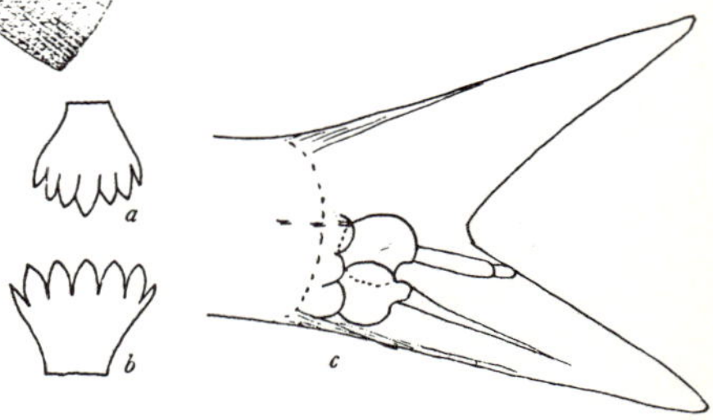

Saccoderma hastata (after Eigenmann, 1915) and *S. melanostigma* with teeth and caudal gland (after Schultz, 1944).

GENUS *OTHONOCHEIRODUS*

The two *Othonocheirodus* species, together with *Monotocheirodon pearsoni*, are adaptations of some *Odontostilbe*-like species, or perhaps of some *Deuterodon*, to a particular diet. Thus the regression of the upper lip, the characteristic form of the mouth and the double row of gill-rakers.

O. eigenmanni, the type species, from the Rio Huallaga in Peru, and *O. lethostigmus*, from Rio Grande do Sul in Brazil (described at first as the type of a new genus, *Odontostoechus*), apparently differ only in the number of their anal rays and pentacuspid teeth; it thus seems that two genera are unnecessary. Both species are 45-50 mm in standard length; they have 34-36 lateral line scales, a conspicuous humeral spot and a lateral band ending at the caudal fin base. *O. eigenmanni* has only 14 anal rays, 4 premaxillary, 2 maxillary, and 6 mandibular teeth. *O. lethostigmus* has 15-18 anal rays, 6 or 7 premaxillary, 4-5 maxillary, and 8-10 mandibular teeth.

Cheirodon axelrodi, the cardinal tetra. Photo by Dr. Herbert R. Axelrod.

A neon tetra, *Paracheirodon innesi* (lower fish), compared with the cardinal tetra, *Cheirodon axelrodi* (upper fish). Photo by Arend van den Nieuwenhuizen.

A group of cardinal tetras, *Cheirodon axelrodi.* Photo by Dr. Herbert R. Axelrod.

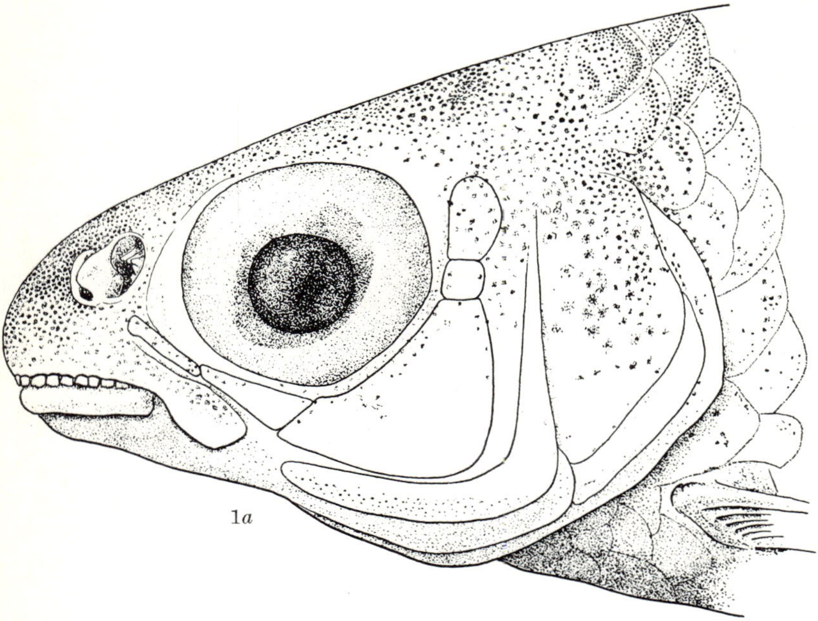

Head of *Othonocheirodus lethostigmus* (after Gomes, 1947, as *Odontostoechus lethostigmus*).

GENUS *CHEIRODON*

The *Cheirodon* species are small tetras, usually 30-45 mm in standard length (except the species from Chile which reach 75 mm). They have pluricuspid compressed teeth resembling a hand, or better a foot, usually with 5-9 points (again with an exception in Chile, where *C. australe* has only three-pointed teeth, possibly by regression), although certain species have up to 12 points. The premaxilla is triangular, without an ascending process. There usually is a conspicuous pseudotympanum (small triangular "window" over the anterior gas-bladder) on both sexes and a strong sexual dimorphism, the males (usually smaller than the females) having the anal lobe thickened, armed with many sexual hooks, and a peduncular keel formed by greatly enlarged interhaemal spines (numbering 4 to 26, depending on the species), except in the subgenera *Lamprocheirodon* and *Pseudocheirodon*.

This genus is the most southern characoid: it is found near Puerto Varas in Chile, south of the 40th parallel. Only *Gymnocharacinus* reaches almost as far south on the eastern side of the continent.

Subgenus *Lamprocheirodon*

Cheirodon axelrodi, the cardinal tetra, is the most colorful characoid, and one of the most decorative fishes. It caused a sensation when it was introduced to the aquarium hobby in 1956. Rumors had already told the aquarium circles that "another neon tetra" lived somewhere in the Amazon basin, but nobody was prepared to see such a gorgeous fish, even more colorful than its cousin. The cardinal tetra is an exception in several respects. Among species of the genus *Cheirodon* (to which it belongs owing to the structure of its jaws and teeth), it occupies a quite separate position, having no interhaemal spines (like *Pseudocheirodon*), yet having the well known color pattern of two unrelated species (*Paracheirodon innesi* and *Hyphessobrycon simulans*), a color which is not shared by any other *Cheirodon* or *Pseudocheirodon*.

Another peculiarity is the fact that it may well be an annual fish. Dr. Rolf Geisler, the well-known limnologist and expert aquarist, does not believe that there is enough food in the waters of the tributaries of the middle Rio Negro (where the cardinals live) to nourish the adults after their first breeding season, and that they probably starve in most places. Only in the aquarium are they able to live longer and spawn several times.

It has been suggested that its peculiar color pattern of a blue-green dorsal iridescence (due to oriented guanine crystals) and bright-red ventral parts may help specific recognition in the poorly lighted black waters which are precisely those of the three neon tetras. It is possible that the similarity of colorations is not coincidental. *Hy. simulans* may be a mimic of *Ch. axelrodi*, with which it is sympatric and which it closely resembles, and it is also much less common than *Ch. axelrodi*. *P. innesi* may have been such a mimic, but it has now been separated by some geologic change.

Cheirodon luelingi (after Gery, 1964).

The albino *Pristella maxillaris.* Photo by Dr. Herbert R. Axelrod.

The normal coloration of *Pristella maxillaris* (formerly called *P. riddlei*). Photo by S. Frank.

Two phases in the spawning sequence of *Pristella maxillaris.* Photos by R. Zukal.

It is perhaps amusing to recall that the correct scientific name of the cardinal tetra was a matter of hours. It was described twice almost at the same time, as *Hyphessobrycon cardinalis* in the *Stanford Ichthyological Bulletin* and as *Cheirodon axelrodi* in the *Tropical Fish Hobbyist*. A judgment of the International Commission for Zoological Nomenclature was necessary, and it was decided that the name *axelrodi* was published Feb. 20, 1956, and the name *cardinalis* the next morning (Feb. 21). Therefore *Ch. axelrodi* must hold forever!

Subgenus *Pseudocheirodon*

This unit, which has been treated as a full genus by certain zoologists, is perplexing. It is supposed to include those *Cheirodon* species which have no interhaemal spines in either sex, like most *Odontostilbe*. Yet the type species, *affinis*, sometimes has a complete lateral line, which is usually the case of a similar form, *Pseudocheirodon terrabae*, and both species fit the definition of *Odontostilbe*! The other two species here associated with *Pseudocheirodon* for convenience may belong to another genus or subgenus. On the other hand, *Cheirodon stenodon* and *C. microdon*, and *Mixobrycon*, are close to the definition of *Pseudocheirodon*, at least concerning the weakness of the interhaemals. As mentioned earlier, *Pseudocheirodon* has been synonymized with *Cheirodon* by Fink and Weitzman.

Cheirodon parahybae (after Eigenmann, 1915).

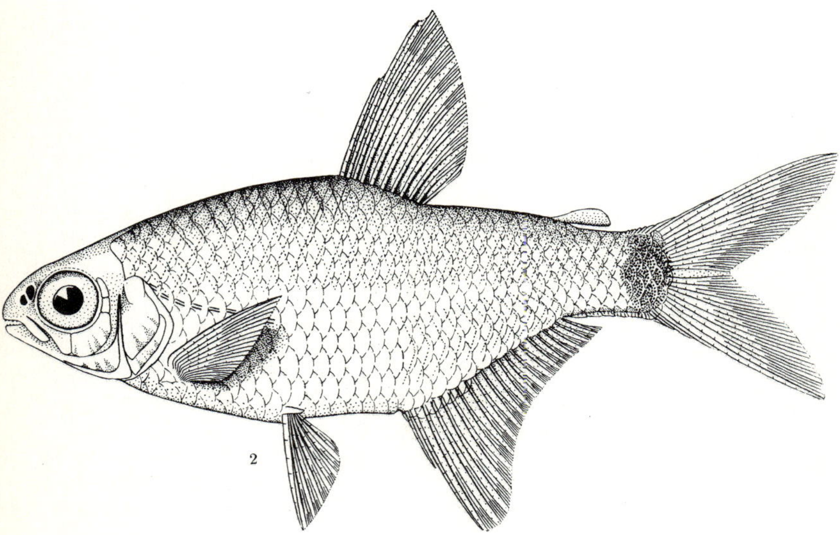

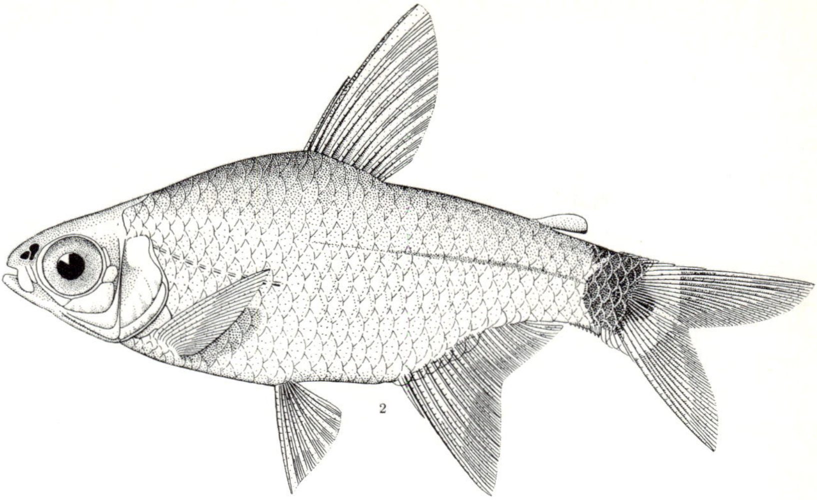

Cheirodon piaba (after Eigenmann, 1915).

The species can be identified by means of the following key:

a. A conspicuous spot on dorsal fin, as in species of the *Hyphessobrycon callistus*-group; iii, 27 anal rays; ((3) 31 lateral scales; body depth about 2.6 in standard length) . . . A species closely resembling *Ch. troemneri* (Rio Madeira basin; the type of the species, of unknown origin, has one small tooth in front of the main, *Cheirodon*-like premaxillary series, which would make it a *Hyphessobrycon*)

aa. No dorsal spot; a medium or large lozenge-shaped caudal spot as in most *Cheirodon*; iv, 16-23 anal rays.

b. Lateral line perforating only 5 or 6 scales; body depth 3.7-4.05 in standard length; a black line above anal fin . . . *Ch. luelingi* (Lower Rio Ucayali in Peruvian Amazon); maybe a "true" *Cheirodon*

bb. (in part after Fink and Weitzman) Lateral line usually perforating 10-34 scales (complete or almost complete in *Ch. terrabae* and certain populations of *Ch. dialepturus*); body depth 2.3-2.7 in standard length; no black line above anal fin.

c. Mandibular teeth with a straight cutting edge; middle premaxillary teeth asymmetrical; no hooks on lower caudal lobe (19-21 branched anal rays) . . . *Ch. affinis* (Panama, often sympatric with *Compsura gorgonae*, which has the same color pattern), and

The black phantom tetra, *Megalamphodus megalopterus.* Photo by Harald Schultz.

A darker phase of the black phantom, *Megalamphodus megalopterus.* Photo by Dr. Herbert R. Axelrod.

Megalamphodus megalopterus. Photo by Hilmar Hansen, Aquarium Berlin.

Male (above) and female (below) *Megalamphodus megalopterus.* Photo by S. Frank.

Ch. terrabae (Costa Rica) (Both forms differ only in certain meristics; they may be subspecies owing to many overlaps and geographical distribution)

cc. Mandibular teeth with the middle cusp enlarged; middle premaxillary teeth not very asymmetrical; upper rays of lower caudal lobe provided with numerous hooklets.

d. 16-19 branched anal rays . . . *Ch. dialepturus*, sympatric with *Compsura gorgonae* (Panama and Costa Rica)

dd. 20-21 branched anal rays . . . *Ch.* species, from Ecuador and Venezuela; could be a subspecies of the preceding

Nominate subgenus *Cheirodon*

This subgenus is a poorly known complex assemblage of relict forms which have been restricted to Chile, and of several smaller, more recently evolved forms, which are most successful in the basin of the Rio de la Plata. The following very imperfect key will suffice here to characterize the species. Needless to say, no attempt can be made to justify the alleged synonymies.

Cheirodon interruptus **(after Eigenmann, 1915) and *C. pisciculus* (after Eigenmann, 1915, as *C. annae*).**

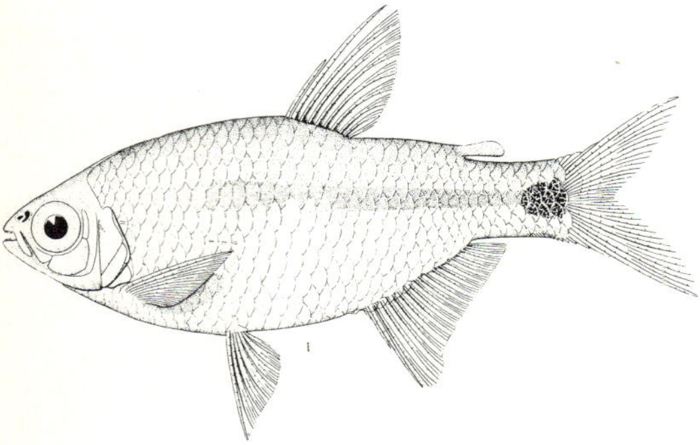

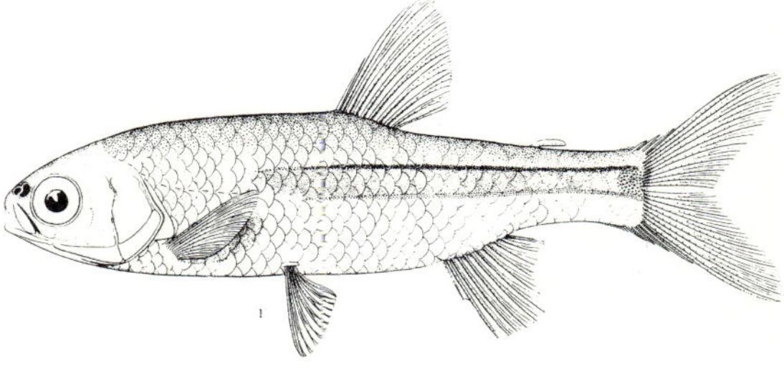

a. Rather small, compressed species from Venezuela to Rio de la Plata, with 18-25 anal rays (rarely 17).

b. Teeth rather narrow and not expanded at tip, the broadest with 5 cusps.

c. Interhaemals weak, 5-10 in number; last mandibular teeth conical; maxilla rather large, about 2/3 of eye diameter (tip of dorsal fin usually dark).

d. Broadest mandibular teeth with a developed middle point and 2 smaller ones on sides; 18-22 anal rays; preventral area rounded; caudal spot large ... *Ch. stenodon* (Rio Parana basin)

dd. Broadest mandibular teeth with 3 equal points and a smaller one on side, as in *Aphyocheirodon*; 23-25 anal rays; preventral area flattened; caudal spot rather faint ... *Ch. microdon* (Upper Paraguay basin; may eventually be considered an *Aphyocheirodon*)

cc. 15-20 interhaemals, prominent in males; mandibular teeth all pluricuspid; maxilla short as in other *Cheirodon* spp., about half eye diameter (a broad caudal spot on entire peduncle) ... *Ch. parahybae* (Rio Paraiba basin)

bb. Teeth broad, expanded at tip, the broadest with 7-12 cusps.

e. 8-13 (rarely 14) interhaemals (*Ch. piaba*-group).

f. 20-25 anal rays.

g. Dorsal and anal fins plain, without black marks.

h. No caudal spot; preventral area rounded (body depth 3 in standard length; 23 anal rays; (10) 34 lateral scales; suborbital leaving a narrow zone behind it) ... *Ch. madeirae* (Rio Guapore basin)

hh. A very conspicuous caudal spot; preventral area flattened.

i. Body depth 2.1-3.1 in standard length; large suborbital covering almost entire cheek; broadest teeth with 7-9 cusps.

j. 1-3 maxillary teeth; belly without black markings. ... *Ch. piaba*; possible synonyms *Ch. calliurus*, *Ch. jaguaribensis*, *Ch. macropterus*, *Ch. micropterus* ? (a common species in Rio Sao Francisco, La Plata basin and Amazon basin)

jj. 0-1 maxillary tooth; belly of the females with a black "gravid" spot, exactly like that of the live-bearing tooth carps ... *Ch. kriegi* (N.E. Paraguay)

ii. Body depth 3.0-3.45 in standard length; great suborbital leaving a broad naked area on cheek; broadest teeth with 10-12 cusps (2 maxillary teeth) ... *Ch. gracilis* (2 subspecies from French Guiana)

gg. First dorsal rays, dorsal base and anal lobe black; caudal spot very large (otherwise much like *Ch. piaba*, with only 4 premaxillary teeth) ... *Ch. notomelas* (Rio Tiete basin)

Unidentified tetra, probably a *Megalamphodus* or *Hyphessobrycon* close to *H. copelandi.* Photo by Dr. Herbert R. Axelrod.

Megalamphodus axelrodi, the calypso tetra. Photo by Dr. Herbert R. Axelrod.

Megalamphodus sweglesi, Swegle's tetra. Photo by S. Frank.

Megalamphodus sweglesi pair, male in front of female. Photo by S. Frank.

ff. 19 anal rays (a medium sized, round caudal spot)
. . . *Ch. felipponei* (a rather poorly known form from Uruguay, possibly synonymous with *Ch. interruptus*)

ee. Usually more than 17 interhaemals, sometimes fewer in immature individuals (always a conspicuous caudal spot; 17-22 anal rays) (*Ch. interruptus*-group)

k. Small size (30 mm); 2 or 3 maxillary teeth; (6) 28-32 lateral scales; 11 predorsal scales; large suborbital entire; head length 4 in standard length
. . . *Ch. insignis*, synonym *Ch. leuciscus* ? (Venezuela and Colombia)

kk. Rather large size for a *Cheirodon* (females up to 60 mm); 1 maxillary tooth; (7-12) 32-36 lateral scales; 13 predorsal scales; large suborbital leaving a broad naked zone behind; head length 4.5-4.66 in standard length
. . . *Ch. interruptus*, possible synonyms *Ch. monodon, Ch. ibicuhiensis, Ch. meinkeni* and *Ch. felipponei* —see above—(S.E. Brazil and La Plata basin)

aa. Rather large (up to 75 mm), less compressed species from Chile, with 12-17 anal rays (rarely 18), and certain regressive characters such as naked predorsal line or loss of tooth cusps (interhaemals strong, projecting; a silvery band; a diffuse, inconspicuous caudal spot).

l. Predorsal line in part not scaled; 10-16 interhaemals; maxilla slightly longer than half an eye diameter; large suborbital very reduced. . . *Ch. pisciculus*, synonym *Ch. annae*, the type species (northern Chile)

ll. Predorsal line fully scaled; 16-26 interhaemals; maxilla smaller than half an eye diameter; large suborbital with a naked zone only behind.

m. Broadest teeth with 5 cusps; body depth 3.1-3.15 and head length 4.35-4.4 in standard length
. . . *Ch. galusdae* (middle Chile)

mm. Broadest teeth with only three cusps; body depth 2.6-3.15 and head length 4.1-4.2 in standard length . . . *Ch. australe* (southern Chile)

GENUS *COMPSURA*

In the cheirodontine group with an incomplete lateral line, *Compsura* takes the place occupied by *Saccoderma* in the preceding group: it is a *Cheirodon* with specialized caudal scales in the ripe males (a gland ?) and

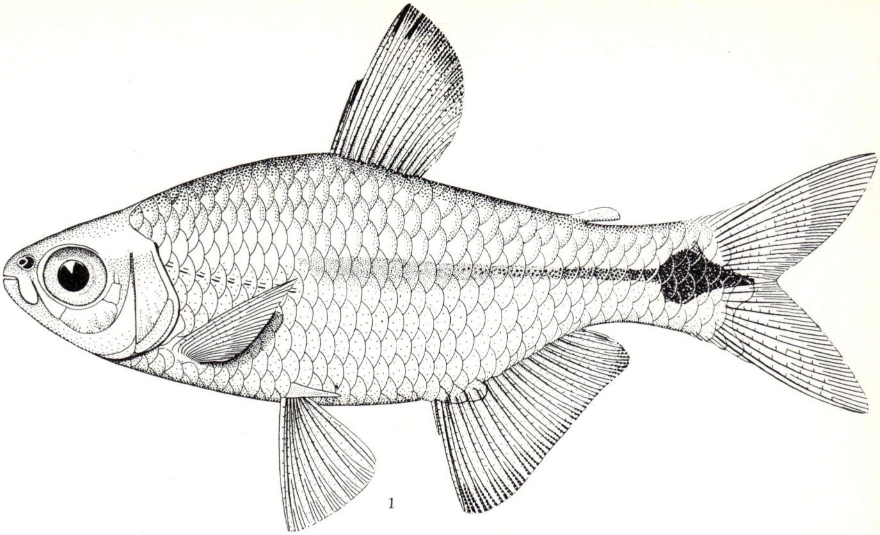

Compsura heterura (after Eigenmann, 1915).

without prominent interhaemals. The two species, the type species *C. heterura* and *C. gorgonae*, are separated by thousands of miles: the former comes from the Rio S. Francisco and the latter from Central America. This poses a phylogenetic problem. It is likely that each species evolved from a different *Cheirodon*-like ancestor and that the genus is unnatural. However, their resemblance in most characters is striking, both attaining about 40 mm in standard length, the body depth 2.5-3 in the standard length, 17-20 anal rays, (5-13) 30-35 longitudinal scales (the lateral line sometimes almost complete in the northern form), the largest teeth with 7-9 cusps and a conspicuous caudal spot.

The two species differ only in the number of premaxillary teeth (4 in the former, 5 in the latter) and in certain markings: *Compsura heterura* has the tip of the dorsal fin black (sometimes also the tip of the anal lobe, in the males only), whereas *C. gorgonae* has a series of V-shaped lines over the anal fin (these lines are useful in separating it from the similar *Ch. affinis*). According to Fink and Weitzman, they also differ in the form of the specialized scales, and *C. gorgonae*, at least, is scarcely separable from a *Cheirodon* at the generic level.

(B) Cheirodontin genera with conical or tricuspid teeth

This polyphyletic group (as stated above) could be divided into a few subgroups, which may have the rank of tribes: *Henochilus* and *Psalidodon* (Henochilini ?), *Probolodus* (Probolodini ?), *Grundulus* and *Phoxinopsis*, together with *Pristella* and several allied genera (Grundulini ?).

The Henochilini are characterized by their large size, flat, trilobed teeth, absence of lips, etc. The Probolodini (with the single species *Probolodus heterostomus*) have a few tricuspid, everted teeth which enable them to

Young *Megalamphodus sweglesi.* Photo by Dr. Herbert R. Axelrod.

Unidentified species of *Megalamphodus.* Photo by R. Zukal.

Axelrodia riesei. Photo by Dr. Herbert R. Axelrod.

Brittanichthys axelrodi. Photo by Dr. Martin Brittan.

scrape scales from other fishes (much as in *Catoprion*). The Grundulini are a complex assemblage of 20-22 genera which again may be split into: (1) *Pristella* and allied genera, (2) *Paracheirodon*, (3) *Aphyodite* and allied genera, and (4) *Grundulus* and *Phoxinopsis*, 2 regressed genera which have lost their adipose fin, lateral line and circumorbital bones.

Tribe Henochilini

Henochilus wheatlandi, from the Rio Mucuri (a coastal river between the States of Bahia and Espirito Santo in Brazil), and *Psalidodon gymnodontus*, from the Rio Iguassu, are large fishes (respectively 190 and 415 mm) vaguely resembling a *Deuterodon* in the body shape as well as in the aspect of the trilobed, spatulate, uncovered upper teeth. *Henochilus* has retained 2 teeth behind the conspicuous main series, which is obviously derived from the outer one of the Tetragonopterinae, and it is clearly intermediate between that group and the Cheirodontinae. Moreover, it has kept some remnants of cusps on certain teeth, which make them more or less pentacuspid.

Psalidodon, apparently derived from it in the past, is decidedly on the cheirodontine "side," having strictly uniserial, trilobed teeth (as well as both lips lacking). Otherwise the two genera, which have the lateral line complete and the circumorbitals well developed, differ in the number of anal rays, scales and maxillary teeth (all more numerous in *Henochilus*). Both are plant-eaters.

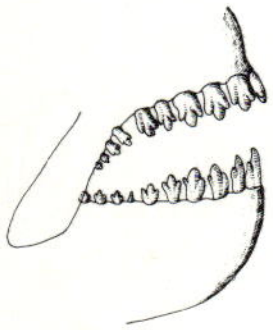

Henochilus wheatlandi (redrawn after Eigenmann, 1921) and jaws of type of *Psalidobon gymnodontus* (right side). Drawings by the author.

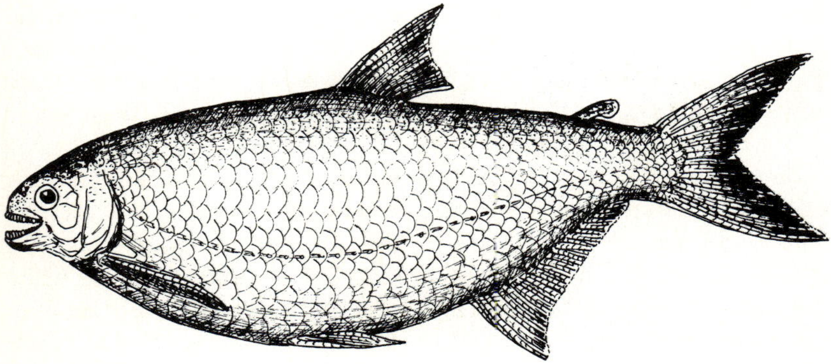

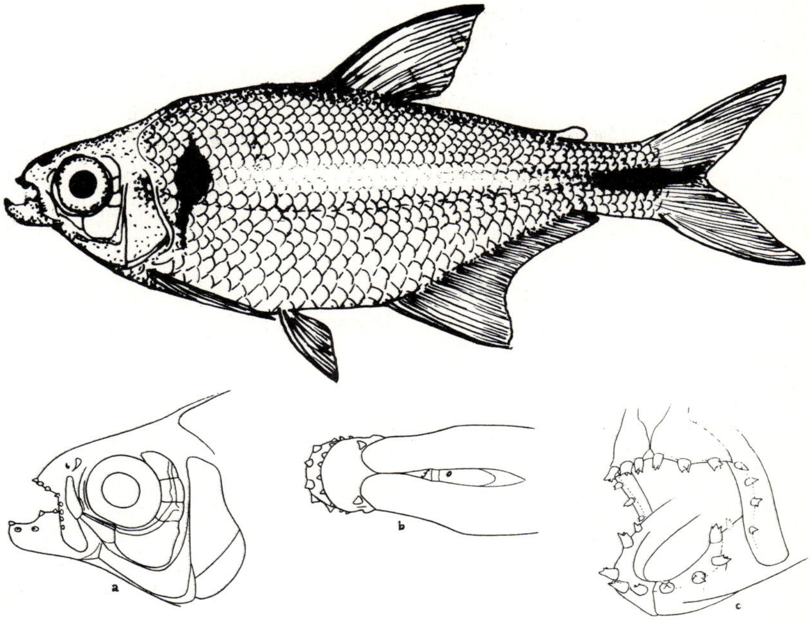

Probolodus heterostomus, type, with head and jaw structure (after Eigenmann, 1915).

Tribe Probolodini

The single species *Probolodus heterostomus*, from coastal streams of S.E. Brazil, is a medium-sized (100 mm) fish strongly resembling *Astyanax fasciatus*, but with a strange dentition which recalls that of *Catoprion*. The tricuspid teeth are few in number (3 on the premaxilla, 2-5 on the maxilla and 3 on the front of the mandible plus 2 on its elevated sides) and protrude out of the jaw, mostly out of the prominent lower jaw as in *Catoprion*. These teeth enable it to take off rows of scales from other fish (possibly from bottom to top as was observed in *Catoprion*).

Another scale-eating Amazonian species has been rather recently collected. It seems to be near *Probolodus* in certain respects, but with two rows of mammilla-like teeth on each jaw. It may be a modification of a *Brycon*-like ancestor, which would raise to 4 the characoid lines in which the scale-eating habit has developed: the Serrasalmidae with *Catoprion*, the Characinae with *Exodon*, *Roeboexodon*, and several *Roeboides*, the Tetragonopterinae-Cheirodontinae with *Probolodus*, and the Bryconinae with the undescribed species referred to above.

From top to bottom: Paracheirodon innesi, Cheirodon axelrodi, Paracheirodon innesi, and *Cheirodon axelrodi.* Photo by Dr. Herbert R. Axelrod.

A school of neon tetras, *Paracheirodon innesi.* Photo by Dr. Herbert R. Axelrod.

Paracheirodon innesi. Photo by Dr. D. Terver, Nancy Aquarium, France.

A spawning pair of neon tetras, *Paracheirodon innesi.* Female below. Photo by S. Frank.

Tribe Grundulini

The study of this complex group began only 20 or 30 years ago, and it will be several more years before the anatomy of each species (certain ones extremely small) is known and the natural evolutionary lines cleared up. At present the group looks much like a Chineze puzzle, and the account that follows can be no more than a caricature of reality. It may be artificially divided into the following sub-units:

a. Body usually compressed; adipose fin present except in 2 *Tyttobrycon* species.

b. Body rather elevated and dorsal fin with a black marking, as in the *Hyphessobrycon bentosi-* or *callistus*-group (lateral line incomplete) . . . *Pseudopristella, Pristella* and *Megalamphodus* (*Aphyocharax axelrodi* probably belongs here)

bb. Body usually not much elevated, dorsal fin without black markings except in a few *Microschemobrycon* and *Axelrodia* (lateral line variable) . . . *Oligobrycon, Atopomesus, Paracheirodon* and the large group of *Aphyodite* and allied genera

aa. Body not much compressed, cyprinodontoid in shape; adipose fin lacking (lateral line very short) . . . *Grundulus* and *Phoxinopsis*

THE *PRISTELLA*-GROUP

This group, with its depth 2.2-3.1 in the standard length, its dorsal spot (acting as a recognition signal?) and its somewhat irregular premaxillary row of teeth (some individuals of certain species may even have one tooth somewhat forward), is very near the *Hy. bentosi*-group, and the limits are often nebulous. However, lumping the two units means a reorganization of most of the tetras (in the large sense), which is improbable in the near future. It is likely that most of the species enter into protective associations, being mimics of other tetras with black dorsal markings. *Pristella* and certain *Megalamphodus* are rather common aquarium fishes, being nicely colored and small (30-50 mm for the whole group). They can be distinguished in the following key:

a. Postorbital part of the cheek well covered; fontanels not very large; anal fin as well as dorsal fin with a black marking.

b. Maxilla with few teeth; some premaxillary or mandibular teeth quinquecuspid (the majority tricuspid); humeral spot quite large; first branched anal ray black . . . *Pseudopristella simulata* (Guianas)

bb. Maxilla with weak conical teeth along its entire margin; broadest teeth all tricuspid, some out of line on the upper jaw in the largest individuals; humeral spot small; anal lobe with a distal black spot . . . *Pristella maxillaris*, synonym *P. riddlei*, the well-known

pristella of aquarists*
(Amazon, Venezuela and Guianas)

aa. Postorbital bones very weak or absent; fontanels usually very large; anal fin usually without black marking (genus *Megalamphodus*, 9 species including *Aphyocharax axelrodi*).

c. 10-20 maxillary teeth (a large humeral bar; dorsal and anal fin dark, very developed in adults) . . . *Megalamphodus megalopterus*, the type species, called the black phantom tetra by aquarists (Rio Guapore basin)

cc. Less than 10 maxillary teeth.

d. Broadest teeth tricuspid; large suborbital usually complete; postorbitals usually not entirely lacking.

e. 27-30 anal rays.

f. Body depth 2.5-2.6 in standard length; a conspicuous humeral bar.

g. About 30 anal rays; anal margined with black . . . *M. eques* (a not well known species from lower-middle Amazon basin)

gg. 27 anal rays; caudal margined with black . . . *M. rogoaguae* (Rio Beni basin in Bolivia)

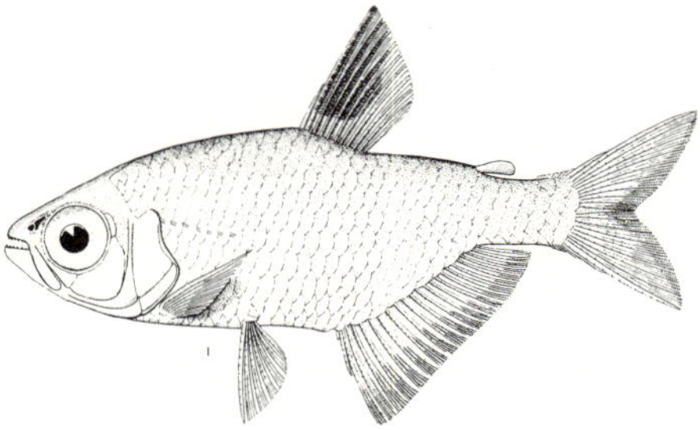

Pseudopristella simulata (photo by the author) and *Megalamphodus micropterus* (after Eigenmann, 1915).

* Dr. J. Böhlke, in his thesis (1954), has shown that *P. maxillaris* and the much better known *P. riddlei* are synonymous, the former antedating the latter.

Two photos in a spawning sequence of the neon tetra, *Paracheirodon innesi*. A shower of eggs can be seen in the right hand page photograph. Photos by J. Elias.

ff. Body depth about 3 in standard length; no humeral markings (27-30 anal rays; about 11 premaxillary teeth and 6-7 maxillary ones) . . . *M. heterestes* ("Brazil," exact locality unknown)

ee. 22-26 anal rays (rarely 27).

g. Body depth 2.4-2.75 in standard length.

h. 7/4½ transverse scales (a narrow anterior humeral bar; 26 anal rays; 9 maxillary teeth) . . . *M. uruguayensis* (Uruguay)

hh. 5/3 or 4 transverse scales.

i. A narrow anterior humeral bar; iii 19 or 20 anal rays (7 maxillary teeth; sometimes a premaxillary tooth out of line) . . . *M. micropterus* (Rio Sao Francisco)

ii. A very large posterior humeral spot; iii, 23 (24) anal rays (4-6 maxillary teeth) . . . *M. sweglesi*, the Swegles tetra (Upper Amazon basin; upper Orinoco basin ?)

gg. Body depth 2.9-3.1 in standard length (small size).

j. A narrow humeral bar; 2 maxillary teeth . . . *M. ecuadorensis* (Pacific slope of Ecuador)

jj. A very large roundish humeral spot; 5 maxillary teeth . . . *M. roseus* (Maroni basin in Guianas)

dd. All teeth conical; large suborbital not complete; postorbitals lacking (24-26 anal rays; body depth 2.7-3.1 in standard length; 3 maxillary teeth) . . . *M. axelrodi*, the calypso tetra (Trinidad; described as an *Aphyocharax*, to which it certainly does not belong)

THE *APHYODITE*-GROUP

Like the preceding, the group comprising *Aphyodite* and 14 other genera (possibly including *Paracheirodon*, the popular neon tetra, that one does not exactly know where to place) is puzzling. Although in the following key it is classified mostly according to the tooth structure, it is probable that such a structure can be obtained from several lines. For example, *Oligobrycon* is possibly evolved from some *Cheirodon*, *Prodontocharax* from *Odontostilbe* (?) and so on. This group abounds in specializations (hooks, interhaemal spines, clupeoid jaws, etc.) and regressions (lack of adipose fin, of lateral line, anal fin short, small to extremely small size, etc.).

The number of monotypic genera (more than 50%) and of genera with 2 species only (more than 30%) is unusually great. A somewhat detailed study of the two largest genera, as well as of *Paracheirodon*, will follow the key.

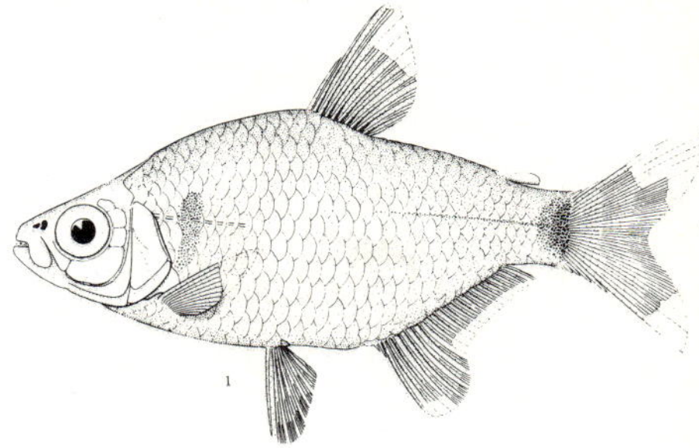

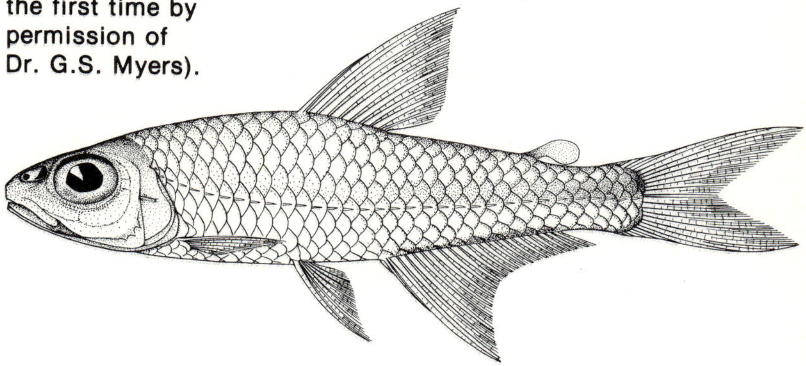

Oligobrycon microstomus, type (after Eigenmann, 1915) and *Atopomesus pachyodus*, type (original drawing by Mr. Pablo Bravo in 1930, published for the first time by permission of Dr. G.S. Myers).

Key to the Aphyoditeina:

a. Teeth heavy, not compressed from front to back.

b. Body deep and compressed; lateral line incomplete; mouth terminal; teeth triscupid, 4 on premaxilla and only 1 or 2 on the small maxilla (color pattern and body shape resembling those of *Cheirodon interruptus*—possibly evolved from some *Cheirodon*) . . . *Oligobrycon microstomus* (Rio Paraiba)

bb. Body elongate (4 times in standard length); lateral line complete; mouth inferior; teeth truncate, somewhat like those of a *Leporinus*, 7 backwardly set ones on premaxilla and 6 on maxilla (an odd species whose affinities are unknown) . . . *Atopomesus pachyodus* (Rio Negro)

aa. Teeth not heavy, tricuspid and compressed from front to back, or small and conical.

c. Teeth mostly tricuspid (except in *Microschemobrycon elongatus* where they are conical and bicuspid).

d. Teeth relatively few in number; premaxilla without ascending apophysis; fontanels very large (as in most *Megalamphodus*); postorbitals completely lacking; lateral line incomplete (body fusiform, not much compressed; peculiar color pattern similar to that of *Cheirodon axelrodi* and *Hyphessobrycon simulans*; affinities unknown) . . . *Paracheirodon innesi*, the popular neon tetra (upper Amazon)

dd. Teeth usually numerous; premaxilla with an ascending prócess; fontanels not very large; postorbitals present; lateral line usually complete (with exceptions).

Unidentified characoid, possibly near *Nannostomus* or *Pyrrhulina,* from the Parima mountains in northern Brazil. Photo by Dr. Herbert R. Axelrod.

A curimatid from the Rio Aguaro. Photo by Dr. Herbert R. Axelrod.

This tetra has the color pattern of *Moenkhausia ceros* or *Hemigrammus levis,* but is much more elongate. Photo by Dr. Herbert R. Axelrod.

A species of *Bryconamericus* or *Knodus.* Photo by Aaron Norman.

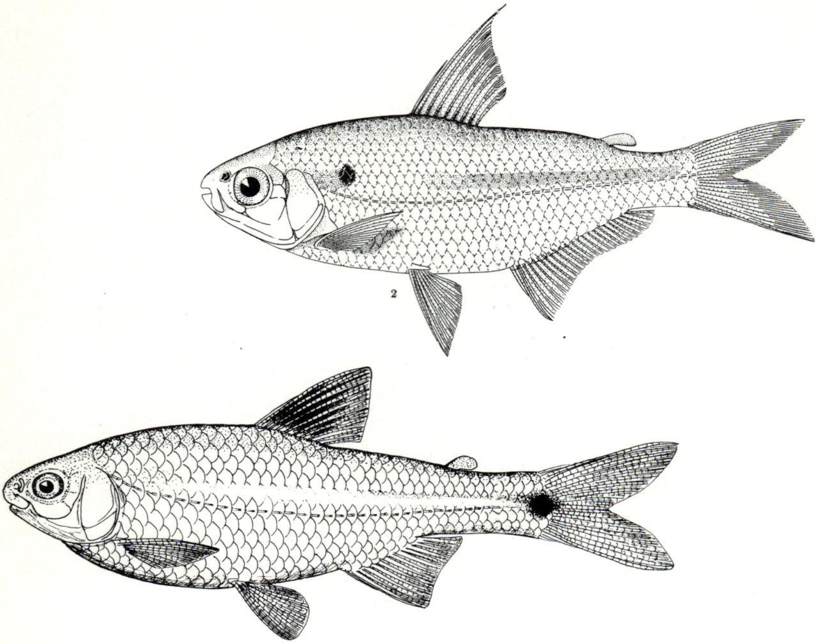

***Parecbasis cyclolepis*, type (after Eigenmann, 1915); and *Prodontocharax melanotus*, type (redrawn from Pearson, 1924).**

e. Caudal fin scaled (see also *Microschemobrycon casiquiare*).

f. Rather large size (75 mm); circumorbital bones well developed; mouth terminal, maxilla not toothed, thickened; sides of mandible raised; probably 2 series of teeth on mandible, as in *Aphyocharacidium* (a small round humeral spot) . . . *Parecbasis cyclolepis* (Rio Madeira)

ff. Small size (30 mm); circumorbital bones rather weak; mouth superior, maxilla with one or two minute teeth, not thickened; sides of mandible not raised; a single mandibular series of teeth (no humeral spot; a black line above anal fin) . . . *Aphyodite grammica* (Guyana, Amazon ?)

ee. Caudal fin naked (20-25 mm in standard length; maxilla not thickened, toothed; usually a pseudotympanum).

g. Mouth inferior; fontanel narrow; sides of mandible raised (lateral line complete, perforating 34-37 scales; 15-26 anal rays; possibly descended from some *Odontostilbe*-like ancestor) . . . *Prodontocharax*, synonym *Amblistilbe* (two species*, *P. melanotus* from Bolivia and *P. alleni* from Rio Ucayali)

* *A. howesi*, the type of *Amblistilbe*, may be a distinct species.

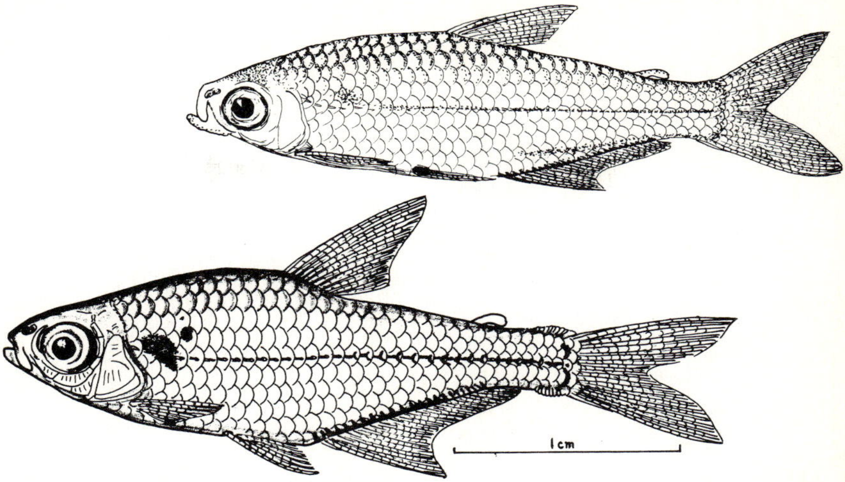

Aphyodite grammica, **type (redrawn from Eigenmann, 1915); and** ***Aphyocharacidium bolivianum,*** **type (after Gery, 1973).**

gg. Mouth terminal or slightly superior; fontanels usually broad; sides of mandible not raised.

h. Two rows of teeth on the mandible; lateral line complete . . . *Aphyocharacidium* (two species, *A. melandetum* from Guianas and *A. bolivianum* from Bolivia)

hh. A single mandibular series of teeth; lateral line complete or incomplete (caudal fin scaled in certain populations of *M. casiquiare*; teeth almost all conical in *M. elongatus*) . . . *Microschemobrycon* (7 species)

cc. Teeth conical except in *Macropsobrycon xinguensis*; lateral line always incomplete; anal fin usually short or rather short; mouth often more or less upturned; small to very small species, 15-45 mm in standard length.

i. Caudal fin scaled (see also *Macropsobrycon xinguensis* (m.)) (anal fin not very short; circumorbital reduced; maxilla developed, somewhat thickened, not toothed; mouth upturned; caudal fin of the male with an ornamented hook) . . . *Brittanichthys* (two species, *B. axelrodi* and *B. myersi*, from Rio Negro)

ii. Caudal fin not scaled.

j. Two rows of teeth on the mandible (lateral line lacking; anal fin very short; maxilla toothed, not thickened; mouth scarcely upturned; sides of mandible raised; a very small, regressed, enigmatic species) . . . *Oxybrycon parvulus* (upper Amazon basin)

jj. A single mandibular series of teeth.

Unidentified tetra from the Rio Negro. Photo by Dr. Herbert R. Axelrod.

Many unidentified tetras turn up in imports of Amazonian aquarium fishes. It is difficult to assign a name to many of them. Photo by S. Frank.

A blue tetra from the upper Rio Arinos, looking somewhat like *Hyphessobrycon vilmae.* Photo by Harald Schultz.

k. Maxilla enlarged, blade-like, clupeoid, its border thickened, with few or no teeth; sides of mandible raised.

l. Mouth clearly upturned; teeth few in number; 20-25 anal rays.

m. Body depth 3.0-3.8 in standard length, which is about 45 mm; maxilla not strongly clupeoid-like . . . *Macropsobrycon* (two species, *M. uruguayanae* from Uruguay and *M. xinguensis*, from Rio Xingu. The latter strongly resembles *Moenkhausia cotinho*; it is atypical in having remnants of cuspids on the broadest teeth and a few scales on caudal fin)

mm. Body depth 4.8-5.3 in standard length, which is about 30 mm; maxilla very clupeoid-like . . . *Thrissobrycon pectinifer* (Rio Negro)

ll. Mouth only slightly upturned; teeth numerous (about 14 on premaxilla); 14 anal rays (body depth about 3.5 in standard length, which is about 30 mm) . . . *Leptobrycon jatuaranae* (Amazon)

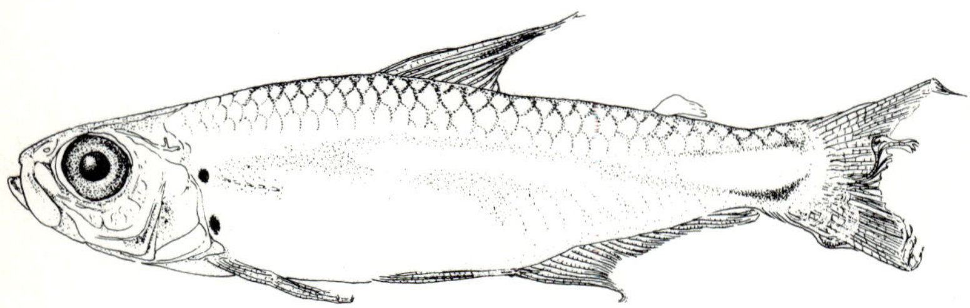

Type of *Brittanichthys myersi*, caudal fin of *B. axelrodi* and of *B. myersi* (after Gery, 1965).

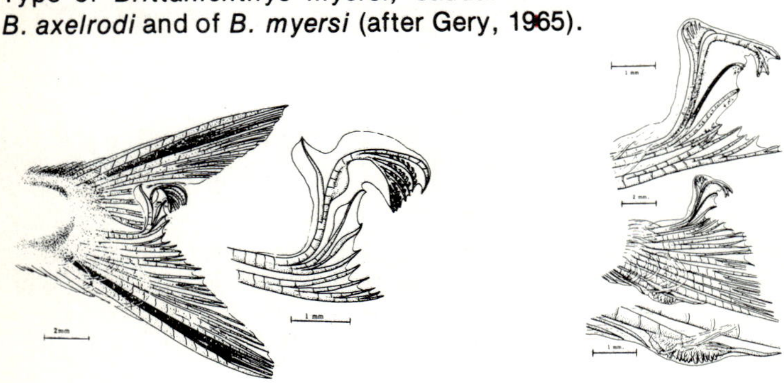

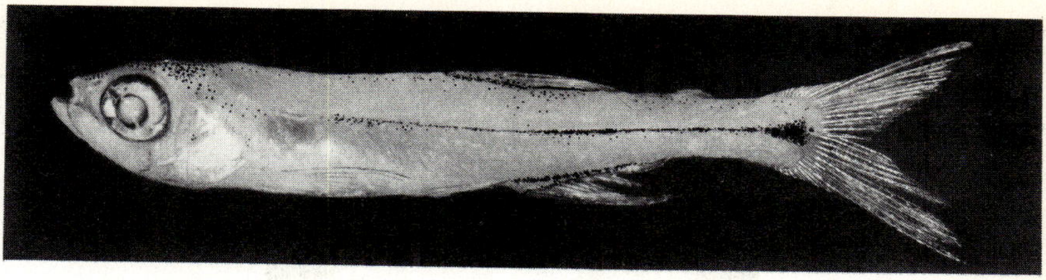

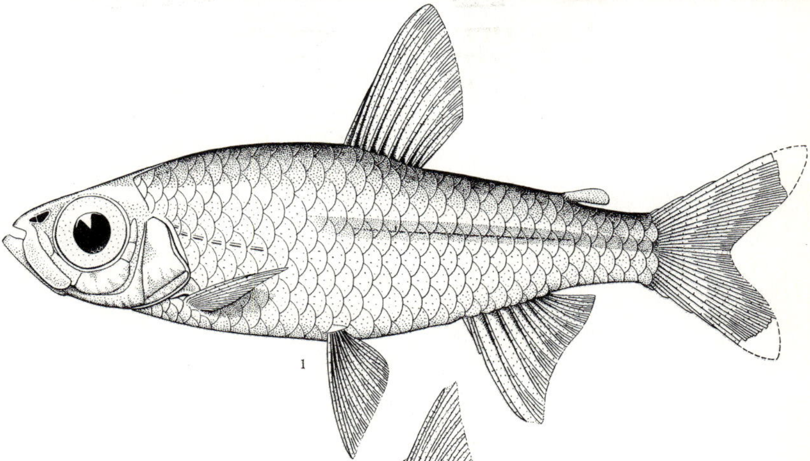

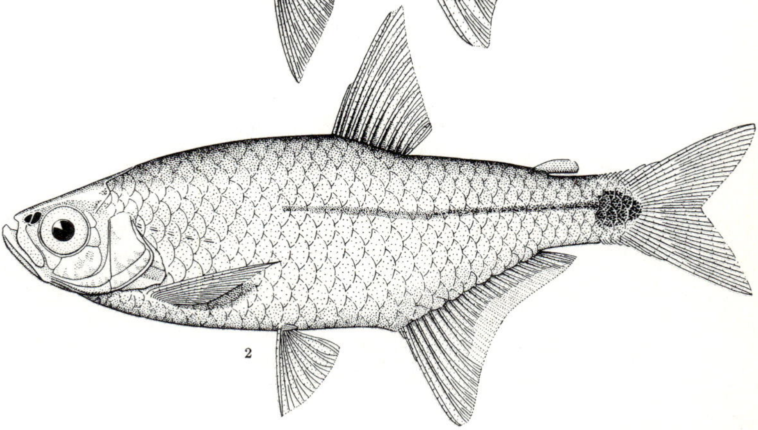

Oxybrycon parvulus*, type (after Gery, 1964); *Leptobrycon jatuaranae*, type and *Macropsobrycon uruguayanae*, type (after Eigenmann, 1915).

kk. Maxilla not enlarged, not thickened, with 0-9 teeth; sides of mandible not raised.

n. Maxilla and side of mandible well toothed; no sexual hooks; no interhaemals

. . . *Axelrodia* (3 species)

nn. Maxilla not toothed (except 1 tooth in *T. spinosus* ?); sides of mandible with few teeth, if any; sexual hooks present on anal fin of adult males; interhaemals present in at least 1 species (which lacks an adipose fin)

. . . *Tyttobrycon* (4 species)

Possibly a species of *Hyphessobrycon.* Photo by Dr. Herbert R. Axelrod.

An interesting new species in the *Moenkhausia*-group. It comes from the Rio Arinos. Photo by Harald Schultz.

Another aspect of the new form from the Rio Arinos. Photo by Harald Schultz.

PARACHEIRODON INNESI, THE NEON TETRA

The well-known neon tetra, which inhabits the tributaries of the upper course of the Rio Solimoes (=Amazon) between Letitia and Iquitos, as well as some places in the Rio Purus, was first described as a *Hyphessobrycon* because its seven tricuspid, premaxillary teeth are set in a somewhat wavy line (which may eventually be understood as a double row; this happens also in *Petitella,* in *Pristella* and in certain *Megalamphodus* and Aphyoditeina). Its correct place in the classification is uncertain. According to its shape, peculiar color pattern and meristics, it would be classified near the cardinal tetra. However, as it was pointed out above, its karyotype (structure of the chromosomes) is not in favor of this hypothesis. *

*Fink & Weitzman (Abstract, ASIH Fifty-seventh Meeting, 1977) placed all three neon tetras in *Paracheirodon.*

GENUS *MICROSCHEMOBRYCON*

The largest unit of the Aphyoditeina, with 7 species, has a provisory status, the whole group being in the process of revision. Most of the species look neotenic, with their small size and larval color pattern (usually a comma-shaped mark at the base of each caudal lobe), and at the same time, rather generalized as compared with other, very specialized, genera. They apparently occupy a central position within the group.

The following key (slightly altered from a previous study) will aid in identifying the species:

a. Lateral line complete or almost complete; 6-12 maxillary teeth.
  b. iii, 15-21 anal rays; a humeral spot; lateral line usually complete.
    c. iii, 15-18 anal rays; humeral spot vertically elongate; a conspicuous crescent-shaped dorsal mark . . . *M. callops* (upper Rio Negro-Orinoco and lower Amazon)
    cc. iii, 19-21 anal rays; humeral spot round, very small; first dorsal ray somewhat darkened . . . *M. geisleri* (lower Amazon)
  bb. iii, 22 anal rays; no humeral spot; lateral line not quite complete . . . *M. guaporensis,* the type species (Rio Guapore)

aa. Lateral line short, perforating 6-23 scales; 3-6 maxillary teeth.
  d. Mouth terminal; iii, 15-18 anal rays; 12-13 premaxillary teeth; no dorsal spot.
    e. Teeth mostly tricuspid; nostrils close together; body depth 3.7-4.6 in standard length . . . *M. casiquiare* (Rio Negro-Orinoco and Rio Madeira)
    ee. Teeth mostly conical or bicuspid; nostrils rather far apart; body depth 4.5-5.05 in standard length . . . *M. elongatus* (Rio Xingu)
  dd. Mouth slightly upturned; iii, 24-27 anal rays; 10-11 premaxillary teeth; a dorsal spot (species looking like elongate *Megalamphodus*).
    f. iii, 24-25 anal rays . . . *M. melanotus* (Guyana, upper Amazon, Rio Negro)
    ff. iii, 27 anal rays . . . *M. hasemani* (Rio Madeira)

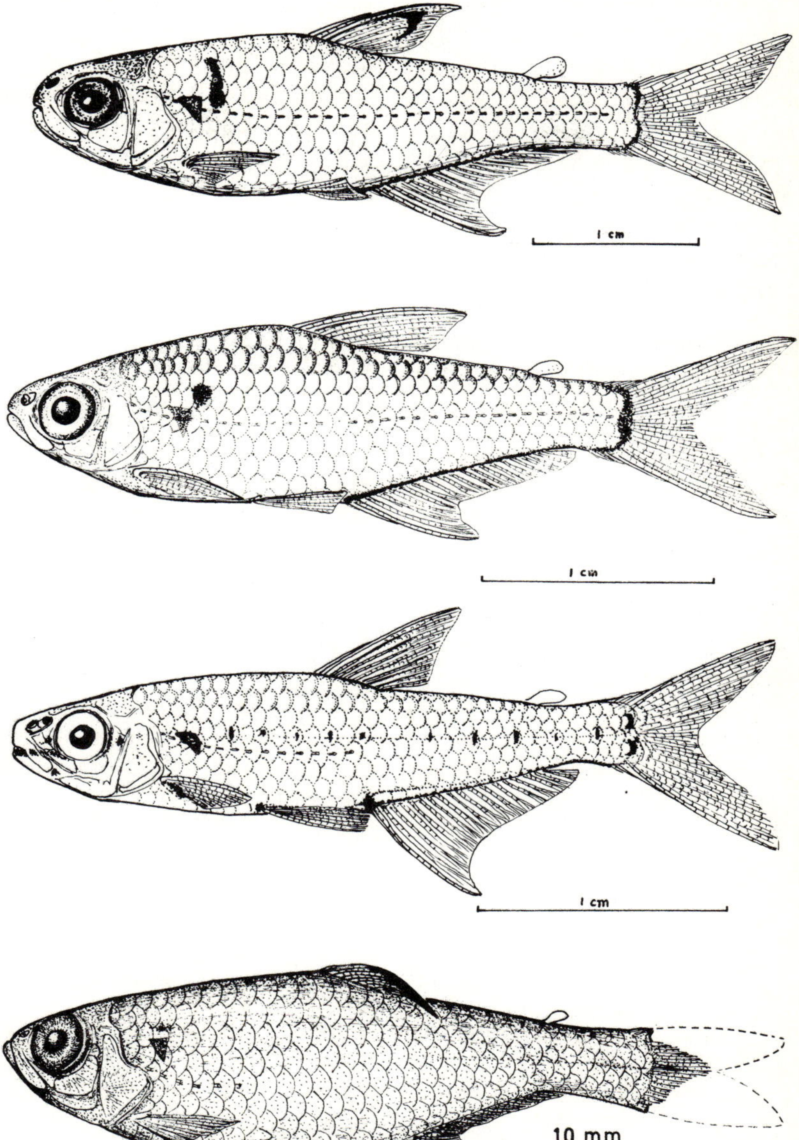

From top to bottom: *Microschemobrycon callops, M. geisleri, M. elongatus* (after Gery, 1973), and *M. melanotus* (after Gery, 1972).

Unidentified tetra from the Amazon basin. Photo by Dr. Herbert R. Axelrod.

Possibly a species of *Hyphessobrycon.* Photo by Dr. Herbert R. Axelrod.

Undescribed *Hyphessobrycon* from Rio Meta, Colombia, strongly resembling *Axelrodia stigmatias.* Photo by Dr. Herbert R. Axelrod.

Unidentified tetra apparently close to *Hyphessobrycon tenuis*. Upper fish a different species, possibly *Hemigrammus schmardae*. Photo by Klaus Paysan.

GENUS *AXELRODIA*

The genus *Axelrodia* was proposed when it was found that the Amazonian *Hyphessobrycon stigmatias* was not a *Hyphessobrycon*, having strictly uniserial, conical teeth. Later on, the minute, intense red *A. riesei* was added; it is one of the most suitable fishes for the aquarium of miniatures. Still later, *A. lindae* (somewhat different from the two others, but also a charming little tetra which would be a prized aquarium fish) was also added.

For anecdotal interest, it may be mentioned that the person to whom the genus was dedicated (the publisher of this book, Dr. Herbert R. Axelrod) is the same one who collected (together with the late William Riese) the second species! It is quite impressive to discover a new species of your "own" genus!

The species can be identified as follows:

a. iii, 14-15 anal rays; 7-10 premaxillary and 4-5 maxillary teeth; a horizontally ovate caudal spot, extending more onto lower part of peduncle than on its upper part.
- b. Body depth 4-5 in standard length; large suborbital bone almost entire; body silvery and yellowish in life; an iridescent spot on upper part of caudal peduncle as in the "head and tail light" . . . *A. stigmatias*, the type species, synonym *A. fowleri* (Amazon basin)
- bb. Body depth 3.1-3.5 in standard length; large suborbital bone reduced; body red in life . . . *A. riesei*, the ruby tetra (upper Rio Meta in Colombia)

aa. iv, 22-25 anal rays; 15 premaxillary and 8-9 maxillary teeth (circumorbital bones very reduced); very unusual color pattern, consisting of a longitudinal band ending below dorsal fin—as in certain *Pyrrhulina* for example—a black line on base of dorsal, anal and caudal fin, and first dorsal and anal rays black . . . *A. lindeae* (Amazon basin, near Obidos and near Porto Velho)

Axelrodia stigmatias (after Gery, 1965).

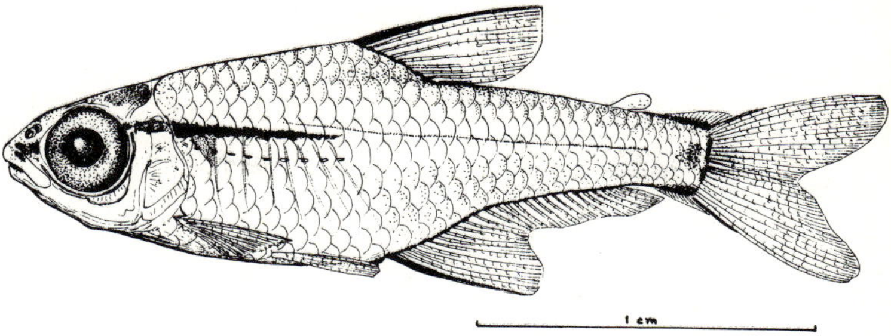

Axelrodia lindeae (after Gery, 1973).

GENUS *TYTTOBRYCON*

The recently described *Tyttobrycon* seems to link *Axelrodia* and *Oxybrycon* with the next group, *Grundulus* et al. Certain of its species are among the smallest vertebrates, the males being fully mature at 14 mm (in standard length). They are very regressed, 2 species having even lost their adipose fin, but are also quite specialized, at least concerning the structures related with sexual activity such as fin-hooks and interhaemal spines.

Tyttobrycon dorsimaculatus and *T. spinosus* (after Gery, 1973).

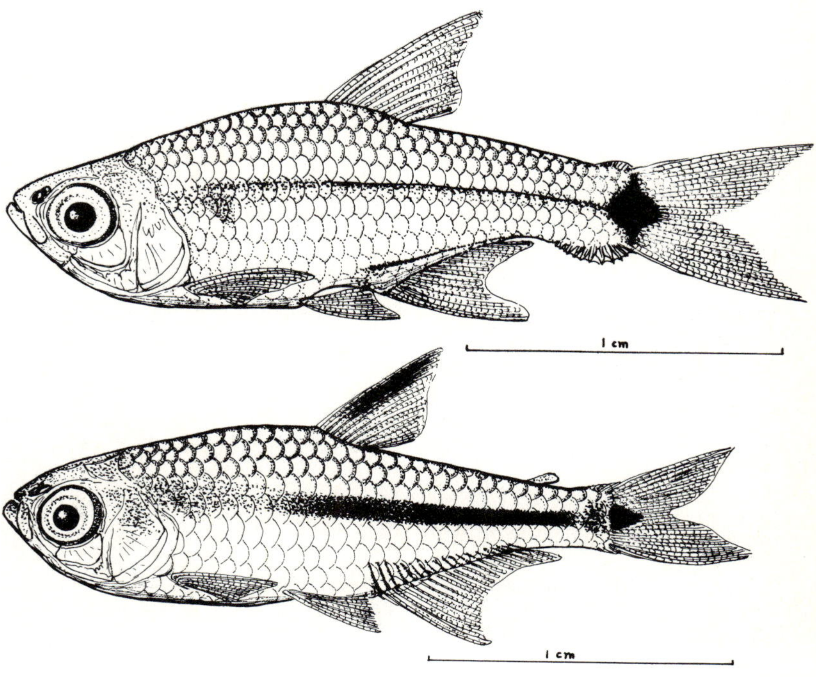

This tetra resembles *Hemigrammus bellottii.* Photo by Dr. Herbert R. Axelrod.

A species very similar to *Hemigrammus rodwayi.* Photo by Aaron Norman.

An unidentified tetra collected back in 1970. Photo by Dr. Herbert R. Axelrod.

Cheirodon sp., a possible new species collected by Dr. Herbert R. Axelrod in 1974. Photo by Dr. Herbert R. Axelrod.

The following key is modified from a recent work:

a. 14-20 branched anal rays; sometimes a maxillary tooth; large suborbital not reduced.

b. Adipose fin present; no interhaemals; dorsal fin with an anterior black marking; a broad lateral band, a triangular caudal spot and a series of transverse lines above anal fin . . . *T. dorsimaculatus* (Rio Chapare in Bolivia)

bb. Adipose fin lacking, interhaemals strong and prominent in the male; anal base forming a sort of keel; no dorsal markings; a large lozenge-shaped caudal spot . . . *T. spinosus* (Rio Chapare basin in Bolivia)

aa. 10-12 branched anal rays; maxilla never toothed; large suborbital bone reduced.

c. Adipose fin present; base of anal forming a sort of keel, the fin armed with several hooklets on each ray in adult male; a broad longitudinal band . . . *T. xeruini*, the type species (Rio Negro basin)

cc. Adipose fin lacking; base of anal fin not keeled, the fin armed with a single hook on each first ray; a longitudinal band interrupted at dorsal fin level, then a longitudinal line somewhat higher; a conspicuous dark band along base of anal fin . . . *T. hamatus* (near Iquitos, Peruvian Amazon)

Tyttobrycon xeruini and *T. hamatus* (both after Gery, 1973).

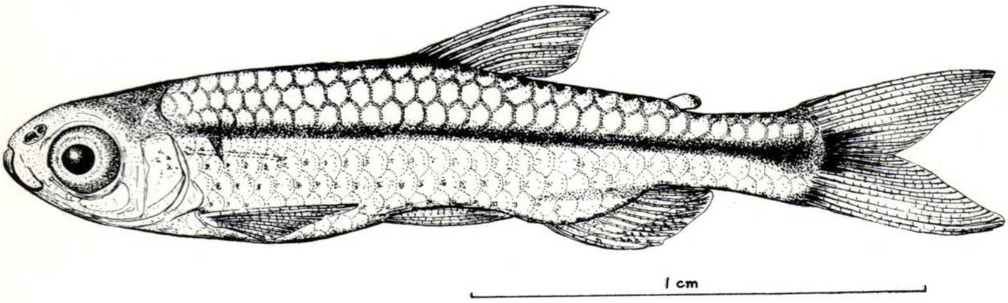

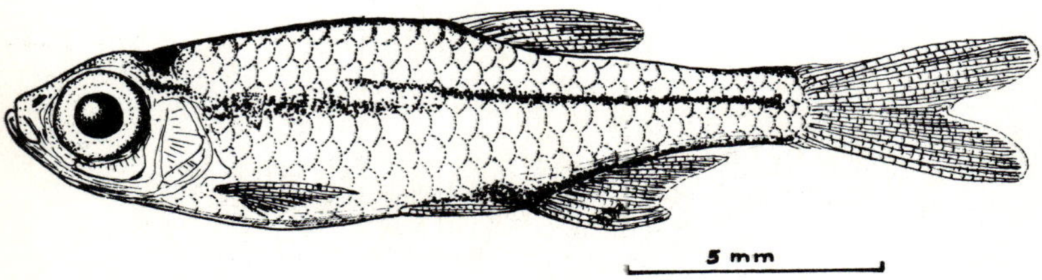

THE *GRUNDULUS*-GROUP

These regressed fishes do not differ much from certain *Tyttobrycon*, as judged by the enumeration of their characters: lateral line perforating only 1-6 scales; adipose fin lacking; anal fin short; a few interhaemals; large suborbital usually reduced, etc. However, the cyprinodontoid shape of their body, much less compressed than any of the Aphyoditeina, suggests that we are dealing with a different—even if convergent—phyletic line. This line must be ancient, because its few actual representatives inhabit the two extremes (or almost) of South America, S.E. Brazil and the plains of Bogota in Colombia.

The following key will be enough to characterize the species:

a. Gill-rakers short but present; teeth strictly conical; predorsal and preventral areas naked; size up to 80 mm . . . *Grundulus bogotensis* (Colombia)

aa. Gill-rakers lacking; teeth partly tricuspid; predorsal and preventral areas scaled; size up to 40 mm.

b. 13-18 anal rays; no pit-lines on head; a dark lateral band along midbody and another one above anal . . . *Phoxinopsis typicus*, probable synonym *Spintherobolus broccae* (surroundings of Rio de Janeiro)

bb. 12 anal rays; a series of conspicuous pit-lines on head; no lateral band . . . *Phoxinopsis papilliferus* (Rio Tiete basin; also known as *Spintherobolus papilliferus*)

Grundulus bogotensis (right) and *Phoxinopsis papilliferus* (after Eigenmann, 1915).

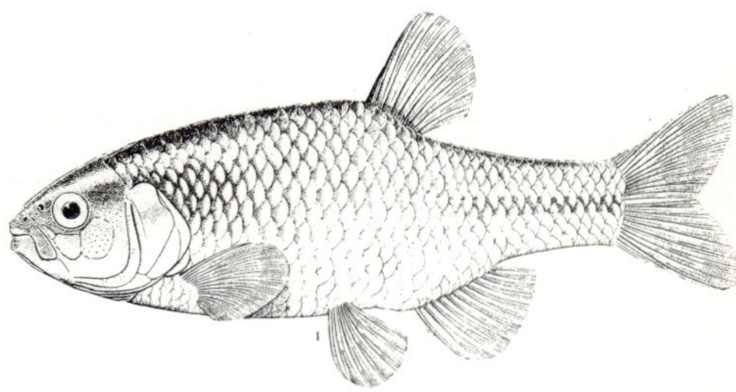

One of the glass characins, possibly a new species. Photo by Dr. Herbert R. Axelrod.

Unidentified *Hemigrammus* or *Moenkhausia*. Photo by Aaron Norman.

UNDESCRIBED OR UNIDENTIFIED SPECIES

Nobody knows exactly how many characoids really do exist. An extrapolation based on the curve of the described species from 1758 to 1970 extended to the year 2000, gives us an estimate of 1160-1220 valid descriptions of South American species, with a limit of 1500 valid species in the more distant future. To that figure must be added the 185-190 African characoids, plus the still unknown species of that area, which should be relatively less numerous than those of South America, maybe 20 or 30 in all. The projected total number of valid characoid species would then be between 1365 and 1440 at the beginning of the next century, with an upper limit of 1700-1750 when (or if) all or almost all species should be known, in the very distant future.

These figures can be compared with the percentage of new species found in modern collections: Almost every collected sample contains a certain percentage of species unknown to science, varying from 3-5% in largely explored regions such as the Middle Amazon River (Rio Solimoes), to more than 10% in places as yet not intensely collected, and up to 40% in very remote, almost unknown areas. To take rather recent examples:

L. P. Schultz collected more than 8,000 specimens of Venezuelan characoid fishes in the Maracaibo Basin (never collected before, or only once very long ago) in 1942. Out of the 42 forms described, 17 (or 40%) were new.

In 1962 the late Harald Schultz (in part unpublished) collected, in the upper reaches of the Arinos-Juruena Rivers in Brazil, some characoid fishes (780 specimens), representing 42 species. Nine (or 21%) species are new, and a number of others, 7 or 8 (16-19%) could represent, at least, new geographical varieties of known species (total again about 40%).

A larger collection from nearby but different basins, that of R. McConnell in 1968 (in part unpublished) in the upper reaches of the Araguaias and Xingu Rivers, includes the same proportions of new species and subspecies.

New species may also be found in old museum collections. Not infrequently species with a peculiar body-shape or color-pattern (such as prominent blotches or bars on flanks) are superficially examined and identified as an old species known to have such a pattern. More careful study may eventually reveal a quite different species (or even genus). Any *Characidium*-species, for example, is likely to be labeled *Characidium fasciatum* if the label is 50-70 years old. That identification has more than a 50% chance of being in error.

The almost endless imports for the aquarium hobby, mostly from the Amazon, are also likely to bring forth, along with popular forms like neon tetras and others, some "trivial" tetras that may be, from time to time, still unknown to science. A study of the chromosomes of numerous characoids also revealed several so-called sibling-species (forms looking exactly alike save for some minor details, noticed only after thorough examination, yet completely reproductively isolated, as judged from their respective incompatible chromosome-set); or even so-called biological or ecological species (i.e. identical in all details of their morphology, yet isolated by certain biological or ecological factors). In these instances, the systematist often has difficulty in determining which of the two (or more) forms correspond to the species already described (represented in a museum by a "type"), owing to the paucity of discriminating characters.

Prospective new species are to be found mostly in the groups of small, very evolutive forms, such as certain pigmy alestids and citharinids, the characidiids (possibly 50 or so yet undescribed species) and chiefly the tetras and allied forms (such as the cheirodontins in the large sense), where at least 100 species would probably be described in the next few years.

Their formal description—which takes most of the time of an ichthyologist, to the detriment of his more important revisional or faunistical duties—necessitates at least two things: (1) good material, well preserved and, if possible, abundant (the description of a single specimen, in the characoids, is more or less a gamble unless it is extremely peculiar—like *Stygichthys* or *Geisleria* for example), and (2) accurate locality. These conditions are quite rarely fulfilled in the last mentioned source of new material, that of the imports of living fishes. Too often, one is forced to be satisfied with some provisory label until more data might be available. That label can be something like "unidentified species of the genus *x-us*" (usually referred to U_1, U_2 and so on), or "close to species *y-us*" (usually shortened into "*aff.y-us*"), or "species of the *x-us* complex" (or group), as it was done in the course of this book.

The worst condition is when only a photograph is available, taken either on the spot or at an import firm. The general shape of the fish, its proportions, the number of its scales and fin-rays (when the picture is sharp enough), together with its color pattern (general shape plus color pattern represent what is called the *habitus*), are often not sufficient for a tentative identification. As it was pointed out several times, the systematics of most characoid groups depends largely on structures not seen in a photograph, such as the teeth, the scale-cover on the caudal base, the completeness of the lateral line, and so on.

As a matter of fact, this characoid book ends up with pictures of a few unidentified species, which may eventually be new to science. It seems convenient to end the review of such a complex fish group with a tone of doubt and hope.

GLOSSARY OF TECHNICAL TERMS

ADIPOSE FIN: Second dorsal fin, usually not rayed, situated above caudal peduncle or just before it, and very similar to that of the trouts.

ADIPOSE LID: Vertical, crescent-shaped membrane which covers (at least partly) the eye of certain characoids such as the hemiodids or the curimatids.

ADVANCED CHARACTERS: Features observed in recently evolved ("modern") animals, and not observed in animals supposed to be very ancient; the opposite of primitive characters.

ALLOMETRY: The growth of certain parts of an animal as compared with its length (or other reference). If the part studied grows faster than the reference dimension, the resulting allometry is called positive, and vice-versa. An increase that is identical for the two parts observed is called isometry.

ALLOPATRIC: Two species are called allopatric when their respective geographic distributions never overlap; the opposite of sympatric.

ANASTOMOSED: Said of two vessels or nerves when they communicate together; by extension, said of the radii of the scales when they are more or less united to form a kind of network.

ANKYLOSED: Said of a joint when it is no longer movable.

APOPHYSIS: A protuberance on the surface of the bones.

ARCHAIC: Of an ancient type, which presents a character of oldness.

AUFWUCHS: German term used to designate the small plants (mosses, lichens and others) growing on the ground or on fallen trees, rocks and so on, and usually containing small animals.

BICUSPID: See teeth.

BIFID: See teeth.

BIOTOPE: The spot into which (where) an animal lives; its surroundings.

BISERIAL: See teeth.

BRUSH-LIKE PATCH OF TEETH: See teeth.

BUCCO-PHARYNGEAL PAPILLAE: Small protuberances on the inner surface of the mouth and beginning of the alimentary canal.

CAUDAL GLAND: A poorly-known structure observed at the base of the caudal fin of the males of certain characoid groups; usually covered with twisted scale(s), and supposed to produce a "smell" to attract females.

CHARACOID, CHARACOID FISH: Teleostean, fresh-water fish belonging to the suborder Characoidei, order Cypriniformes.

CHARACOID TYPE: Having characoid characters (for example adipose fin, Weberian ossicles, peculiar teeth, and so on).

CANINE: See teeth.

CANINIFORM: See teeth.

CILIATE SCALES: Scales with comb-like, smooth "teeth" along their free, so-called apical, border; very similar to pectinate scales.

CONICAL TEETH: See teeth.

CONTACT ORGANS: The ensemble of organs permitting contact during mating, including tubercles, etc.

CONTACT TUBERCLES: Small outgrowths usually appearing on the snout and head of certain male cyprinids during the breeding season; often called pearl organs or nuptial tubercles.

CONVERGENCY (EVOLUTIONARY): When two independent lines, during their evolution, have ended in the same adaptive character(s). Example *Salmo* (trouts) and *Salminus* (characins).

CRENULATE SCALES: A somewhat intermediary state between ciliate and very strongly ctenoid scales (crenate is perhaps better).

CRYPTIC: See superspecies.

CTENOID SCALES: Scales with more or less hard teeth along their free border (apical border), and sometimes on their surface.

CUSPS (or CUSPIDS): See teeth.

CYCLOID SCALES: Scales with smooth borders and surface.

CYCLO-CTENOID SCALES: Intermediate state between cycloid and ctenoid scales, the teeth being smooth, small and not numerous; very close to the crenate (crenulate) type.

DERMAL FLAP: Usually employed in the sense of a bridge or lappet of skin.

DETRITIVOROUS: Eating mostly or solely waste.

ECTOPTERYGOID: See teeth.

FIN FORMULAE: Of the three unpaired fins (dorsal, anal and caudal), only the formula of the first two is usually given (the number of caudal rays being constant in the characoids, with few exceptions). The first two or three, sometimes more, unbranched (or simple) rays are numbered in small Roman numbers: ii,iii,iv and so on; in certain very rare cases there exists a strong, spinous ray, which is numbered in Roman capitals: "I." The branched rays are numbered in Arabic numbers. The formula of the two paired fins (pectoral and ventral, or pelvic) is of systematic importance only in the Characidiidae and certain other groups. In most families there is usually one simple ray (rarely spinous) usually followed by 6-9 branched rays.

FLUTED TENON: The form of a specific kind of joint between two jaws-bones.

FONTANEL: Lozenge-shaped or triangular hole in the top of the skull, the brain at this point being covered only by the skin.

FUSIFORM BODY: Spindle-shaped body, like that of the tuna for example, that gives to an aquatic animal lowered resistance in its passage through the water allowing it to reach a greater velocity.

GILL-RAKERS: Small, soft spines along the inner ridge of the gill-arches, probably acting as a filter; not to be confounded with the much longer and numerous gill-filaments outside the arches, which are the respiratory organs.

GONORHYNCHIFORMES: Order of teleostean fishes formerly known as the Chanoidei, supposedly more primitive than the Cypriniformes and having given rise to them. Mostly African at the present time: *Cromeria, Kneria, Phractolaemus*, for example.

HERBIVOROUS: Eating herbs (more generally plants).

HUMERAL HIATUS: See pseudotympanum.

HUMERAL SPOT: A black or dark spot in the humeral region, on the flank behind opercle, occurring in many characids and a few other groups of characoids. It is supposed to be a recognition signal.

HYALINE: Transparent, without marks.

ICHTHYOPHAGOUS: Fish eating.

INCISIFORM TEETH: See teeth.

INFRAORBITAL CANAL: The sensorial canal of the lateral line system that runs around the eye within the small, flat bones, called infraorbitals or circumorbitals, that border the orbit.

INTERHAEMAL SPINES: Small spines protruding out of the lower border of the caudal peduncle in the males of certain Cheirodontinae.

ISTHMUS: Triangular area situated on the lower part of the fish, just behind the head, between the gills and the base of the pectoral fins.

JAWS; Composed, in the characoids, of three bones on each side; the premaxilla in the middle and the maxilla laterally, on the upper jaw, and the mandible (or dentary bone, that actually is a part of the mandibulary bone). These bones are usually toothed. However, on the maxilla the teeth are, more often than not, crowded on the upper part of the bone near its joint with the premaxilla.

KARYOTYPE: The number and form of the chromosomes of an animal or plant, as seen and photographed under the microscope, from a stained preparation on a slide (of a regeneration tissue or gonad).

KEELED: In the form of a keel, with an acute angle like that of a hull.

LIMNIVOROUS: Mud eating.

MANDIBULARY TEETH: See teeth.

MAXILLARY TEETH: See teeth.

MICROPREDATOR: Predator (i.e. living on living animals) eating only very small prey.

MOLARIFORM TEETH: See teeth.

MONOTYPIC (GENUS): A genus with only one species.

MONOTYPIC (SPECIES): A species without subspecies.

MULTICUSPID TEETH: See teeth.

NEOTENY: Evolutionary process, by which an animal becomes sexually mature while still in its larval stage, or has retained larval or juvenile characters when adult (which is almost the same thing). This process, also called paedomorphosis by Garstang, is believed to have played a large role in the evolution of several groups of invertebrates as well as vertebrates (including man) according to certain zoologists.

NEUROCRANIUM: The part of the skull that is in connection with the brain, excluding the bones which mostly form the jaws and their suspension, partly called splanchnocranium.

NICHE: Abstract notion of the milieu on which the animal depends, not in terms of space at least in the sense used by modern ecologists, but as concerns its food. For example a fish feeding mostly on a certain copepod occupies a different niche from that of another species feeding mostly on another copepod species. According to most ecologists each species has its own niche and no two species can occupy the same niche.

NOMINATE: A taxon (subgenus or subspecies for example) that bears the same name as the equivalent higher taxon (genus or species).

NUPTIAL TUBERCLES: See contact tubercles.

OSTARIOPHYSINE CHARACTER: Character shared by an ostariophysan; see Weberian ossicles.

PAEDOMORPHOSIS: Term coined by Garstang to designate an evolutionary process very similar to neoteny.

PALATINE TEETH: See teeth.

PEARL ORGANS: See contact tubercles.

PECTINATE SCALES: Scales with a comb-like free edge.

PENTACUSPID: See teeth.

PHARYNGEAL BAGS: Pockets in the mucosa, situated in the hinder part of the mouth.

PHARYNGEAL TEETH: See teeth.

PIT-LINES: Very tiny holes in the skin, arranged in parallel, wavy lines and depending on the lateral sensory system. The pit-lines are usually situated on the top and sides of the head.

PLURICUSPID TEETH: See teeth.

POLYMORPHISM: Phenomenon of genetic origin, causing the individuals of a single species to be not all of the same coloration or, less often, of the same form.

POLYTYPIC: Said of a genus with more than one species, or of a species with subspecies; the opposite of monotypic.

PREMAXILLARY TEETH: See teeth.

PRIMITIVE CHARACTER: Usually said of a character possessed by an ancient form such as a fossil, which is still to be seen in a living form, often together with more "modern" characters.

PROCHARACOID: Unofficial term coined to designate the hypothetical ancestor of the characoids.

PSEUDOTYMPANUM: Translucent, triangular area in the humeral region, on the side behind the opercle where the muscles are missing and the anterior part of the gas-bladder is directly in contact with the skin, which is supposed to improve the hearing (see Weberian ossicles). Quite often present in very young characoids, the pseudotympanum, also called humeral hiatus, persists in several species, when adult, chiefly among the Cheirodontinae.

PTERYGOID: See teeth.

QUADRICUSPID: See teeth.

QUINQUECUSPID: See teeth.

RADII (OF SCALES): Furrows in the scales, usually radiating from their center (or nucleus); not to be confounded with the circuli, which are thinner, much more numerous and usually concentric, being related to the growth of the fish.

RUGOSE (SCALE): In certain species the surface of the scale is not smooth but spread with tiny tubercles or spines; this is usually in relation to a ctenoid (dented) apical border.

SCALE FORMULAE: Most of the species descriptions give the following data concerning the number of scales on the body (none are found on the head in characoid fishes): (1) Longitudinal: Range of the number of scales along lateral line from end of opercle to beginning of caudal fin (example: Sq.L.30-33); when the lateral line is not complete the number of perforated scales is given between parentheses and followed by the total number (example: Sq.L. (7-8)30-33). (2) Transverse: Range of the number of scales in a transverse line usually between dorsal and ventral fins, the scale of the lateral line not being counted (example: Sq. tr. 4-5 / 3); when there is no lateral line the corresponding row is treated as if there might be a lateral line (as in the preceding example), or the number of transverse scales is given in full (example: Sq.tr. 7-8); frequently the two formulae are joined: first the scales above lateral line, then the longitudinal scales, then the scales below (example: Sq. 4-5 / (7-8)30-33 / 3). (3) Predorsal: Range of the number of scales along the predorsal line from the occipital process to the dorsal fin or, if the line is irregular, along a side-line. (4) Circumpeduncular: Number of scales around the middle part of the caudal peduncle counting both sides plus one scale above and below. (5) Preventral and postventral scales: These formulae are rarely used.

SERRAE (ABDOMINAL): Also called abdominal spines, the serrae are formed by the scales of the ventral region, which became spinous forming a saw along the trenchant belly of certain groups. The number of serrae in front and behind the ventral fins is of systematic importance in the Serrasalmidae.

SEXUAL DIMORPHISM: External difference in the form (or coloration) between the two sexes of a species; the characters showing this difference are called secondary sexual characters.

SEXUAL HOOKLETS: Tiny hooks or spines along the foremost rays of the ventral and anal fins of the males of most Characidae (rarely in other groups), sometimes on the whole fin or even on caudal and dorsal fins. These hooks, which may belong to the so-called contact organs, are supposed to act as a coupling device between the two sexes during mating to insure the best fertilization.

SIBLING SPECIES: Two sympatric species which resemble each other very closely, like twins. They are different species because they cannot interbreed, owing to some isolation mechanism (different seasons, different behavior, different chromosome structure, etc.) that prevents the obtaining of fertile hybrids in nature (they are said to be "reproductively isolated"). The present author favors the term "cryptic" to designate species which are sibling but not sympatric.

SPERMATOPHORE: Microscopic sack that holds the spermatozoids in several animal groups. It is deposited by the male into the genital tract of the female and allows fertilization to occur long after mating.

SUPERSPECIES: A step in the classification of living beings which is situated between the species and the subgenus. It is not expressed with a latin name like the two last-named taxa, and it usually indicates a group of species that are very closely related, often "cryptic" and allopatric. The *Artenkreis* (meaning circle of species) of Rensch has the same meaning.

SYMPATRIC: Sympatry or sympatricity is an important concept in the study of the evolution of the species. It means that two forms live together in the same biotope, each being sympatric with the other. It follows that these forms, even if only slightly different (sometimes only in the structure of their chromosomes), have to be reproductively isolated (see sibling species), hence good species. Otherwise their characters would have been mixed and one would have been dealing with a single species or a hybrid. On the other hand, allopatry or allopatricity does not give such an insurance, and two slightly different allopatric forms may be of the same species (races or subspecies), because one can not usually prove that, if they might come into contact, they would be reproductively isolated. The following discrimination table, from Mayr, Linsley and Usinger (Methods and Principles of Systematic Zoology, McGraw-Hill, ©1953), is of daily use to the systematist:

	Reproductively isolated or not	Living together (sympatric)	Geographically separated (allopatric)
Identical or almost identical forms	not isolated	same population	same subspecies
	isolated	sibling species	cryptic species
Slightly different forms	not isolated	individual variant (same population)	different subspecies
	isolated	different species	different species

It is usually difficult to tell whether two allopatric forms are sexually isolated or not, even with experiments in aquariums where the sexual behavior is not the same as in nature.

SYMPHYSIS: A kind of joint that is only slightly movable.

SYSTEMATICS: The part of biology that deals with the classification of living things and their evolution, with the following three categories: Nomenclature, or naming the entities in accordance with the International Code; Taxonomy, or classifying them in answer to our innate need of order; and Systematics in the strict sense dealing with the relationships of species and higher categories, mostly from an evolutionary point of view.

TAXON (plural Taxa): A taxonomic unit, not prejudging its rank.

TEETH: The characoids are the most heterodont fish group, that is, a group with very different teeth from one taxon to the other. As these teeth play the largest role in their classification (but not in the study of their evolution), the present section is given in more detail. The teeth are usually present on the jaws (premaxillary, maxillary and mandibular teeth), in the throat (pharyngeal teeth, not well known in most families) and sometimes on the palate (palatine teeth) where they are situated mostly on the part of the palate called the ectopterygoid (sometimes simplified into pterygoid which includes three palatine bones). When the fish is large enough, the teeth can be observed under a stereomicroscope, or even a simple magnifying lense. A useful tool is a minute dryer that blows air, preferably hot, on the lips in order to dry them without damaging the whole fish. When the fish is very small, as in the Tetragonopterinae and Cheirodontinae for example where the teeth are of great systematic importance, the dissection of the jaws on one side is necessary; the soft tissues are destroyed with a drop of diluted chloride bleach (Clorox), while watching under the microscope, and simply rinsed before the chemical has attacked the tiny bones. The jaws and teeth can be observed, drawn or photographed, with a magnification usually between 25x and 80x. When there is sufficient biological material (say more than 2 or 3 specimens), it is possible to stain one specimen with alizarin (which stains only the bone red) after having cleared the soft tissues, usually with diluted potash (KOH). The process is rather difficult and long (several days) but gives splendid preparations where the whole skeleton, including the jaws, is to be seen to the finest detail.

Tooth formula: The teeth are usually counted on the premaxilla first, then on the maxilla and on the mandible, occasionally on the palate, and their number is given, by most ichthyologists, from one side only (although both sides are counted for verification). When there is more than one row (teeth said to be biserial, instead of uniserial), which occurs mostly in the Characidae, the two (eventually three) rows are separated by an oblique bar or other system (example of a tetragonopterine tooth formula: pmx 2-3 / 5 (rarely 6); mx 2-3; mdb (or dn for dentary) 5 + several (smaller teeth on side)).

Form of the teeth: The teeth may have only one point or cusp (more correctly cuspid), or several, in which case they are called pluricuspid or multicuspid (more correctly multicuspidate); with two points, bicuspid(ate) or bifid, with three points tricuspid, then quadricuspid, quinquecuspid (or pentacuspid), etc. With more than five cusps one usually gives, for brevity, the range of the number of points, which may be as many as 9-12, or sometimes even much more in certain "hand-like," flat teeth of the Cheirodontinae or Hemiodinae. Independent of the number of cusps, the tooth may be more or less conical and pointed, in which case it is called caniniform (without cusps), or even canine (like that of a dog) when very big; it may be compressed with the cutting edge straight or denticulated, like our incisors (incisiform tooth), or broad with the crushing part more or less flat like our molars (molariform tooth). In rare cases, the conical teeth can be very small, numerous, sometimes arranged in a patch like a brush.

TRENCHANT (BELLY): With an acute keel.

TRICUSPID: See teeth.

UNISERIAL: See teeth.

VALVULA (plural VALVULAE): Kind of fold in the mucosa that prevents the backward flow of some liquid or substance, usually in a vessel, or in the intestine.

WEBERIAN OSSICLES: Chain of small bones (the tripus, intercalarium, scaphium and claustrum), derived from the apophyses of the first three vertebrae. These ossicles connect the anterior part of the gas-bladder with the inner ear. The possession of Weberian ossicles characterizes an ostariophysan (superorder Ostariophysi, orders Cypriniformes and Siluriformes).

XANTHOCHROMIC: Of a yellow or golden color; xanthochromism is among the less rare mutations in aquarium fishes together with melanism (black mutation), albinism (without pigment, which gives to the fish a pale rosy color) and long finnage.

BIBLIOGRAPHY I: Africa

AHL, E., 1922.—*Phago maculatus* sp. n.—*Blatter f. Aquar. Terr. Kunde:* 23-25.

———, 1923.—*Phago maculatus* sp. n.—*Zool. Anz., Leipzig,* **56**, pl. 23, fig. 1.

———, 1928.—Drei neue westafrikanische Fische.—*Id.,* **78** (9/10): 205-208 (*Neolebias landgrafi:* 206).

ANONYMOUS, 1965.—Fisheries Research Bull. 1963-1964.—*Game and Fisheries Dept., Minist. of Lands and Natur. Resources,* Lusaka, 226 pp, 13 figs.

ARNOULT, J., 1965.—Contribution a la faune du Congo (Brazzaville) (Mission A. Descarpentries & A. Villiers). XI. Poissons dulcaquicoles.—*Bull. I.F.A.N.,* **27**, ser. A N°3: 1188-1191 (Char.: 1188-1189).

AZIZ, I.A., 1965.—Absorption of some parts of the chondrocranium among teleostean fishes during later stages of development.—*Bull. Fac. Sci. Cairo Univ.,* N°**39**: 135-142, 4 figs.

———, 1965.—The development of the hyoid arch in some Nile bony fishes larvae.—*Id.,* N°**39**: 149-157, 3 figs.

BADENHUIZEN, T.R., 1967.—Some notes on the population dynamics of *Hydrocyon vittatus* (Castelnau) in Lake Kariba.—*Hydrobiologica,* **30**: 527-540, 6 figs.

BARNARD, K.H., 1948.—Report on a collection of fishes from the Okovango River, with notes on Zambesi fishes.—*Ann. S. Afr. Mus.,* **36**: 407-455 (Char.: 413-417).

BELL-CROSS, G., 1965.—Additions and amendments to the checklist of the fishes of Zambia.—*The Puku, Occ. Pap. Dept. Game and Fisheries,* Zambia, 3: 29-43.

———, 1967.—Preliminary observations on *Hydrocynus vittatus* in the upper Zambesi River system.—*Fish. Res. Bull.,* Zambia, 1965-66: 21-28.

BERTMAR, G., 1962.—On the ontogeny and evolution of the arterial vascular system in the head of the African characidean fish *Hepsetus odoe.*—*Acta Zool.,* **43**: 255-295, 12 figs.

BILHARZ, Th., 1852.—*Alestes macrolepidotus,* ein neuer Nilfish.—*Sitz. ber. Akad. Wiss. Wien,* **9**: 469-472, pl. 37.

BLACHE, J., 1964.—Les poissons du bassin du Tchad et du bassin adjacent du Mayo Kebbi. Etude systematique et biologique (avec la collaboration de F. Miton, A. Stauch, A. Iltis et G. Loubens).—*O.R.S.T.O.M.,* Paris, 483 pp, 147 figs.

BLACHE, J. & F. MITON, 1960.—Poissons nouveaux du bassin du Tchad et du bassin adjacent du Mayo Kebbi.—*Bull. Mus. Nat. Hist. Nat.,* Paris, **32**(1): 100-107.

______, 1961.—Sur le statut de quelques especes de poissons du bassin du Tchad et du bassin adjacent du Mayo Kebbi.—*Id.*, **32**(5): 395-400 (Char.: 396).

BLEEKER, P., 1963.—*Nat. Verh. Holl. Matsch. Wetensch.*, **1863**: 1-136.

BLOCH, M.E., 1794.—Naturgeschichte der auslandische Fische, Berlin, 1785-95 (*Salmo odoe*, Vol. 8: 122, pl. 386).

BOESEMAN, M., 1957.—On a collection of fishes from Stanley Pool (Belgian Congo).—*Zool. Mededel. Leiden*, **35**(11): 139-151 (Char.: 141).

______, 1963.—An annotated list of fishes from the Niger delta.—*Zool. Verh., Leiden*, N° **61**: 5-48, pls. 1-6 (Char.: 6).

BORODIN, N.A., 1936.—On a collection of freshwater fishes from lakes Nyassa, Tanganika and Viktoria in Central Africa.—*Zool. Jahrb.*, Jena (Syst.), **68**: 1-34 (non vidi).

BOULENGER, G.A., 1897.—*Ann. Mag. Nat. Hist.*, London, (6) **20**, Char.: 424-425 (non vidi).

______, 1898.—Report on the collection of fishes made by Mr. J.E.S. Moore in lake Tanganika during his expedition 1895-1896.—*Trans. Zool. Soc., London*, **15**: 1-30, pls. (Char.: 25).

______, 1898.—Materiaux pour la faune du Congo, fasc. 2: Elopes, Characins, Cyprins.—*Ann. Mus. Congo*, Zool. i: 21-38 (Char.: 23-29).

______, 1899.—Poissons nouveaux du Congo. IV. Polypteres, Clupes, Mormyres, Characins.—*Ann. Mus. Roy. Congo Belge*, Zool. (1) 1, 4: 59-96, 9 pls. (Char.: 75-94, pls. 34-39).

______, 1900.—Materiaux pour la faune du Congo, fasc. 6: Mormyres, Characins, Cyprins, Silures, Acanthopterygiens, Dipneustes.—*Id.*, Zool. i: 129-164, 9 pls. (Char.: 130-131, pl. 48).

______, 1901.—Les Poissons du bassin du Congo.—*Publ. Etat independant du Congo*, Bruxelles, **62**, 532 pp. (Char.: 139-204).

______, 1901.—List of the fishes of the characinid genus *Alestes* M. & T., with a key to their identification.—*Ann. Mag. Nat. Hist. London*, (7)**8**: 486-490.

______, 1901.—List of the fishes of the characinid genus *Distichodus* M. & T., with a key to their identification.—*Id.*, (7)**8**: 510-513.

______, 1902.—Contribution to the ichthyology of the Congo, II. On a collection of fishes from the Lindi River.—*Proc. Zool. Soc., London*, **1**: 265-271, 3 pls.

______, 1902.—Description of a new characinid fish (*Citharidium ansorgii* n. g. et sp.) discovered by Dr. W.J. Ansorge in southern Nigeria.—*Ann. Mag. Nat. Hist., London*, (7)**9**: 144-145, pl. 3.

______, 1902.—Materiaux pour la faune du Congo, additions a la faune ichthyologique du bassin du Congo.—*Ann. Mus. Congo*, 1, ser. 2, fasc. 2: 19-57, 10 pls. (Char.: 27-28).

______, 1903.—Descriptions of new freshwater fishes from southern Cameroon.—*Ann. Mag. Nat. Hist., London*, (7)**12**: 435-441 (Char.: 435-437).

———, 1903.—Description of a new West African fish of the genus *Alestes*. —*Id.*, (7)**12**: 594-595.
———, 1903.—On the fishes collected by Mr. G.L. Bates in southern Cameroon.—*Proc. Zool. Soc., London*, pt. 1: 21-29, 5 pls.
———, 1903.—Report on the fishes collected by Mr. Oscar Neumann and Baron Carlo von Erlanger in Gallaland and southern Ethiopia.—*Id.*, pt. 2: 328-334, pls. (Char.: 329).
———, 1904.—Description of a new fish of the genus *Alestes* from Natal.—*Ann. Mag. Nat. Hist., London*, (7)**14**: 155.
———, 1905.—Descriptions of four new freshwater fishes discovered by Dr. W.J. Ansorge in Angola.—*Id.*, (7)**15**: 457-459.
———, 1905.—A list of freshwater fishes of Africa.—*Ibid.*, (7)**16** (Char.: 40).
———, 1905.—A series of fishes from lake Chad and the Shari River.—*Proc. Zool. Soc., London*, i: 151 (non vidi).
———, 1906.—Descriptions of new fishes discovered by Mr. E. Degen in lake Victoria.—*Ann. Mag. Nat. Hist., London*, (7)**17**: 433-452 (Char.: 435).
———, 1906.—Fourth contribution to the ichthyology of lake Tanganyika. Report of the collection of fishes made by Dr. W.A. Cunnington during the third Tanganyika expedition 1904-1905.—*Trans. Zool. Soc., London*, **17**: 537-619, 2 pls. (Char.: 547, pl. 30, fig. 3).
———, 1907.—Description of three new fishes from Central Africa.—*Ann. Mag. Nat. Hist., London*, (7)**20**: 485-489.
———, 1907.—Zoology of Egypt; the fishes of the Nile.—2 vols., *London*, li + 578 pp., pls. A-F, 1-97, in-4° (Char.: 98-156, pls. 16-26).
———, 1909.—Catalogue of the fresh-water fishes of Africa in the British Museum (Natural History), vol. I, Selachii-Teleostomi.—*London*, xi + 373 pp. (Char.: 174-298).
———, 1909.—Descriptions of new freshwater fishes discovered by Mr. G.L. Bates in south Cameroon.—*Ann. Mag. Nat. Hist., London*, (8)**4**: 186-188.
———, 1910.—On a large collection of fishes made by Dr. W.J. Ansorge in the Quanza and Bengo rivers, Angola.—*Id.*, (8)**6**: 537-561.
———, 1911.—L'etat present de nos connaissances sur la faune ichthyologique du Congo.—*Rev. Zool. Bot. Afric.*, **1**(1): 12-21.
———, 1911.—Descriptions of new freshwater fishes discovered by Dr. W.J. Ansorge in Portuguese Guinea.—*Ann. Mag. Nat. Hist., London*, (8)**7**: 373-377.
———, 1912.—Description of two new fishes from the Nile system.—*Ibid.* (8)**10**: 601-602.
———, 1912.—Poissons recueillis dans la region du bas Congo par M. le Dr. W.J. Ansorge.—*Ann. Mus. Congo* (*Ser. Zool.*), Bruxelles, ii, **3**: 1-25, 6 pls. (Char.: 7-14).
———, 1913.—Descriptions of three new fishes discovered in the Gold

Coast by Dr. H.G.F. Spurell.—*Proc. Zool. Soc., London*: 51-53, pl.
______, 1913.—Sur une petite collection de poissons recueillis dans l'Uele par la mission dirigee par M. Hutereau.—*Rev. Zool. Bot: Afric.*, **2**: 155-161.
______, 1915.—Descriptions of new freshwater fishes from Sierra Leone.—*Ann. Mag. Nat. Hist., London,* (8)**15**: 202-204.
______, 1916.—Catalogue of the fresh-water fishes of Africa in the British Museum (Natural History), vol. IV, Acanth. (continued)-Plectognathi-Addenda.—*London*, xxvii + 392 pp. (Add. Char.: 175-197).
______, 1917.—Description of new fishes from Lake Tanganyka forming part of the collection made by the late Dr. C. Stappers.—*Ann. Mag. Nat. Hist., London,* (8)**20**: 363-368 (Char. 363-364).
______, 1919.—A list of the freshwater fishes of Sierra Leone.—*Id.*, (9)**4**: 34-36.
______, 1920.—Poissons de la mission Stappers (1911-1913) pour l'exploration hydrographique et biologique des lacs Tanganyka et Moero.—*Rev. Zool. Bot. Afric.*, **8**(1): 1-57 (Char.: 10-12 and 34-35).
______, 1920.—Poissons recueillis au Congo belge par l'expedition du Dr. C. Christy.—*Ann. Mus. Congo, Zool. Ser 1,* Materiaux pour la faune du Congo, **2**(4): 1-39, pls. 23-25.
______, 1921.—Description de deux poissons nouveaux de l'Ituri.—*Rev. Zool. Bot. Afric.*, **8**: 148-150.
BOWMAKER, A.P., 1961.—Research results, lake Bangweulu.—*Rep. Fish. Res. Org. N. Rhod.* Nyasaland, N°**11**: 6-20.
BURCHARD, J.E. Jr., 1968.—Observations on *Phago loricatus*, an ectoparasitic fish from the Niger Delta region of West Africa.—*Journ. Zool., London,* **155**: 1-4.
CAREY, T.G., 1967.—Feeding habits of some fishes in the Kafue River.—*Fish. Res. Bull.*, Zambia, **4**(1965-1966): 105-109, 2 tabs.
CASTELNAU, F.L. de, 1861.—Memoire sur les poissons de l'Afrique australe.—*Paris*, in-8°, vii + 78 pp (Char.: 65-66).
CHABANAUD, P., 1934.—Mission J. Thomas en Afrique equatoriale francaise-Poissons.—*Faune des Colonies francaises,* **5**(4): 195-231 (non vidi).
COCKERELL, T.D.A., 1910.—The scales of the African characinid fishes. —*Smithson. misc. Coll.*, **56**: 1-10, 2 pls.
COULTER, G.W., 1966.—The occurrence of *Hydrocynus goliath* (Blgr) in Lake Tanganyika.—*The Puku, Occ. Pap. Dept. Game and Fisheries,* Zambia, **4**: 87-90, fig.
CUVIER, G., 1817.—Le Regne animal, distribue d'apres son organisation . . . *ii, Poissons.*—Paris, 532 pp. (Char.: 167-168).
______, 1818.—Sur les Poissons du genre Myletes.—*Mem. Mus. Hist. Nat.*, Paris, **4**: 444-456.
______, 1819.—Sur les Poissons du sous-genre *Hydrocyon*, sur deux nouvelles especes de *Chalceus*, sur trois nouvelles especes de Serrasalmes, et sur l'*Argentina glossodonta* de Forskahl, qui est l'*Albula gonorhynchus* de Bloch.—*Id.*, **5**: 351-379, 3 pls.

———, 1829.—Le Regne animal distribue d'apres son organisation. . ., 2nd ed., *ii, Poissons.*—Paris (Char.: 313).
DAGET, J., 1948.—La collection de Poissons d'eau douce de l'I.F.A.N.—*I.F.A.N.*, Dakar, *Catalogues*, iii: 3-59 (Char.: 15-20).
———, 1951.—Poissons d'eau douce nouvellement entres dans les collections de l'I.F.A.N.—*Bull. I.F.A.N.*, **13**(4): 1141-1151.
———, 1952.—La Reserve naturelle integrale du mont Nimba. XV. Poissons.—*Mem. I.F.A.N.*, **19**: 311-334 (Char.: 311-313).
———, 1954.—Les Poissons du Niger superieur.—*Id.*, **36**: 1-391 (Char.: 107-173).
———, 1957.—Les Poissons, in BLANC, M. & J. DAGET, Les eaux et les Poissons de Haute Volta. *Ibid.*, **50** (Mel. biol. 3): 113-168, figs. 10-30 (Char.: 125-131).
———, 1959.—Les Poissons du Niger superieur (lere note complementaire). —*Bull. I.F.A.N.*, **21** A, 2: 664-668.
———, 1959.—Note sur les *Distichodus* (Poissons characiformes) de l'Ouest africain.—*Id.*, **21** A: 1275-1303, 14 figs.
———, 1960.—Poissons de la Volta noire et de la haute Comoe (Mission d'Aubenton-Arnoult, Oct.-Dec. 1959).—*Bull. Mus. Natnl. Hist. Nat.*, Paris, **32**(4): 320-330.
———, 1961.—Contribution a la connaissance de la faune du fleuve Senegal. Poissons du Baoule et du Bakoy.—*Id.*, **32**(6): 506-512.
———, 1961.—Faune relicte sur le revers oriental du plateau de Bandiagara (Afrique occidentale).—*Hydrobiologia*, **18**(1-2): 95-108.
———, 1961.—Un Poisson nouveau de Cote d'Ivoire: *Citharinus eburneensis* n. sp. (Citharinidae).—*Bull. Mus. Natnl. Hist. Nat.*, Paris, **33**(5): 485-486.
———, 1961.—Poissons du Niari-Kouilou recoltes par MM. Ch. Roux, J. Ducroz et J.P. Troadec (Afrique noire, region Gabon-Congo).—*Id.*, **33** (6): 577-586.
———, 1961.—Le Parc National du Niokolo-Koba (2e fasc.). XXXV. Poissons.—*Mem. I.F.A.N.*, **62**: 325-362 (Char: 330-333).
———, 1961.—Note sur les *Nannocharax* (Poissons Characiformes) de l'Ouest africain.—*Bull. I.F.A.N.*, **23**, Ser. A, 1: 165-181.
———, 1962.—Les Poissons du Fouta Dialon et de la basse Guinee.—*Mem. I.F.A.N.*, **65**: 210 pp, 13 pls. (Char.: 60-67).
———, 1962.—Le genre *Citharinus* (Poissons, Characiformes).—*Rev. Zool. Bot. Afric.*, **66**(1-2): 81-106.
———, 1962.—Le genre *Citharidium* (Poissons, Characiformes).—*Bull. I.F.A.N.*, **24**, Ser. A, 2: 505-522.
———, 1963.—La Reserve naturelle integrale du Mont Nimba. XXVII. Poissons (2e note).—*Mem. I.F.A.N.*, **66**: 573-600 (Char.: 575-577).
———, 1963.—Sur plusieurs cas probables d'hybridisation naturelle entre *Citharidium ansorgi* et *Citharinus distichodoides.*—*Id.*, **68**: 81-83, fig.
———, 1965.—Les genres *Nannaethiops* et *Neolebias* (Poissons characifor-

mes).—*Rev. Zool. Bot. Afric.*, **72**(1-2): 1-24.
______, 1965.—Poissons nouveaux de Cote d'Ivoire.—*Bull. Mus. Natnl. Hist. Nat.*, Paris, **36**(5): 590-595.
______, 1966.—Note sur les *Micralestes* du bassin tchadien (Pisces, Characidae).—*Id.*, **38**(2): 129-131.
______, 1967.—Poissons de Republique centrafricaine recoltes par M.A. Stauch en 1962.—*Cah. O.R.S.T.O.M., Ser. Hydrob.*, **1**(1-4): 3-5.
______, 1967.—Le genre *Ichthyborus* (Poissons, Characiformes).—*Id.*, 1(1-4): 139-154.
______, 1969.—Poissons d'eau douce du Senegal recoltes par MM. A. Villiers et C. Reizer.—*Bull. Mus. Natnl. Hist. Nat.*, Paris, **40**(6): 1114-1115.
DAGET, J. & A. ILTIS, 1965.—Poissons d'eau douce de Cote d'Ivoire (eaux douces et eaux saumatres).—*Mem. I.F.A.N.*, **74**: 1-385, 4 pls. (Char. 58-92).
DAGET, J. & P. PLANQUETTE, 1967.—Sur quelques poissons de Cote d'Ivoire avec la description d'une espece nouvelle, *Clarias lamottei* n. sp. (Pisces, Siluriformes, Clariidae).—*Bull. Mus. Natnl. Hist. Nat.*, Paris, **39**: 278-281, fig.
DAGET, J. & A. STAUCH, 1963.—Poissons de la rive droite du Moyen-Congo. Mission A. Stauch (fev.-avr. 1961).—*Bull. Inst. Rech. Sci. Congo*, **2**: 41-48 (Char.: 43-44).
______, 1963.—Poissons de la partie camerounaise du bassin de la Benoue.—*Mem. I.F.A.N.*, **68** (Melanges ichthyologiques dedies a Achille Valenciennes): 85-107 (Char.: 89-91).
DAVID, L., 1935.—Poissons de Kapanga (Katanga).—*Rev. Zool. Bot. Afric.*, **27**: 293-294 (non vidi).
______, 1936.—Contribution a la faune ichthyologique du lac Tanganyika. —*Id.*, **28**: 149-160 (non vidi).
______, 1937.—Poissons de l'Urundi.—*Ibid.*, **29**: 413-420 (non vidi).
DAVID, L. & M. POLL, 1937.—Contribution a la faune ichthyologique du Congo belge.—*Ann. Mus. Congo belge*, Zool. (I), 3(5): 189-294, pl.
DUMERIL, A., 1856.—Ichthyologie analytique ou classification des Poissons. . . —*Mem. Acad. Sci.*, Paris, **27**(1): 1-511 (Afric. Char.: 455-458).
EHRENBERG, C.G., 1828.—Symbolae physicae. . . ex itinere per Africam . . . Berolini.—pages ? (non vidi).
EIGENMANN, C.H. & F. OGLE, 1907.—An annotated list of characin fishes in the United States National Museum and the Museum of Indiana University, with descriptions of new species.—*Proc. U.S. Natnl. Mus.*, Washington, **33**: 1-36 (Afric. Char.: 30).
FANG, P.W., 1944.—Poissons d'Afrique recueillis par Henri Lhote, charge de la mission Lebaudy.—*Bull. Soc. Zool. France*, **69**: 191-192 (non vidi).
FORSKAL, P., 1775.—Descriptiones animalium. . . quae in itinere orientali observavit.—*Niebuhr, Havniae,* in-4°, 164 pp, 43 pls, map (Char.: xiii & 66).
FOWLER, H.W., 1906.—Further knowledge of some heterognathous

fishes.—*Proc. Acad. Nat. Sci., Philadelphia*, **58**: 293-351, 431-483, 60 figs. (Afric. Char.: 319, 445, 467).
———, 1919.—The fishes of the United States Eclipse Expedition to West Africa.—*Proc. U.S. Natnl. Mus.*, Washington, **35**: 195-292.
———, 1925.—Fishes from Natal, Zululand, and Portuguese East Africa.—*Proc. Acad. Nat. Sci., Philadelphia*, **77**: 187-268 (Char.: 197-198).
———, 1930.—The freshwater fishes obtained by the Gray African Expedition—1929. With notes on other species in the Academy collection.—*Id.*, **82**: 27-83, 26 figs. (Char.: 28-32 and 54-60).
———, 1931.—The Fishes obtained by the De Schauensee South African Expedition, 1930.—*Ibid.*, **83**: 233-249 (Char.: 236).
———, 1934.—Fishes obtained by Mr. H.W. Bell-Marley chiefly in Natal and Zululand in 1929 to 1932.—*Ibid.*, **86**: 405-514, 53 figs. (non vidi).
———, 1935.—Scientific results of the Vernay-Lang Kalahari Expedition, March to September 1930. The freshwater fishes.—*Ibid.*, **88**: 243-335, 152 figs. (Char.: 255-262).
———, 1935.—South African fishes received from Mr. H.W. Bell-Marley in 1935.—*Ibid.*, **87**: 361-408, 39 figs. (non vidi).
———, 1935.—Notes on fishes obtained in East Africa in 1935.—*Proc. New England Zool.*, **15**: 45-56.
———, 1936.—Zoological results of the George Vanderbilt African Expedition of 1934. Part III—The Freshwater Fishes.—*Proc. Acad. Nat. Sci., Philadelphia*, **88**: 243-335 (Char.: 252-272).
———, 1949.—Results of the two Carpenter African Expeditions, 1946-1948. Part II-The Fishes.—*Id.*, **101**: 233-275 (Char.: 244-251).
———, 1951.—Some new or emended names of fishlike vertebrates.—*The Fish Cult.*, **30**(10): Supp.: 1-4.
———, 1958.—Some new taxonomic names of fishlike vertebrates.—*Not. Nat. Acad. Nat. Sci., Philadelphia*, **N°310**: 1-16.
FRYER, G. et al., 1965.—Predation and its effect on migration and speciation in African fishes: a comment.—*Proc. Zool. Soc., London*, **144**: 301-322.
GARMAN, S., 1895.—The Cyprinodonts.—*Mem. Mus. Comp. Zool.*, Harvard, 19(1): 1-179, 12 pls. (*Neolebias unifasciatus*, a characoid, page 39).
GERY, J., 1968.—*Ladigesia roloffi*, a new genus and species of African characoid fishes.—*Trop. Fish Hobb.*, **16**(8): 78-87.
———, 1968.—Un essai de systematique quantitative: la classification des especes du genre *Alestes* s. lat. (Pisces, Characoidei).—*Vie et Milieu*, Ser. A, **19**(1-A): 171-194.
———, 1970.—Salmler, Zitter-und Messeraale.—Kap. 14: 287-319, 8 pls., in *Grzimeks Tierleben*, Kindler Verlag, Munchen.
GEOFFROY-SAINT-HILAIRE, G., 1809.—Poissons du Nile, de la Mer Rouge et de la Mediterranee.—*Description de l'Egypte. . . Histoire naturelle*, vol. 1, pt. 1: 1-52, 27 pls., Paris in-4° (Char.: 40-49, pls. 4-5).

GIDEIRI, Y., 1967.—Fishes of the Blue Nile between Khartoum and Roseires.—*Rev. Zool. Bot. Afric.*, **76**: 132-136, fig.

GILTAY, L., 1929.—Un Characide nouveau du Congo belge (*Belonophago hutsebouti* n. g., n. sp.), de la sous-famille des Ichthyoborinae.—*Id.*, **18** (2): 271-276.

______, 1930.—Notes ichthyologiques. VI. *Ichthyoborus besse congolensis*, nov. subsp. du Nyonga, Congo belge.—*Ibid.*, **19**(1): 393-396.

GMELIN, J.F., 1788.—Caroli a Linne. . . Syst. Nat. . . . —*Fishes*, vol. **1**: 1126-1516 (*Salmo aegypticus*: 1386).

GOSSE, J.-P., 1963.—Le milieu aquatique et l'ecologie des Poissons dans la region de Yangambi.—*Ann. Mus. Roy. Afr. Centr.*, Ser. in-8°, Zool. N° **116**: 111-270, 10 pls. (Char.: 180-197, pls. 6-7).

______, 1966.—Remarques systematiques sur quelques especes de la faune ichthyologique congolaise.—*Rev. Zool. Bot. Afric.*, **73**(1-2): 186-200 (Char.: 191-193).

______, 1968.—Les Poissons du bassin de l'Ubangi.—*Mus. Roy. Afr. Centr., Docum. Zool.* N° **13**: 1-56 (Char.: 15-26).

GRAS, R., 1960.—Liste des Poissons du Bas Dahomey faisant partie de la collection du laboratoire d'Hydrobiologie du Service des Eaux, Forets et Chasse du Dahomey.—*Bull. I.F.A.N.*, **23** ser. A N° 2: 572-586 (non vidi).

______, 1962.—Etude biometrique des *Alestes baremose* du bassin du Bas-Chari.—*Id.*, **24** ser. A: 245-261, 2 figs.

GREEN, J., 1967.—The distribution and variation of *Daphnia lumholtzi* (Crustacea: Cladocera) in relation to fish predation in lake Albert.—*East Africa Journ. Zool.*, London, **151**: 181-197, 4 figs.

GREENWOOD, P.H., 1958.—Fishes of Uganda.—*Kampala: Uganda Society* (non vidi).

______, 1959.—The characin fishes of lakes Victoria and Kyoga.—*Ann. Mag. Nat. Hist., London,* (13)**1**: 41.

______, 1963.—A collection of fishes from the Aswa River drainage system, Uganda.—*Proc. Zool. Soc., London,* **140**(1): 61-74 (Char.: 63).

______, 1972.—New fish fossils from the Pliocene of Wadi Natrun, Egypt.—*Journ. Zool., London,* **168**: 503-519 (descr. of the fossil sp. *Alestes deserti*).

GROENEWALD, A.A.V.J., 1964.—The role of *Eutropius depressirostris* Peters (Pisces: Schilbeidae) as a predator in a lowveld impoundment in the Transvaal.—*Hydrobiologia,* **23**: 267-273, 2 figs.

GUERIN-MENEVILLE, F., 1830-1844.—Iconographie du Regne animal de G. Cuvier.—Paris, 70 pls., vol. iv, *Poissons.* (pl. 56, fig. 1, *Myletes*).

GUNTHER, A., 1864.—Catalogue of the fishes in the British Museum. Vol. 5, London: v-xxii, 1, 455 (Characinidae, fam. 2: 278-380).

______, 1865.—Description of a new characinoid genus of fish (*Phago loricatus*) from West Africa.—*Ann. Mag. Nat. Hist., London,* (3)**15**: 209-210, pl. 5.

______, 1867.—New fishes from the Gabon and Gold Coast.—*Id.*, (3)**20**: 110-117 (Char.: 112-114, pl. 3).

______, 1869.—The fishes of the Nile.—in *Petherick J. & B.H. Travels in central Africa.* Vol. ii, Append. C, London in-8° (Char.: 240-250, pls. 2-3).
______, 1871.—Report on several collections of fishes recently obtained for the British Museum.—*Proc. Zool. Soc., London*: 652-675, 18 pls. (Char.: 669-670, pl. 65, fig. C).
______, 1873.—New fishes from Angola.—*Ann. Mag. Nat. Hist., London* (4)**12**: 142-144.
______, 1893.—Second report of the reptiles, batrachians and fishes transmitted by Mr. H.H. Johnston, C.B., from British central Africa.—*Proc. Zool. Soc., London:* 616-628, 5 pls.
______, 1894.—Report on the collection of reptiles and fishes made by Dr. J.W. Gregory during his expedition to Mount Kenia.—*Id.*: 84-91, 4 pls. (Char.: 90).
______, 1896.—Report on a collection of fishes made by Dr. A Donaldson Smith during his expedition to lake Rudolf.—*Ibid.*: 217-224, pl. (Char.: 223).
______, 1896.—Report on a collection of reptiles and fishes made by Miss M.H. Kingsley during her travels on the Ogowe River and old Calabar.—*Ann. Mag. Nat. Hist., London,* (6)**17**: 261-285, 3 pls. (Char.: 279, pl. 15).
______, 1899.—An account of a collection of fishes made by Mr. R.B.N. Walker, CMCZ, on the Gold Coast.—*Proc. Zool. Soc., London*: 716-732, 5 pls. (Char.: 731, pl. 45, fig. B).
______, 1903.—Last account of fishes collected by Mr. R.B.N. Walker, CMCZ, on the Gold Coast.—*Id.*, 1902 pt.` 2: 330-339, 4 pls. (Char.: 338).
HARDING, D., 1961.—Research on Lake Kariba.—*Rep. jt. Fish. Res. Org. Nth. Rhod.* Nyasaland, N°**11**: 25-51.
HECKEL, J., 1849.—Russeger's Reise Egypt, iii (Char.: 307-308—non vidi).
HEIM, W., 1935.—Ueber die Rachensacke der Characinidien und uber verwandte akzessorische Organe bei andern Teleosteern.—*Zool. Jahrb.* (*Anat.*), **60**: 61-106, 23 figs., pls. 1-2.
HERRE, A.W., 1936.—A new cyprinid genus and a new Characin from Portuguese East Africa.—*Proc. Biol. Soc., Washington*, **49**: 99-102.
HILGENDORF, F., 1894.—Neue Characinidengattung, *Petersius*, aus dem Kingani-Flusse in Deutsch-Ostafrika.—*Sitzb. Ges. naturf. Fr., Berlin:* 172-173.
HOEDEMAN, J.J., 1951.—Studies on African characid Fishes. The tribe Alestidi.—*Beaufortia*, 3: 1-8.
______, 1954.—Aquariumvissen-encyclopaedie.—*De Bezige Bij*, Amsterdam, 527 pp.
______, 1956 (1958?).—Aanvulling Aquariumvissen-encyclopaedie.—*De Bezige Bij*, Amsterdam: 529-619 (Char.: 543-572).
______, 1956.—*Nannaethiops geisleri*, eine neue Art aus dem Kongo, nebst einer ubersicht uber die Arten der Sippe Nannaethiopidi.—*Die Aqu.*

Terr. Zeitschr., **9**(10): 259-261, fig.

______, 1969.—Aquariumvissen Encyclopaedie, vols. 1-6.—*Elsevier*, Amsterdam (Char.: vol. 3, pp. 59-189.)

HOLDEN, M.J., 1963.—The population of fish in dry season pools of the river Sokoto.—*Colon. Off. Fish. Publ.*, 58 pp., 10 figs. (non vidi).

HOLLY, M., 1926.—Einige neue Fischformen aus Kamerun.—*Anz. Akad. Wiss. Wien:* 155-157.

______, 1926.—Zwei neue Siluriden und ein neuer Characinide aus Kamerun.—*Id.*: 157-159.

______, 1927.—Mormyriden, Characiniden und Cypriniden aus Kamerun. —*Sitzb. Akad. Wiss. Wien* (I), 136: 115-150.

______, 1927.—Fische aus dem Niger, gesammelt von Herrn A. Weidholz.—*Zool. Anz.*, **73**: 129-132 (non vidi).

______, 1927.—Weitere Fische aus Kamerun.—*Sitzb. Akad. Wiss. Wien* (I), **136**: 421-427 (non vidi).

______, 1928.—Uber die Berechtigung der Characiniden-Art *Alestes senegalensis* Steindachner.—*Zool. Anz., Leipzig*, **76**: 310-311, 2 figs.

______, 1928.—Fische aus dem Scharifluss, gesammelt von Herrn. A. Weidholz, darunter drei noch nicht beschriebene Formen.—*Id.*, **79**: 183-190.

______, 1929.—Einige neue afrikanische Fischformen.—*Sitzb. Anz. Akad. Wiss. Wien:* 1-4 (non vidi).

______, 1930.—Synopsis der Susswasserfische Kameruns.—*Sitzb. Akad. Wiss. Wien*, 139: 195-281, 2 pls. (Char.: 225-233).

HUBBS, C.L., 1939.—*Hepsetus* to replace *Hydrocyonoides* and *Sarcodaces* for a genus of African fresh water Fishes.—*Copeia*, N°**3**: 168.

HUBRECHT, A., 1881.—On a collection of fishes from the St. Paul's river, Liberia, with description of three new species.—*Notes Leyden Mus.*, **3**: 66-71.

HULOT, A., 1950.—Premiers apercus sur la repartition des Poissons dans les environs de Yangambi.—*C.R. Conf. pisc.*, Elisabethville, 1949, Min. Col., Bruxelles, pp. ?

______, 1956.—Contribution a la connaissance de la biologie des poissons du genre *Citharinus* (Citharinidae) de l'Afrique centrale.—*Bull. Afr. Congo belge*, **47**(4): (sep.) 1-12, 7 figs.

ISBRUCKER, I.J.H., 1970.—*Lepidarchus adonis signifer*, a new subspecies of characoid fish from Liberia (Pisces, Cypriniformes, Characidae).—*Beaufortia*, **18**(234): 133-140.

JACKSON, P.B.N., 1961.—The fishes of the middle Zambesi.—*Kariba Stud., Ichthyology* (Manchester): 1-36, pls. 1-4. (Char.: 17-22).

______, 1961.—The fishes of the northern Rhodesia. A check-list of indigenous species.—*Govern. Printer*, Lusaka: v-xv, 1-140 (Char.: 7).

JOANNIS, L. de, 1835.—Observations sur les Poissons du Nil, et description de plusieurs especes nouvelles.—*Mag. Zool.*, iii, 5e annee, N° 1-14, 15 pls. (Char.: pls. 6-10).

JOHNELS, A.G., 1954.—Notes on fishes from the Gambia River.—*Ark. f. Zool., Stockholm*, **2**(6), 17: 327-411, 19 figs.

JUBB, R.A., 1958.—A preliminary report on the collections of freshwater fishes made by the Bernard Carp Expedition to the Caprive Strip, 1949, the lower Sabi River, 1950, and to Barotseland, 1952.—*Occ. Pap. Nat. Mus. South Rhodesia*, **22**B: 177-189, fig.

______, 1961.—An illustrated guide to the freshwater fishes of the Zambezi, river, lake Kariba, Pungwe, Sabi and Limpopo rivers.—*Stuart Mauning, Bulawayo*: i-ix, 1-171, 106 figs., 26 pls. (Char.: 79-83).

______, 1963.—A revised list of the freshwater fishes of southern Africa.—*Ann. Cape Prov. Mus.*, iii: 5-39.

______, 1967.—Freshwater fishes of southern Africa.—*A.A. Balkema,* Cape Town, 248 pp., 57 pls. (Char.: 77-83).

KNER, R., 1859.—Zur Familie der Characinen. I.—*Denkschr. Akad. Wiss. Wien*, 17 (Afr. Char.: 152 and 159).

______, 1860.—*Id.*, II.—*Ibid.*, 18 (Afr. Char.: 18-20).

LACEPEDE, B., 1803.—Hist. Nat. Poissons, vol. 5 (Afr. Char.: 274).

LINNAEUS, C., 1758.—Systema naturae. . ., Tom. i. *Regnum animale,* ii, 824 pp. (Afr. Char.: 531).

— in HASSELQUIST, 1762.—Reise Palaest. (Char.: 395, 422) (non vidi).

LADIGES, W., 1964.—Beitrage zur Zoogeographie und Oekologie der Susswasserfische Angolas.—*Mitt. Hamburg. Zool. Mus. Inst.*, **61**: 221-272, pls. 9-10.

LAMBERT, J.G., 1960.—Les Poissons d'aquarium du bassin du Congo.—*Bull. Soc. Roy. Anvers.*, N° **19**: 1-50, 27 pls.

______, 1961.—Contribution a l'etude des poissons de foret de la cuvette congolaise.—*Ann. Mus. Roy. Afr. Centr.*, Tervuren, Ser. in-8°, Zool. N° **93**: 1-40, pls. 1-5 (Char.: 10-20).

______, 1961.—Sur une petite collection de Poissons de l'Ituri-Uele.—*Rev. Zool. Bot. Afric.*, **64**(3-4): 309-314.

LOHBERGER, 1930.—Fische aus dem Wamifluss.—*Zool. Anz. Leipzig*, **89**: 88-94, fig.

LONNBERG, A.J.E., 1903.—On a collection of fishes from the Cameroon, containing new species.—*Ann. Mag. Nat. Hist., London,* (7)**12**: 37-46 (Char.: 37-38).

______, 1907. Fishes.—in SJOSTEDT, *Zool. Kilimandjaro-Meru Expedition* 1905-1906, **5**: 1-8 (Char.: 2).

LONNBERG, E. & H. RENDAHL, 1920.—On some freshwater fishes from lower Congo.—*Ann. Mag. Nat. Hist., London,* (9)**6**: 167-176.

MARLIER, G., 1953.—Etude biogeographique du bassin de la Ruzizi, basee sur la distribution des Poissons.—*Ann. Soc. Roy. Zool. Belgique,* **84**(1): 177-224.

______, 1958.—Recherches hydrobiologiques du lac Tumba (Congo belge, Province de l'Equateur).—*Hydrobiologia,* **10**: 352-385, 8 figs.

MATTHES, H., 1964.—List of the types of African freshwater fishes in the Amsterdam Zoological Museum, with notes on their synonymy.—*Beaufortia,* **10**(122): 177-182 (Char.: 178-179).

______, 1964.—Les Poissons du lac Tumba et de la region d'Ikela. Etude systematique et ecologique.—*Ann. Mus. Roy. Afr. Centr.*, in-8°, Sci. Zool., N° **126**: tab., pls. 1-6 (Char.: 40-79, pls. 1-3).

______, 1965.—Quelques poissons nouveaux du Congo.—*Rev. Zool. Bot. Afric.*, **71**(1-2): 177-193, 4 figs.

______, 1968.—The food and feeding habits of the tiger-fish, *Hydrocyon vittatus* (Cast. 1861) in lake Kariba.—*Beaufortia*, **15**: 143-153, 2 figs.

MEINKEN, H., 1957 sequ.—in HOLLY, MEINKEN, RACHOW, Die Aquarienfische in Wort und Bild, *Stuttgart*, 7. Fam.: Characidae (Afr. Char.: 1089/91, 1154-1157). See also Rachow, A.

MONARD, A., 1940.—Resultats de la mission scientifique du Dr. Monard en Guinee portuguaise, 1937-1938. vii-Poissons.—*Arq. Mus. Boc. Lisboa*, 11: 91-104.

MONOD, Th., 1950.—Notes d'Ichthyologie ouest-africaine.—*Bull. I.F.A.N.*, Dakar, **12**: 1-71, 110 figs.

MULLER, J. & F.H. TROSCHEL, 1844.—Synopsis generum et specierum familiae Characinorum (Prodromus descriptionis novorum generum et specierum).—*Arch. Naturgesch.*, **10**(1): 81-99 (Afr. Char.: 84-88).

______, 1845.—Horae ichthyologicae. Beschreibung und Abbildung neuer Fische; erstes und zweites Heft. Die Familie der Characinen.—*Berlin:* 1-40, pls. 1-11, tab. (Afr. Char.: 8-9, 11-13, 18).

MURAMOTO, J.C., S. OHNO & N.B. ATKIN, 1968.—On the diploid state of the fish order Ostariophysi.—*Chromosoma*, **24**: 59-66, 7 figs., 4 tabs.

MYERS, G.S., 1924.—Mutanda ichthyologica. *Neoborus* Boulenger and *Barbus rubripinnis* Nichols and Griscom.—*Rev. Zool. Afric.*, **12**(3): 397.

______, 1926.—Two new genera of African characin fishes.—*Id.*, **13**(3-4): 174-175.

______, 1929.—Cranial differences in the African characin Fishes of the genera *Alestes* and *Brycinus*, with notes on the arrangement of related genera.—*Amer. Mus. Novit.* N° **342**: 1-7, 2 figs.

______, 1949.—The family name of the characid Fishes.—*Copeia* N° **3**: 195-204.

______, 1950. On the characid Fishes called *Hydrocynus* and *Hydrocyon* by Cuvier.—*Proc. Calif. Zool. Club*, **1**(9): 45-47.

______, 1960.—Review of: Poll, 1957, Les genres des Poissons d'eau douce de l'Afrique.—*Copeia* No. 1: 78-79. (*Ra* and *Ichthyoborus*).

NAWAR, G. & E.G. YOAKIM, 1963.—A study on the fecundity of the Nile Characid *Alestes baremose* (Joannis, 1835).—*Ann. Mag. Nat. Hist., London* (13)**6**: 157-160.

NICHOLS, J.T., 1923.—New African fishes.—*Amer. Mus. Novit.*, N° **57**: 1-3.

NICHOLS, J.T. & R. BOULTON, 1927.—Three new minnows of the genus *Barbus*, and a new Characin from the Vernay Angola Expedition.—*Id.*, N° **264**: 1-8, 4 figs.

NICHOLS, J.T. & L. GRISCOM, 1917.—Fresh-water fishes of the Congo basin obtained by the American Museum Congo Expedition, 1909-1915.

—*Bull. Amer. Mus. Nat. Hist.*, **37**, Art. xxv: 653-756, pls. 64-83 (Char.: 677-689, pls. 66-71, 80).
NICHOLS, J.T. & F.R. LA MONTE, 1933.—A new catfish, *Amphilius pictus*, and a discussion of a small Liberian collection of fishes.—*Amer. Mus. Novit.*, N° **626**: 3 pp., fig.
______, 1933.—New fishes from the Kasai district of the Belgian Congo.—*Id.*, N° **656**: 6 pp., 4 figs. (non vidi—Char. ?).
______, 1934.—More new fishes from the Kasai district of the Belgian Congo.—*Ibid.*, N° **723** (non vidi—Char. ?).
______, 1938.—Two new fishes from lake Victoria.—*Ibid.*, N° **1001**: 1-3, 2 figs.
NORMAN, J.R., 1932.—A collection of fishes from Sierra Leone.—*Ann. Mag. Nat. Hist. London*, (10)**10**: 180-185, 3 figs.
PAPPENHEIM, P., 1911.—Zoologische Ergebnisse der Expedition des Herrn G. Tessmann nach Sud-Kamerun und Spanisch-Guinea-Fische.—*Mitt. Zool. Mus., Berlin*, **5**: 505-528 (Char.: 510-513).
______, 1914.—Fische—in *Wissenschaftliche Ergebnisse Deutsch. Zentral-Afika Expedition* 1907-1908, 5 (Zool. 3): 225-259, 10 pls. (Char.: 232-234, pl. 4, fig. 5).
______, 1926.—Fische aus Ostkamerun.—*Sitz. ber. Gesell. Naturf. Fr., Berlin:* 133-134.
PELLEGRIN, J., 1900.—Poissons nouveaux du Congo francais.—*Bull Mus. Hist. Nat.*, Paris, **6**: 98-101.
______, 1900.—Poissons nouveaux ou rares du Congo francais.—*Id.*, **6**: 177-182; 348-354.
______, 1901.—Poissons nouveaux ou rares du Congo francais.—*Ibid.*, **7**: 328-332.
______, 1903.—Poissons recoltes dans l'Ouelle par la Mission du Bourg de Bozas.—*Ibid.*, **9**: 326-328.
______, 1904.—Characinides nouveaux de la Casamance.—*Ibid.*, **10**: 218-221.
______, 1904.—Poissons du Chari et du Tchad, recoltes par la Mission Chevalier-Decorse.—*Ibid.*, **10**: 309-313.
______, 1904.—in A. CHEVALIER, Mission Chari-lac Tchad, 1902-1904.—*L'Afr. centr. franc.*, pp. (non vidi).
______, 1905.—La faune ichthyologique du lac Tchad et du Chari.—*C.R. 6e Congres Intl. Zool., Berne* 1904: 608-609.
______, 1905.—Mission scientifique de Ch. Alluaud en Afrique orientale, juin 1903-mai 1904. Systematique.—*Mem. Soc. Zool. France*, **17**: 174-185, pl. 16 (Char.: 181).
______, 1905.—Les Poissons d'eau douce de l'Afrique tropicale francaise.—*Bull. Soc. Acclim., Paris*, **51**: 209-225; and Rev. colon., Paris: 400-408.
______, 1906.—Catalogue des Characinides africains des collections du Museum d'Histoire naturelle de Paris.—*Bull. Mus. Hist. Nat., Paris*, **12**: 474-480.
______, 1906.—Characinides nouveaux du Congo.—*Id.*, **12**: 366-368.

______, 1906.—Collections recueillies par M.E. Haug, dans l'Ogooue. Poissons.—*Ibid.*, **12**: 467-471.
______, 1906.—Poisson nouveau du Niger appartenant au genre *Distichodus.*—*Bull. Soc. Zool. France,* **31**: 109-110.
______, 1907.—Poissons du Sud-Cameroun recueillis par la Mission Cottes. —*Bull. Mus. Hist. Nat. Paris,* **13**: 319-332.
______, 1907.—Poisson nouveau du genre *Alestes* (*A. erythropterus*).—*Bull. Soc. Zool. France,* **32**: 139-140.
______, 1907.—Poissons de la cote occidentale d'Afrique, part II.—*Acta Soc. Linn. Bordeaux,* 32 pp., 17 figs. (non vidi).
______, 1907.—Les Poissons du lac Tchad.—*Rev. Scient., Paris,* **5**, ser. 7: 614-618.
______, 1907.—Sur une collection de Poissons recueillis par M.E. Haug a Ngomo (Ogooue).—*Bull. Soc. Philom., Paris,* (9)**9**: 17-42 (Char.: 21-25, pl. 1).
______, 1908.—Collections recueillies par M.E. Haug, dans l'Ogooue. Liste des Poissons et description d'une espece nouvelle. 2e note.—*Bull. Mus. Hist. Nat. Paris,* **14**: 347-349.
______, 1908.—Sur une seconde collection de poissons recueillis par M.E. Haug a Ngomo (Ogooue).—*Bull. Soc. Philom. Paris,* Ser. 9, **11**: 184-190 (Char.: 186).
______, 1909.—Collections recueillies pas M.E. Haug dans l'Ogooue. Poissons.—*Bull. Mus. Hist. Nat. Paris,* **15**: 66-68.
______, 1909.—Poissons de la Komadougou et du lac Tchad recoltes par la mission Tilho-Gaillard.—*Id.*, **15**: 240.
______, 1909.—Sur la faune ichthyologique du lac Tchad.—*C.R. Acad. Sci., Paris,* May 17, 3 pp.
______, 1909.—Sur la faune ichthyologique du lac Victoria.—*Id.*, July 12 (Char.: 1 sep.).
______, 1909.—Sur une seconde collection recueille par M.E. Haug a Ngomo (Ogooue).—*Bull. Soc. Philom., Paris,* Ser. 10, **1**: 17 (non vidi).
______, 1909.—Note complementaire sur une seconde collection de Poissons recueillis par M.E. Haug a Ngomo (Ogooue).—*Id.*, Ser. 10, **1**: 45-53, 3 figs.
______, 1910.—Mission scientifique de Ch. Alluaud en Afrique orientale (1908-1909). Poissons.—*Mem. Soc. Zool. France,* **22**(3-4): 281-298, pl. 14.
______, 1911.—Poissons du Sud-Cameroun recueillis par la Mission Cottes. —in *La Mission Cottes au Sud-Cameroun* (1905-1908): 230-234, pl 34, figs. 1-3.
______, 1911.—Description d'un poisson nouveau de l'Ogooue appartenant au genre *Nannocharax.*—*Bull. Soc. Zool. France,* **36**: 180-181.
______, 1911.—Poissons de l'Afrique occidentale francaise. Mission de M. Gruvel, 3e note.—*Id.*, **36**: 182-186.
______, 1913.—Sur une collection de poissons du Moyen-Niger recueillie par

M. le Dr. G. Bonet.—*Bull. Mus. Hist. Nat., Paris,* **5**: 1-2.
______, 1913.—Poissons du lac Victoria. Voyage de Ch. Alluaud et R. Jeannel en Afrique orientale (1911-1912).—*Mem. Soc. Zool. France,* **26**: 57-71, pl. 5.
______, 1914.—Poissons nouveaux du Haut-Zambeze recueillis par M.V. Ellenberger.—*Bull. Soc. Zool. France,* **39**: 24-28.
______, 1914.—Les Poissons du bassin du Tchad, Mission Tilho (1906-1909) —III. *Documents scientifiques*: 131-282, pls. 1-11 (Char.: 193-212).
______, 1914 or 1915.—Les Poissons du bassin de l'Ogooue.—*C.R. Ass. fr. Avanc. Sci., Congres du Havre:* 500-505 (Char.: 502).
______, 1919.—Poissons du Gribingui recueillis par M. Baudon. Description de sept especes nouvelles.—*Bull. Soc. Zool. France,* **44**: 201-214.
______, 1920.—Poissons nouveaux du Haut-Zambeze recueillis par M.V. Ellenberger-2e note.—*Id.,* **45**: 148.
______, 1920.—Poissons de l'Onham et de l'Oubanghi recueillis par M. Baudon. Description de deux Cyprinides nouveaux.—*Ibid.,* **45**: 245-248 (Char.: 246-247).
______, 1921.—Les Poissons des eaux douces de l'Afrique occidentale.—*C.R. Ass. fr. Avanc. Sci., Congres de Rouen:* 633-638.
______, 1921.—Poissons du Gribingui et de la M'Loko recueillies par M. Baudon. Description d'un Characinide nouveau.—*Bull. Soc. Zool. France,* **46**: 47-50.
______, 1922.—Poissons d'Oubanghi-Chari recueillis par M. Baudon. Description d'un genre, de cinq especes et d'une variete.—*Id.,* **47**: 64-76.
______, 1922.—Poissons nouveaux ou rares du Musee du Congo.—*Rev. Zool. Afric.,* **10**(2): pages ? (non vidi).
______, 1923.—Les poissons d'eau douce de l'Afrique occidentale (du Senegal au Niger).—*Larose Ed., Paris,* 373 pp. (Char.: 89-120).
______, 1923.—Poissons—in Mission Rohan-Chabot (Angola et Rhodesia 1912-1914).—*Impr. natle. Paris*: 171-176, pl. (Char.: 172).
______, 1924.—Poissons de la region de Loango (Gabon) recueillis par M. Baudon. Description de deux especes et de deux varietes nouvelles.—*Bull. Soc. Zool. France,* **49**: 284-291.
______, 1924.—Poissons du Niger recueillis par M. Jean Thomas. Description de deux especes nouvelles.—*Bull. Mus. Hist. Nat., Paris:* 457- (non vidi).
______, 1925.—Poissons du nord du Gabon et de la Sangha recueillis par M. Baudon. Description de deux especes et d'une variete nouvelle.—*Bull. Soc. Zool. France,* **50**: 97-106 (Char.: 98).
______, 1926?.—Les Characinides du Congo.—*C.R. Ass. fr. Avanc. Sci., Congres de Grenoble* 1925: 420-421.
______, 1926.—Contribution a l'etude de la faune ichthyologique du Niger et de la Guinee francaise d'apres les envois de M. Jean Thomas.—*Bull. Com. Et. Hist. Sci. A.O.F.,* jan-mars 1926, 26 pp., figs. (Char.: 6-10).
______, 1926.—Description de Characinides nouveaux recoltes au Congo

belge par le Dr. Schouteden.—*Rev. Zool. Afric.*, **13**(3-4)(1925): 157-164.

______, 1926.—Poissons de l'Afrique orientale anglaise recueillis par le R.P. Bernhardt. Description de trois especes nouvelles.—*Bull. Soc. Zool. France*, **51**: 384-390.

______, 1926.—Characinide nouveau du Cameroun recueilli par M. Th. Monod.—*Id.*, 51: 390-392.

______, 1927.—*Distichodus vexillifer* n. sp.—*Ibid.*, **52**: 139-?? (non vidi).

______, 1927.—A propos du *Distichodus vexillifer* Pellegrin.—*Ibid.*, **52**: 557-558.

______, 1927.—Mission Guy Babault. Poissons du lac Tanganyka.—*Bull. Mus. Hist. Nat. Paris*: 499-501 (Char.: 499).

______, 1928.—Mutanda ichthyologica: *Clupeocharax* Pellegrin.—*Rev. Zool. Bot. Afric.*, **16**: 80-82, and *Ann. Mus. Congo*, **3**(1): 20-21.

______, 1928.—Characinide et Cyprinide du Cameroun recueillis par M. Th. Monod. Description de deux especes et d'une variete nouvelle.—*Bull. Soc. Zool. France*, **53**: 309-316.

______, 1928.—Poissons etranges du Cameroun.—*Le monde colonial Illustre*, and *Togo-Cameroun Mag. trim.*, Paris, Apr., 1932: 141-145, 9 figs.

______, 1928.—Poissons du Chiloango et du Congo recueillis par l'expedition du Dr. Schouteden (1920-1922).—*Ann. Mus. Congo belge, Zool.* Ser. 1, Mater. pour la Faune du Congo, **3**(1): 1-50.

______, 1928.—Poissons du Kasai (Congo belge). Description d'un genre nouveau et de quatre especes nouvelles.—*Bull. Soc. Zool. France*, **53**: 103-113 (Char. 292-293).

______, 1928.—Sur une collection de Poissons du Cameroun recueillie par M. Chamaulte.—*Ann. Soc. Sci. Nat. Charente-inferieure*, **1**(1): 1-11, fig. (Char.: 2-4).

______, 1928.—Poissons de la region des lacs Kivu et Edouard.—*Bull. Mus. Hist. Nat., Paris*: 82-86, 3 figs.

______, 1929.—Mission saharienne Augerias-Draper, 1927-28. Poissons.—*Id.*, 2e ser., **1**(2): 134-139 (Char.: 135-136).

______, 1929.—Les Poissons des eaux douces du Cameroun.—*Acad. Sci. Colon.*, 12: 369-375, fig.

______, 1929.—Les Poissons des bassins cotiers du Cameroun.—*C.R. Ass. fr. Avanc. Sci., Congres du Havre:* 450-453 (Char.: 451).

______, 1930.—Poissons de l'Ogooue, du Kouilou, de l'Alima et de la Sangha, recueillis par M.A. Baudon. Description de 5 especes et 5 varietes nouvelles.—*Bull. Soc. Zool. France*, **55**: 196-210.

______, 1930.—Poissons du Kouilou recueillis par M.A. Baudon. Description d'une espece et d'une variete nouvelles.—*Id.*, **55**: 334-336, fig.

______, 1931.—Poissons du Kouilou et de la Nyanga recueillis par M.A. Baudon.—*Ibid.*, **56**: 205-211 (Char.: 207-208).

______, 1931.—Poissons de la Louesse (Kouilou) recueillis par M.A. Baudon. Description d'une variete nouvelle.—*Ibid.*, **56**: 219-221.

______, 1934.—Voyage de Ch. Alluaud et P.A. Chappuis en Afrique occidentale francaise (Dec. 1930-Mars 1931).—*Arch. Hydrobiol.*, **26**: 101-120, 6 figs. (non vidi).

______, 1934.—Sur une collection de poissons du Moyen-Congo recueillis par Madame Petit-Renaud.—*Ann. Soc. Sci. Nat. Charente-Inferieure*, N.S., **1**(7): 121-127, 2 figs.

______, 1935.—Poissons de la region du Kivu recoltes par M. Guy Babault. —*Rev. Zool. Bot. Afric.*, **27**(3): 376-385 (Char.: 377).

______, 1938.—Poissons de l'Afrique equatoriale francaise de J. Thomas.—*Bull. Soc. Zool. France,* **63**: 369-378 (Char.: 372-373).

PELLEGRIN, J. & J. ROUX, 1928.—Poissons du Kasai (Congo belge) du Musee de Bale.—*Rev. suisse Zool.*, **35**(17): 291-302.

PERUGIA, A., 1891.—Intorno ad alcuni pesci raccolti al Congo dal Capitano Giacomo Bove.—*Ann. Mus. Civ. Storia Nat. Genova*, (2)**10**: 967-977 (Char.: 975).

PETERS, W., 1852.—Diagnosen von neuen Flussfischen aus Mossambique. —*Monatsber. Akad. Wiss. Berlin:* 275-276 (char.), 681-685.

______, 1868.—Reise nach Mossambique; *iv. Flussfische.*—:1-116, 20 pls. (Char.: 64-74, pls. 12-13).

______, 1877.—Ueber die von Dr. Reinhold Buchholz in Westafrika gesammelten Fische.—*Monatsber. Akad. Wiss. Berlin* fur 1876: 244-252 (Char.: 845).

PFAFF, J.R., 1933.—Report on the Fishes collected by Mr. Harry Madsen during Pr. O. Olufsen's expedition to French Sudan in the years 1927-28. —*Vidensk. Medd.*, 94: 273-315, 13 figs.

PFEFFER, G., 1893.—Ostafrikanische Fische gesammelt von Herrn Dr. F. Stuhlmann im Jahre 1888 u. 1889.—*Jahrb. Hamburg. Wiss. Anst.*, **10** (2): 1-49, pls. 1-3 (Char.: 39).

______, 1894.—Fische, in K. Mobius, *Deutsch-Ost-Afrika*, vol. iii, 72 pp., 24 figs.—Hamburg u. Berlin, in-8° (non vidi—Char.: 41-44).

PIENAAR, U. de V., 1968.—The freshwater fishes of the Kruger National Park.—*Nat. Parks Board Trust. Rep. S. Africa:* 1-82, figs. 1-46 (Char.: 22-25).

POLL, M., 1933.—Contribution a la faune ichthyologique du Katanga.—*Ann. Mus. Congo belge, C. Zool.*, Ser. 1, **3**(3): 101-152 (Char.: 119-124).

______, 1938.—Poissons du Katanga (bassin du Congo) recoltes par le Professeur P. Brien.—*Rev. Zool. Bot. Afric.*, **30**(4): 389-423, 15 figs.

______, 1939?.—Les Poissons d'eau douce du Congo belge.—*Encycl. Congo belge:* 259-282, 8 pls. (Char.: 266-268, pls. 2-3).

______, 1939.—Exploration du Parc National Albert—Mission H. Damas (1935-1936)—Poissons. *Inst. Parcs. Nat. Congo belge,* **6**: pages ? (non vidi).

______, 1939.—Exploration du Parc National Albert-Mission G.F. de Witte-Poissons.—*Id.*, **24**: 1-81, 11 pls. (Char.: 13).

______, 1939.—Poissons du Sankuru. Poissons de l'Ubanghi.—*Bull. Cercle zool. Congo*, **16**: 42-44, 2 figs.

______, 1939.—Les Poissons du Stanley-Pool.—*Ann. Mus. Congo belge, C Zool.*, Ser. 1, **4**(1): 1-60.
______, 1941.—Poissons nouveaux de la Cote d'Ivoire.—*Rev. Zool. Bot. Afric.*, **34**: 133-143 (non vidi).
______, 1942.—Les poissons du lac Tumba, Congo belge.—*Bull. Mus. Roy. Hist. Nat. Belgique*, **18**(36): pages ?
______, 1945.—Note complementaire sur les Poissons du Stanley-Pool.—*Rev. Zool. Bot. Afric.*, **38**(3-4): 290-297, pls. 18-20.
______, 1945.—Descriptions de Mormyridae et de Characidae nouveaux du Congo belge, avec une etude du genre *Stomatorhinus* et des genres de Characidae nains africains.—*Id.*, **39**(1): 36-77, 9 figs.
______, 1946.—Revision de la faune ichthyologique du lac Tanganika.—*Ann. Mus. Roy. Congo belge, Zool.* (1)**4**(3): 141-364, 87 figs., 3 pls.
______, 1948.—Note sur une serie de types de Mormyridae, de Characidae et de Citharinidae du Museum de Paris.—*Bull. Mus. Natl. Hist. Nat. Paris*, **20**(1): 75-81.
________, 1948.—Poissons recueillis au Katanga par H.J. Bredo.—*Bull. Mus. Roy. Hist. Nat. Belgique* 24(21): 1-24.
______, 1953.—Poissons non Cichlidae.—Explor. hydrobiol. lac Tanganika (1946-1947). *Resultats scientifiques*, 3, 5A: 1-251, 34 figs.
______, 1953.—Les Poissons d'aquarium du Congo belge.—*Bull. Soc. Roy. Zool. Anvers.*, N°2.
______, 1954.—Poissons de foret des environs de Yangambi (Stanleyville) recueillis par A. Hulot.—*Ann. Mus. Congo*, Tervueren, in-4°, Zool., 1: 56-68 (Miscell. Zool. H. Schouteden).
______, 1957.—Les genres des Poissons d'eau douce de l'Afrique.—*Ann. Mus. Roy. Congo belge*, N°**54**: 1-191, 425 figs. (Char.: 93-103).
______, 1959.—Recherches ecologiques sur la faune ichthyologique du Stanley-Pool.—*Ann. Soc. Roy. Belgique*, **89**(1): 183-201.
______, 1959.—Recherches sur la faune ichthyologique de la region du Stanley-Pool.—*Ann. Mus. Congo, Zool.*, **71**: 77-174n pls. 12-26 (Char.: 81-86, 126-131).
______, 1966.—Geographie ichthyologique de l'Angola.—*Bull. Acad. Roy. Sci. Outre-Mer*, seance du 22.2 1966: 355-365 (Char.: 359).
______, 1967.—Revision des Characidae nains africains.—*Ann. Mus. Roy. Afr. Centr.*, Ser. in-8°, Sci. biol. N°**162**: 1-158.
______, 1967.—Contribution a la faune ichthyologique de l'Angola.—*Museu do Dundo. Subsid. para o Estudo Biol. Lunda*, Publ. cult. N°**75**: 1-381 (Char.: 91-131, figs. 33-34).
______, 1969.—Le prolongement caudal de la vessie hydrostatique des Poissons actinopterygiens.—*Bull. Acad. Roy. Belgique*, Cl. des Sci., 5e ser., **55**: 486-505 (Char.: 494-495).
POLL, M., & J. DAGET, 1968.—Description d'*Hemistichodus lootensi* (Pisces, Citharinidae).—*Bull. Mus. Natnl. Hist. Nat.*, 39(6): 1060-1065, figs. 1-3.

POLL, M. & J.P. GOSSE, 1963.—Revision des genres *Nannaethiops* Gunther 1871 et *Neolebias* Steindachner 1894, et description de trois especes nouvelles (Pisces, Citharinidae).—*Ann. Mus. Roy. Afr. Centr.*, Tervueren, Ser. in-8°, Sci. zool., N° **116**: 7-38, pls. 1-2.

______, 1963.—Contribution a l'etude systematique de la faune ichthyologique du Congo central.—*Id.*, N° **116**: 41-104, pls. 1-4 (Char. 46-51, pls. 1-2, 70-74, 93-94).

POLL, M. & J. LAMBERT, 1964.—*Congocharax gossei* sp. n. du Gabon (Ogooue) (Pisces, Citharinidae).—*Rev. Zool. Bot. Afric.*, **70**(3-4): 336-340 (fig. p. 407).

POLL, M. & B. ROMAN, 1967.—Poissons nouveaux de la Haute-Comoe.—*Id.*, **75**(1-2): 179-187 (Char.: 179-184).

POLL, M. & L. TIHON, 1945.—Note complementaire sur les poissons du Stanley-Pool.—*Ibid.*, **38**(3-4) pages ? (non vidi).

RACHOW, A., 1936 seq.—in Holly, Meinken, Rachow, Die Aquariumfische in Wort u. Bild, 7. Fam., Characinidae, *Stuttgart* (Afr. Char.: 191, 232/33, 459, 633-634, 809/10, 814-815, 822-823).

REICHENOW, A., 1875.—Ueber die ichthyologische Sammlungen der deutschen Expedition nach der Loango-Kuste.—*Sitzber. Ges. Naturf. Fr. Berlin:* 146-148.

RICARDO, C.K., 1939.—The fishes of lake Rukwa.—*Journ. Linn. Soc., London*, **40**: 625-656.

______, 1939.—Report on the fishes and fisheries of lake Rukwa in Tanganyika territory, and the Bangweulu region in northern Rhodesia.—*London* (non vidi).

RICARDO-BERTRAM, C.K., 1940.—The fishes of the Bangweulu region. —*Journ. Linn. Soc., London*, **41**: 183-217, 2 figs.

ROBERTS, T.R., 1966.—Description and osteology of *Lepidarchus adonis*, a remarkable new characid fish from West Africa.—*Stanford Ichth. Bull.*, 8(3): 209-227, 11 figs.

______, 1967.—Tooth formation and replacement in characoid Fishes.—*Id.*, 8(4): 231-247.

______, 1967.—*Virilia*, a new genus of sexually dimorphic characid fish from West Africa, with remarks on Characoids having an incomplete lateral line.—*Ibid.*, 8(4): 251-259.

ROMAN, B., 1966.—Les Poissons des hauts bassins de la Volta.—*Ann. Mus. Roy. Afr. Centr.*, Zool. N° 150, 191 pp., 7 pls.

______, 1971.—Peces de Rio Muni, Guinea Ecuatorial (Aguas dulces y salobres).—*Fundacion La Salle de Ciencias naturales.*

______, 1973.—Une nouvelle espece d'*Alestes* (Characidae) dans le fleure Niger.—*Notes et Documents Voltaiques*, **6**(3): 51-55.

ROUX, Ch., 1958.—Notes sur la riviere Malela et le lac Sinda dans la region de Loudima (Moyen Congo).—*Bull. Mus. Natl. Hist. Nat., Paris*, **29**: 467-468 (non vidi).

———, 1960.—La faune des poissons du Niari (Moyen Congo), etude du milieu et comparaison avec les bassins voisins.—*C.R. Soc. Biogeogr.*, **37**, N°324: 42-46, map.

RUPPEL, W., 1829.—Beschreibung und Abbildung mehrerer neuer Fische im Nil entdeckt.—*Frankfurt/Main* 1829-35, in-4° (Char.: 5).

———, 1832.—Forsetzung Beschreibung Fische Nil (Char.: 12, pl. 2, fig. 3 —non vidi).

SANDON, H., 1950.—An illustrated guide to the freshwater fishes of the Sudan.—*Sudan Notes and Records*, London: 1-61, 14 pls.

SAUVAGE, H.E., 1879.—Notice sur la faune ichthyologique de l'Ogooue.—*Bull. Soc. Philom. Paris*, (7)**3**: 90-103.

———, 1880.—Etude sur la faune ichthyologique de l'Ogooue.—*Nouv. Arch. Mus. Hist. Nat. Paris*, 2e ser., **3**: 5-56.

———, 1882.—Notice sur les Poissons du territoire d'Assinie (Cote d'Or).—Mission scientifique de M. Chaper.—*Bull. Soc. Zool. France*, **7**: 313-325 (Char.: 319-321).

———, 1883.—Description de quelques poissons de la collection du Museum d'Histoire naturelle.—*Bull. Soc. Philom. Paris*, (7)**7**: 156-161.

———, 1884.—Note sur les Poissons de Franceville (Haut-Ogooue).—*Bull. Soc. Zool. France*, **9**: 193-198, pl.

———, 1884.—Notice sur une collection de reptiles et de poissons recueillie a Majumba, Congo.—*Id.*, **9**: 199-208, pl. and figs.

SCHILTHUIS, L., 1890-1892.—On a collection of fishes from the Congo, with description of some new species.—*Tijdschr. Ned. Dierk. Ver.* (2), **3**: 83-92, pls. (Char.: 89-90).

SCHULTZ, L.P., 1942.—The fresh-water fishes of Liberia.—*Proc. U.S. Natl. Mus.*, **92**: 301-348 (Char.: 311-316).

STARKS, E.C., 1926.—Bones of the ethmoid region of the fish skull.—*Stanford Univ. Publ., Biol. Sci.*, **4**(3): 137-338, 58 figs.

STEINDACHNER, F., 1870.—Zur Fischfauna des Senegal, III.—*Sitz. ber. Akad. Wiss. Wien*, **61**(1): 533-583, 8 pls. (Char.: 538-549, pl. 3).

———, 1881.—Ichth. Beitrage X.—*Id.*, **83**: 179-219 (Char.: 200).

———, 1894.—Die Fische Liberia's.—*Notes Leyden Mus.*, **16**: 1-96, 4 pls. (Char.: 62-64, 78, pl. 4, fig. 3-4).

———, 1906.—Uber *Momopholis erlangeri* sp. nov., aus Abissinien, *Alestes sadleri* Blg. aus dem Viktoria-Nyansa und *Varicorhinus tornieri*, sp. nov., aus Deutsch-Kamerun.—*Ann. Naturh. Hofmus. Wien*, **21**: 149-155, pl. (Char.: 151).

———, 1913.—Zur Fischfauna des Dscha, eines sekundaren Nebenflusses des Kongo, im Bezirke Molundu, Kamerun.—*Denkschr. k. Akad. Wiss. Wien*, Math-Naturw. Kl., **89**: 1-64, pls. 1-9 (Char.: 12-23, pl. 2) (Summary published in 1912 in: *Anz. k. Akad. Wiss. Wien*, **49**: 443-449—same title).

———, 1915.—Bericht uber die ichthyologischen Aufsammlungen der Bruder Adolf und Albin Horn wahrend einer im Sommer 1913 ausgefuhrten Reise nach Deutsch-Ostafrika.—*Id.*, **92**: 59-86, 5 pls. (Char.: 64-69, pls. 1-2).

SVENSSON, G., 1933.—Freshwater fishes from the Gambia River.—*K. Svensk. Vet. Akad. Handl.*, **12**(3): 1-102, pls. 1-8, 28 figs. (Char.: 48-63, pls. 2-3).

SWAINSON, W., 1838 or 1839.—Natur. Hist. and Classif. of Fishes, Amphibians and Reptiles, I (pp. 259-260, *Hepsetus*, a Characoid).

THOMINOT, A., 1886.—Sur quelques Poissons nouveaux appartenant a la collection du Museum d'Histoire naturelle.—*Bull. Soc. Philom. Paris*, (7)**10**: 161-171 (Char.: 167).

THYS van den AUDENAERDE, D.F.E., 1968.—Addendum to the "Freshwater fishes of Fernando Po"—On some fishes collected by Col. J.J. Scheel.—*Rev. Zool. Bot. Afric.*, **78**: 123-128.

TRAVASSOS, H., 1951-52.—Catalogo dos generos e subgeneros da subordem Characoidei.—*Dusenia*, **2**(3-6) and **3** (2-4).

TREWAVAS, E., 1936.—Dr. Karl Jordan's expedition to South West Africa and Angola: The freshwater fishes.—*Novit. Zool.*, **40**: 63-74, 2 pls.

TREWAVAS, E. & F.R. IRVINE, 1947.—Freshwater Fishes, in Irvine, F.R., *The Fishes and Fisheries in Gold Coast*, London: 221-282, figs. 135-190.

VAILLANT, L., 1893.—*Bull. Soc. Philom. Paris*, (8)**5** (non vidi).

VALENCIENNES, A., in Cuvier & Valenciennes, 1849.—Histoire Nat. Poiss., **22** (Afr. Char.: 94-95, 157, 172-175, 180-190, 255, 345, pl. 639).

VANDERPLANK, F.L., 1939.—Some interesting fishes from the Mayonga River.—*The Aquarist*, London, **9**: 275 (non vidi).

VINCIGUERRA, D., 1895.—Esplorazione del Giuba e dei affluenti compiuta dal Capitano V. Bottego durante gli anni 1892-93 sotto gli auspicii della Societa geographica italiana (Pesci).—*Ann. Mus. Civ. Storia Nat., Genova*, (2)**15**: 19-60 (Char.: 28).

______, 1897.—Pesci raccolti da Don Eugenio dei Principi Ruspoli durante l'ultimo suo viaggio nelle regioni dei Somali e dei Galla; nota.—*Id.*, (2) **17**: 24-31 (Char.: 28-29).

______, 1897.—Pesci raccolti dal Cap. V. Bottego durante la sua seconda spedizione nelle regioni dei Somali e dei Galla; nota.—*Ibid.*, (2)**17**: 343-364 (Char.: 355).

______, 1898.—I pesci dell'ultimo spedizione del Capitano Bottego.—*Ibid.*, (2)**19**: 240-260 (Char.: 257-258).

______, 1928.—Pesci raccolti dal March. de Saverio Patrizi nel bacino del Congo.—*Ibid.*, **53**: 1-25, pl. 1.

WEBER, M., 1897.—Beitrage zur Kenntniss der Fauna von Sud-Afrika.—*Zool. Jahrb.*, (*Syst. Abth.*), **10**: 135-199, pl., 2 figs. (Char.: 155).

WERNER, F., 1906.—Beitrage zur Kenntniss der Fishfauna des Nils.—*Sitzber. Akad. Wiss. Wien, math.-Nat. Kl.*, **115**: 1097-1158, 4 pls., 6 figs., and *Anz. Akad. Wiss Wien*, 43: 325-327 (Char.: 1126-1130, pl. 3).

WHITLEY, G.P., 1931.—New names for Australian fishes.—*Austr. Zool., Sydney*, 6: 334.

WORTHINGTON, E.B., 1932.—Scientific results of the Cambridge Expedition to the East African lakes, 1930-31. Fishes other than Cichlidae.—*Journ. Linn. Soc. London*, **38**: 121-134 (Char.: 123).

______, 1933.—The fishes (other than Cichlidae) of lake Bangweulu and adjoining regions, including description of three new species.—*Ann. Mag. nat. Hist., London*, (10)**12** (Char.: 40).

BIBLIOGRAPHY II: South America

Addenda to the 1972 list
(Gery, 1972, Poissons Characoides des Guyanes: 82-133).

ANONYMOUS, 1964.—Eradication of Piranhas in inland waters.—*Trop. Fish Hobb.*, Oct., 1964: 36-49 (extract from an article in *Commercial Fisheries Review*, March, 1963).

ASTORQUI, date ?.—Peces de la cuenca de los grandes lagos de Nicaragua (non vidi).

AUBRY, J.C., 1969.—Contribution a la systematique des Characidae—Etude de l'anatomie viscerale.—*Ann. Soc. Roy. Zool. Belgique*, **99**(4): 253-254.

BERKENKAMP, H.O., 1972.—Unsere Beilbauchfische.—*Aquarien Magazin:* 196-203.

BEZERRA E SILVA, J.W., & O.F. Dourado, 1975.—Curva de rendimento da Traira, *Hoplias malabaricus* Bloch, do acude publico "Pereira de Miranda" (Pentecoste, Ceara, Brasil).—*Centre techn. for. trop., Notes Doc. Peche et Pisc.*, N.S. N° 10: 77-83.

______, 1975.—Curva de rendimento da Curimata commun, *Prochilodus cearensis* Steindachner, do acude publico "Pereira de Miranda" (Pentecoste, Ceara, Brasil).—*Centre tech. for. trop., Notes Doc. Peche et Pisc.*, N.S. N° 10: 69-76.

BRAGA, R.A., 1961.—Erradicacao de piranhas no acude publico "Poco da Cruz" (Inaja, Pernambuco). Desenvolvimento da bacia hidrografica.—*Bol. Mus. Nac., N.S. Zool.* N° 6, 22 pp.

______, date ?.—*Id.*, 2. Tinguijamento das colecoes d'agua.—*Ibid.* (non vidi).

______, 1972.—Ecologia e etologia de Piranhas no Nordeste do Brasil.—Resumen in *Deutsche Nachrichten*, Sao Paulo.

BUSSING, W.A., 1974.—*Pterobrycon myrnae*, a remarkable new glandulocaudine characid fish from Costa Rica.—*Rev. Biol. Trop.* 22(1): 135-159.

BUSSING, W.A., and T.R. ROBERTS, 1971.—Rediscovery of the glandulocaudine fish *Pterobrycon* and hypothetical significance of its spectacular humeral scale (Pisces: Characidae).—*Copeia* N°1: 179-181.

CALA, P., 1970.—Sobre la explotacion y control de la fauna amazonica.—*II Simposio y foro de Biol. trop. amaz.:* 145-151.

CHARDON, M., 1967.—Reflexions sur la dispersion des Ostariophysi a la lumiere de recherches morphologiques nouvelles.—*Ann. Soc. Roy. Zool. Belgique*, **97**(3): 175-186.

CHIRICHIGNO, NORMA F., 1963.—Estudio de la fauna ictiologica de los esteros y parte baja de los rios del Departamento de Tumbes (Peru), Lima. Serv. Pesq. Peru.—*Ser. Divulg. Cient.* No. 22, Lima, Characoidei: 19-24.

DAGET, J., 1970.—A propos des "Poissons aveugles" de G. Thines.—*Rev. Quest. Sci.* **141**(2): 287-293 (Char.: 290-291).

DAHL, G., 1955.—An ichthyological reconnaissance of the Sinu River.—*Rev. Linn.* N°1: 11-19, Sincelejo.

______, 1958.—Los Peces del rio Sinu. Informe preliminar.—*Secr. Agricult. Monteria*, Jun. (non vidi).

______, 1960.—New fresh-water Fishes from Western Colombia.—*Caldasia*, 8(39): 451-484, 12 figs.

______, 1960.—Los peces en los rios colombianos.—*Rev. nac. Agr.*, **662**: 22-24 (non vidi).

______, 1971.—Los Peces del Norte de Colombia.—*Instituto de Desarrollo de los Recursos Naturales Renovales* (INDERENA), Bogota.—Characoidei: 100-155.

EIGENMANN, C.H., 1919-1920.—Addenda and Corrigenda-Cheirodontinae.—*Mem. Carnegie Mus.*, 7: xix and xxi-xxii.

FERNANDEZ YEPEZ, A., 1972.—Analisis ictiologico del complejo hidrografico (04) "Rio Yaracuy."—*Dir. Obras hidraulicas,.Min. Obras publicas, Rep. Venezuela.*

FINK, W.L., & S.H. WEITZMAN, 1974.—The so-called cheirodontin fishes of Central America with description of two new species (Pisces: Characidae).—*Smiths. Contrib. to Zool.* N°172: 1-46, 26 figs., 15 tables.

FONTENELE, O., 1959.—Ha uma guerra contra as Piranhas nos rios de Pernambuco.—*Diaro de Pernambuco, Recife*, edit. 26. 03. 1959 (non vidi).

______, 1960.—Relatorio dos Servicos executados em 1959 pelo Servico de Piscicultura do Depto. nac. Obras contra as Secas.—*Fortaleza* N°219, Ser. II-M, 132 pp.

______, 1960.—A erradicacao da piranha nos acudes do Nordeste.—*Bol. Depto. nac. Obras contra Secas, Rio de Janeiro*, **21**(8): 299-304.

______, 1961.—Relatorio dos Servicos executados em 1960 pelo Servico de Piscicultura do Depto. nac. Obras contra as Secas.—*Fortaleza*, N°235, Ser. II-M, 158 pp.

GANNON, R., 1961.—Piranhas banned by the Government, a good regulation?.—*Trop. Fish Hobb.*, May: 48-51.

GEISLER, R., H.A. KNOPPEL & H. SIOLI, 1971.—Okologie der Susswasserfische Amazoniens. Stand und Zukunftsaufgaben der Forschung.—*Naturwiss.*, 58(6): 303-311.

GERY, J., 1971.—On certain fresh-water fishes from British Honduras.—*The Aquarium*: 48-52, 2 figs., 1 table.

______, 1971.—Corrected and supplemented descriptions of certain characoid fishes described by Henry W. Fowler, with revisions of several of their genera.—*Stud. Neotrop. Fauna*, **7**(1): 1-35, 6 figs.

______, 1972.—Contribution a l'etude des Poissons characoides de l'Equateur, avec une revision du genre *Pseudochalceus* et la description d'un nouveau genre endemique du rio Cauca en Colombie.—*Acta Humboldtiana, ser. geol., paleont., biol.* N°2 (110), 112 pp., 8 pls., 20 ills.

______, 1972.—Remarques sur quelques poissons characoides de la Colombie et de l'Equateur, avec la description d'une nouvelle espece de *Pseudochalceus.*—*Rev. Suisse Zool.*, **79**(2): 931-945.

______, 1972.—Poissons characoides des Guyanes. I. Generalites.—II. Famille des Serrasalmidae.—*Zool. Verh. Leiden* N°122, 250 pp., 16 pls.

______, 1972.—Une sous-famille nouvelle de Poissons characoides sudamericains: les Geisleriinae.—*Vie et Milieu,* 22(1), Ser. C: 153-166.

______, 1973.—Sur la nomenclature et les types de livree de *Carnegiella strigata* (Gunther).—*Aquarama*, 7(21): 16-19.

______, 1973.—New and little-known Aphyoditeina (Pisces, Characoidei) from the Amazon basin.—*Beitr. Neotrop. Fauna*, **8**(1): 81-137.

______, 1974.—Notes sur quelques Anostomidae (Pisces, Characoidei) du bassin amazonien.—*Vie et Milieu,* **23**(1), Ser. C.: 143-175.

GODOI, M.P. de, date ?.—Estudos osteologicos da traira *Hoplias malabaricus malabaricus* (Bloch 1974) (Pisces, Cypriniformes, Erythrinidae) (non vidi).

______, date ?.—Nova especie de *Astyanax* Baird & Girard, 1854 (Pisces, Characidae, Heterognathi) (non vidi).

ISBRUCKER, I.J.H., 1971.—*Axelrodia stigmatias* (Fowler, 1914), een Zvidamerikaanse Dwergkarperzalm.—*Het Aquar.*, **42**(5): 112-114.

KNOPPEL, H.A., 1972.—Zur Nahrung tropischer Susswasserfische aus Sudamerica (Anostomidae, Curimatidae, Hemiodidae, Characidae).—*Amazoniana,* **3**(2): 231-257, 3 figs., 6 tables.

LAHILLE, F., 1922.—Nombres vernaculares de algunos de nuestros peces de agua dulce.—*Publ. Diec. Labor. Investig. agricolo-ganaderas, Lab. Zool.* N°37 (non vidi).

LOPEZ S., M.I., 1972.—Contribucion al estudo de los peces Cheirodontinae (Familia Characidae) de Costa Rica y Panama: Generos *Compsura y Pseudocheirodon.*—*Rev. Biol. Tropic.*, **20**(1): 93-129.

LUENGO, J.A., 1973.—Expedicion uruguaya al Mato Grosso, 1955. I. Pisces (Resumen).—*Bol. Soc. Zool. Uruguay,* **2**: 22.

MAGO LECCIA, F., S.D. (1971?).—La ictiofauna del Casiquiare.—*Rev. Def. Naturaleza, Caracas,* Venezuela, 1(4): 5-11.

———, 1972.—Consideraciones sobre la sistematica de la familia Prochilodontidae (Osteichthyes, Cypriniformes), con una sinopsis de las especies de Venezuela.—*Acta Biol. Venez.*, **8**(1): 35-96, 15 figs., 13 tables.

MARKL, H., 1971.—Schallerzeugung bei Piranhas (Serrasalmidae, Characoidei).—*Zeit. Vgl. Physiol.*, **74**(1): 39-56.

MAYER, R., date ?.—Uber das vorkommen von Astacin im Neonfish (non vidi—cit. by AXELROD, H.R., What makes a Neon Tetra so colorful, *Trop. Fish Hobb.*, June, 1964: 20-23).

MEINKEN, H., 1975.—*Microschemobrycon meyburgi* n. sp. aus dem Rio Xeruini (Pisces: Characoidei: Aphyoditeina).—*Senckenbergiana Biol.*, **56**(4/6): 217-220.

MELVILLE, R.V., 1975.—*Chilodus* and *Caenotropus*—Revised proposals —Z.N. (S) 1502.—*Bull. Zool. Nomencl.*, **32**(1): 45-50.

MENEZES, N.A., 1974.—Redescription of the genus *Roestes* (Pisces, Characidae).—*Pap. Avuls. Zool. S. Paulo*, **27**(17): 219-225.

MENEZES, N.A., 1976.—On the Cynopotaminae, a new subfamily of Characidae.—*Arq. Zool. S. Paulo*, **28**(2): 1-191.

MYERS, G.S., 1932.—Some notes on the characin, *Astyanax mexicanus*, in Texas.—*Aquatic Life, Baltimore*, 16(3) (non vidi).

———, 1972.—A Monograph on the Piranha (reprinted from The Aquar. Journ., N° 2-3, 1949, with important additions: Supplementary notes on the classification of Piranhas; How dangerous are Piranhas? etc.).—*The Piranha Book, T.F.H. Edit.:* 7-51, 54-63, 97-102.

NOMURA, date ?.—Biologia de piava, *Leporinus copelandi* Steindachner, 1875, do rio Mogi Guassu, Estado de Sao Paulo (non vidi).

OLIVEIRA & SILVA, date ?.—Aspectos morfo-estructurais da mucosa digestiva de *Curimata* sp. (non vidi).

RINGUELET, R.A., R.H. ARAMBURU & A.A. de ARAMBURU, 1967.—Los Peces argentinos de agua dulce.—*La Plata*, 602 pp., 37 figs., 10 pls.

ROBERTS, T.R., 1972.—Ecology of fishes in the Amazon and Congo basins. —*Bull. Mus. Compar. Zool., Harvard Univ.*, **143**(2): 117-147 (Char.: 119).

———, 1973.—The Glandulocaudine Characid Fishes of the Guayas Basin in Western Ecuador.—*Bull. Mus. Compar. Zool., Harvard Univ.*, **144** (8): 489-514.

———, 1973.—Osteology and relationships of the Prochilodontidae, a South American Family of Characoid Fishes.—*Bull. Mus. Compar. Zool., Harvard Univ.*, **145**(4): 213-235.

———, 1973.—Interrelationships of Ostariophysans, in Interrelationships of fishes, ed. by P.H. Greenwood, R.S. Miles and C. Patterson.—*Supp. N° 1 to Zool. Journ. Linn. Soc*, **53**: 373-395.

———, 1974.—Dental polymorphism and systematics in *Saccodon*, neotropical genus of fresh-water fishes (Parodontidae, Characoidei).—*Journ. Zool., London*, **137**: 303-321.

———, 1974.—Osteology and classification of the Neotropical Characoid

Fishes of the Families Hemiodontidae (including Anodontinae) and Parodontidae.—*Bull. Mus. Compar. Zool., Harvard Univ.*, **146**(9): 411-472.

ROSEN, D.E., 1970.—A new tetragonopterine characid fish from Guatemala.—*Am. Mus. Novit.*, N° 2435: 1-17.

______, 1972.—Origin of the characid fish genus *Bramocharax* and a description of a second, more primitive species in Guatemala.—*Am. Mus. Novit.*, N° 2500: 1-21.

SCHEEL, J.J., 1972.—The chromosomes of the third Neon Tetra.—*Trop. Fish Hobb.*, July: 60-65.

STEINDACHNER, F., 1887.—Ichthyologische Beitrage XIV.—*Sitz. ber. Akad. Wiss. Wien, M-N. Kl. 1*, **96**: 56-68, pls. 1-4 (Char.: *Elopomorphus orinocensis:* 66-67, pl. II, figs. 2-2a).

THOMERSON & GREENFIELD, date ?.—Preliminary key to the freshwater fishes of Belize (British Honduras) (non vidi).

WEITZMAN, S.H., 1975.—Der Flugelschuppensalmler, *Pterobrycon myrnae*, ein bezaubernder Aquarienfish der Zukunft aus Costa Rica.—*Die Aquar. Terr. Zeitschr.*, **28**(12): 407-410.

WEITZMAN, S.H., & W.L. FINK, 1971.—A new species of characid fish of the genus *Nematobrycon* from the rio Calima of Colombia (Pisces, Characoidei, Characidae).—*Beaufortia* V. 19, N° 248: 57-77.

WEITZMAN, S.H., & J.S. COBB, 1975.—A revision of the South American Fishes of the genus *Nannostomus* Gunther (Family Lebiasinidae).—*Smithson. Contrib. Zool.*, N° 186: 1-36, 34 figs.

WILKENS, H., 1971.—Genetic interpretation of regressive evolutionary Processes: Studies on hybrid eyes of two *Astyanax* cave populations (Characidae, Pisces).—*Evolution*, **25**(3): 530-544.

ZBINDEN, K., 1973.—Verhaltenstudien an *Serrasalmus nattereri.*—*Rev. Suisse Zool.*, **80**(2): 521-542.

INDEX TO LATIN NAMES

The page numbers in ordinary type refer to the principal page(s) on which the species occurs; page numbers in *italic type* refer to *secondary references;* and page numbers in **bold type** refer to **illustrations.**

Valid scientific names are in ordinary type. Synonyms are in italic type; higher taxa (families, tribes, etc.) are in capital letters.

The following abbreviations are used for those ichthyologists who have published more than 20 valid species. Appearance of a zoologist's name and a date within parentheses means that the genus name has been changed subsequent to the original publication.

A = Agassiz
B = Boulenger
C = Cope
E = Eigenmann
F = Fowler
G = Gery
Gu = Gunther
K = Kner
M = Myers
S = Steindachner
V = Valenciennes

A

B

F

G

N

O

P

T

U

V

W

X

Y

Z